Immersed Methods for Coupled Continua

Frontier Research in Computation and Mechanics of Materials and Biology

ISSN: 2315-4713

Published:

Vol. 6 *Immersed Methods for Coupled Continua*
by Sheldon Wang & Lucy Zhang

Vol. 5 *Gradient Smoothing Methods with Programming: Applications to Fluids and Landslides*
by G R Liu & Zirui Mao

Vol. 4 *A Practical Guide to OpenSees*
by Quan Gu & Surong Huang

Vol. 3 *Multiscale Optimization and Materials Design*
by Jun Yan & Gengdong Cheng

Vol. 2 *Advanced Continuum Theories and Finite Element Analyses*
by James D Lee & Jiaoyan Li

Vol. 1 *Introduction to Practical Peridynamics: Computational Solid Mechanics Without Stress and Strain*
by Walter Herbert Gerstle

Frontier Research in Computation and Mechanics of Materials and Biology – Vol. 6

Immersed Methods for Coupled Continua

Sheldon Wang
Midwestern State University, USA

Lucy Zhang
Rensselaer Polytechnic Institute, USA

World Scientific

NEW JERSEY • LONDON • SINGAPORE • BEIJING • SHANGHAI • HONG KONG • TAIPEI • CHENNAI • TOKYO

Published by

World Scientific Publishing Co. Pte. Ltd.
5 Toh Tuck Link, Singapore 596224
USA office: 27 Warren Street, Suite 401-402, Hackensack, NJ 07601
UK office: 57 Shelton Street, Covent Garden, London WC2H 9HE

Library of Congress Control Number: 2024062247

British Library Cataloguing-in-Publication Data
A catalogue record for this book is available from the British Library.

Frontier Research in Computation and Mechanics of Materials and Biology — Vol. 6
IMMERSED METHODS FOR COUPLED CONTINUA

ISBN 978-981-323-450-5 (hardcover)
ISBN 978-981-323-451-2 (ebook for institutions)
ISBN 978-981-323-452-9 (ebook for individuals)

For any available supplementary material, please visit
https://www.worldscienti ic.com/worldscibooks/10.1142/10833#t=suppl

Desk Editors: Soundararajan Raghuraman/Steven Patt

Typeset by Stallion Press
Email: enquiries@stallionpress.com

To Our Families and Friends

Preface

This compendium provides a broad introduction to the original Immersed Boundary Methods and current extensions in the computational mechanics community. It highlights a comprehensive review of immersed methods in the context of computational fluid and solid mechanics and fluid–structure interactions. This invaluable title will provide useful reference materials and an interesting read for academics, researchers, professionals, and students.

The rapid development of computational mechanics and computer technology during the last two decades has brought compelling computational tools to the disposal of scientists and engineers. With the recent advances of petaflop and GPU/CPU processors that enabled and enhanced mathematical computation capabilities, simulation-based engineering processes are gaining more acceptance and are rapidly becoming viable design alternatives. Moreover, the traditional classification of various disciplines of mechanics has been blurred, and interaction problems involving two or more fields are often encountered in ever more challenging engineering practices. One of these interacting disciplines is the so-called fluid–structure or fluid–solid interaction (FSI) modeling, which lies at the intersection of fluid mechanics, heat transfer, solid mechanics, and dynamics. The new research frontiers in the quantitative understanding of biological systems compel researchers to grasp fundamentals with depth in the fields of physics, chemistry, and mathematics. In addition, complex system dynamics derived from interacting systems is essential for understanding phenomena pertaining to multiple temporal and spatial scales. In many aspects, the study of FSI systems is becoming a case study of modeling of complex systems, and the computational procedures developed and implemented might have broader applications.

The key question is how to acquire, within a reasonable period of time, a knowledge base in relevant disciplines with both depth and breadth. It is obviously not feasible for students to wade through the traditional curricula of applied mathematics, solid mechanics, continuum mechanics, theory of elasticity, structural mechanics, fluid mechanics, vibration, and dynamics.

In this book, we endeavor to present a unified exposition of fundamental principles and solution techniques based on the Immersed Boundary Methods for problems involving fluid–solid interactions. Through the publication of this work, we hope to bridge the gap between applied mathematics and computational science & engineering communities and start the transition of mechanics education to a more integrated model. The benefit of this approach is that it will provide students with a convenient entry into complicated modeling methodologies stemmed from the Immersed Boundary Method and required applied mechanics and applied mathematics and help them quickly acquire sufficient background knowledge in preparation for subsequent focus on special areas which are often interdisciplinary in nature.

This book intends to serve as the basis of a course for senior undergraduate or first-year graduate students. The comprehensive presentation of the fundamentals of fluid–solid systems will also provide applied scientists and mathematicians with recent references and an overview of the subject.

In this book, we extract the unique physical and mathematical aspects of various FSI problems, lay a foundation for computational approaches based on the concepts of the Immersed Boundary Methods for steady or stable transient analysis, and explore numerical approaches for static and dynamic instability analysis. It is important for us to point out that this book is not all-encompassing and certainly not written as an encyclopedia of all applications and subfields related to FSI problems. Although many examples derived from various engineering fields are introduced for didactic purposes, it is not our intention to focus on problems in any specific industry. A complete set of working Fortran 90 codes named as IFEM–ICM for two-dimensional immersed continuum method with mixed finite element formulations for both fluid and solid domains and implicit matrix-free Newton–Krylov iterative solution procedures are available on https://github.com/lucy-t-zhang/IFEM-ICM.git.

We would like to acknowledge a few individuals without whom we would not have had the knowledge or time to prepare this book. Foremost is Professor Charles Peskin in the Courant Institute of Mathematical Sciences at New York University, our wonderful role model and good friend. Both authors are also grateful for their respective Ph.D. advisors Professor Klaus Jürgen Bathe at Massachusetts Institute of Technology and Professor Wing Kam Liu at Northwestern University. It is really a blessing to have worked with many friends and professionals over the course of preparing this book. The financial support from the National Science Foundation (NSF) is also gratefully acknowledged. Finally, we would like to dedicate this book to our patient and loving families and dear friends, without whom this book would not be possible.

Sheldon Wang and Lucy Zhang

About the Authors

Sheldon Wang is a Professor of the McCoy School of Engineering at Midwestern State University (MSU), a member of the Texas Tech University System, Wichita Falls, TX. He is a fellow of ASME. He received his B.S. in Naval Architecture and Ocean Engineering from Shanghai Jiao Tong University (1988), his M.S. in Ocean Engineering (1993), and his Ph.D. in Applied Mechanics (1995) from the Massachusetts Institute of Technology (MIT). He has worked in the Institute of Paper Science and Technology (currently, Renewable Bioproducts Institute at Georgia Tech), Polytechnic University (currently, Tandon School of Engineering at New York University), and Department of Mathematical Sciences at New Jersey Institute of Technology (NJIT) as a faculty member before joining MSU in 2009 as the Chair and Professor of the McCoy School of Engineering. His research interests include computational fluid and solid mechanics, fluid-solid interactions, and quantitative modeling of complex dynamical systems. He is the author of over 50 well-cited journal papers, the research monograph *Fundamentals of Fluid-Solid Interactions: Analytical and Computational Approaches*, published by Elsevier Science in 2008, and the textbook *Essential Mathematical Tools for Engineers*, published by Sentia Publishing in 2022. Polytechnic University selected him as one of the five founding Othmer Junior Faculty Fellows from 2002 to 2004. He also received awards from the US Air Force Summer Fellowship Program in 2008 and 2009 and the National Science Foundation Summer Institutes Fellowship Program from 2003 to 2012.

Lucy Zhang is a Professor of Mechanical, Aerospace, and Nuclear Engineering at Rensselaer Polytechnic Institute (RPI), Troy, NY. She is a Fellow of ASME. She received her B.S. from Binghamton University in December of 1997 and obtained her M.S. and Ph.D. from Northwestern University, IL, respectively. She joined the Mechanical Engineering Department at Tulane as an Assistant Professor. She moved to Rensselaer Polytechnic Institute in 2006 and was later promoted to Associate Professor and Professor. Her research interests are building advanced and robust computational tools and software for accurate and efficient multiphysics and multiscale simulations that can be used for engineering applications in biomechanics, micro- and nano-mechanics, medicine, and defense projects. From 2021 to 2024, she served as a Program Director in the Mechanics cluster: Mechanics of Materials and Structures (MoMS) and Biomechanics and Mechanobiology (BMMB) programs within the CMMI Division at NSF. She is also the co-host of *This Academic Life* podcast series.

Contents

List of Figures

List of Tables

Chapter 1

Essential Mathematics

Although Newton's laws retain their importance throughout the modern study of mechanics, the mathematical tools required in various disciplines of mechanics have evolved into something far more complicated and sophisticated than the original calculus.

The analysis of fluid–structure interaction (FSI) problems requires the idealization of both fluid and solid domains in a continuum form that can be solved systematically and repetitively. In order to coherently present the formulation of such mathematical models as well as the solution procedures, we aim in this chapter to establish first a mathematical foundation for kinematic and dynamic descriptions of fluids and solids and their respective continua.

In this chapter, we present basic mathematical concepts, notations employed in this book, and some fundamental concepts of linear algebra, complex analysis, and advanced calculus. Kinematics for mass points, rigid bodies, and continuous media are also presented along with Reynolds transport theorems and variation principles, from which both integral and differential equilibria of mass, momentum, and energy conservation laws can be derived. Some useful mathematical tools relevant to immersed methods are also discussed in this chapter. We expect that these fundamentals will not only streamline the presentation of this book but also enables the reader to quickly pick up fundamental concepts crucial for building up necessary tools to tackle FSI problems without wading through numerous traditional books on mathematics and mechanics.

1.1 Fundamentals and Notations

The purpose of this section is to form a foundation for elementary mathematical concepts and to present notations employed throughout this book.

1.1.1 *Mappings, spaces, and tensors*

A mapping (function, transformation) is a connection from a set $\mathcal{X}$ to a set $\mathcal{Y}$, denoted as $f : \mathcal{X} \to \mathcal{Y}$, which assigns for every $x \in \mathcal{X}$, a unique element $y \in \mathcal{Y}$, and is usually written as $y = f(x)$ or (x, y). Moreover, $\mathcal{X}$ is called the domain of f and the set of elements in $\mathcal{Y}$ which are images of elements in $\mathcal{X}$ is called the range of f. In the case, where the domain and range are the same, f is also called an operator. We introduce an operation $\circ$ between two mappings $f : \mathcal{X} \to \mathcal{Y}$ and $g : \mathcal{Y} \to \mathcal{Z}$, such that $\forall x \in \mathcal{X}$, $g \circ f(x) = g(f(x))$. Furthermore, a mapping $f : \mathcal{X} \to \mathcal{Y}$ is called a linear mapping, if for $x_1, x_2 \in \mathcal{X}$ and $\lambda \in \mathcal{R}$, where $\mathcal{R}$ is a set of real numbers,

$$f(x_1 + x_2) = f(x_1) + f(x_2),$$
$$f(\lambda x_1) = \lambda f(x_1).$$

There are quantities in physics characterized by magnitudes only, such as mass, length, time, and temperature. Such quantities are often called scalars. Apart from their physical units, scalars are just real numbers and are denoted in this book by upper or lower-case ordinary letters. To describe physical quantities characterized by both magnitude and direction, such as force, velocity, and heat flux, we introduce the concept of vectors. The geometric representation of a vector is a directed line segment. Since the translation of a vector in space will not change its magnitude or direction, we can always assign the origin O and a spatial point P as the initial point and the terminal point of a vector, respectively, and a vector always corresponds to a position vector $\overrightarrow{OP}$. It is clear that a vector from a point $\mathbf{x}_1$ to a point $\mathbf{x}_2$ can be expressed as $\mathbf{x}_2 - \mathbf{x}_1$. As in number systems, we also introduce rules, so-called linear algebra, for the manipulation of vectors.

To represent three-dimensional space in nature, we often introduce a rectangular coordinate frame $oxyz$, the so-called Cartesian coordinate system, named after René Descartes (1596–1650), every point of which is uniquely defined with three real numbers x_1, x_2, and x_3. Such a space is denoted as $\mathcal{R}^3$, and called the Euclidean space, named after Euclid (~300 BC). Thus, a spatial point p can be represented by (x_1,x_2,x_3), where x_1, x_2, and x_3 stand for the distances relative to the origin $o(0,0,0)$ at the intercept of x_1-, x_2-, and x_3-axes. Likewise, we can introduce a real n-dimensional Cartesian space (hyperspace, or manifold) $\mathcal{R}^n$, as a set of all n-tuples of real numbers. A typical element in $\mathcal{R}^n$ is called a point (or vector) and denoted as $\mathbf{r}$, $\mathbf{x}$, or $(x_1,\ldots,x_n)$, where x_i is the ith coordinate of that point, or the ith component of that vector. Conventionally, $(x_1,\ldots,x_n)$ refers a row vector. Moreover, in mathematics and mechanics, a typical vector $\mathbf{x}$ refers to a column vector. To delineate $(x_1,\ldots,x_n)$ from the column vector $\mathbf{x}$, we introduce $\mathbf{x} =< x_1,\ldots,x_n >$. Naturally, for two elements $\mathbf{x}_1$ and $\mathbf{x}_2$ in $\mathcal{R}^n$ and $\lambda \in \mathcal{R}$, we have $\mathbf{x}_1+\mathbf{x}_2 = (x_{11}+x_{21},\ldots,x_{1n}+x_{2n})$ and $\lambda\mathbf{x}_1 = (\lambda x_{11},\ldots.,\lambda x_{1n})$. Furthermore, the magnitude of vector $\mathbf{r}$ is defined as $|\mathbf{r}| = \sqrt{x_i x_i}$. Note that, without specification, in indicial notations, we adopt both the Einstein summation convention, i.e., a dummy index (superscript or subscript) repeated in any expression is summed throughout its range, and the range convention, i.e., a dummy index repeated in a term takes all values in its range. In this book, vectors are denoted as boldface lowercase letters. A null or zero vector with a magnitude of 0, and a unit vector with a magnitude of 1, are denoted as $\mathbf{0}$ and $\mathbf{e}$, respectively.

Denoting the unit vectors along the x_1, ..., x_n-axes as $\mathbf{e}_1$, ..., $\mathbf{e}_n$, e.g., $\mathbf{e}_i$ is a vector with 1 as the ith coordinate and 0 elsewhere. We can then express a vector $\mathbf{r}$ as $\mathbf{r} = x_i\mathbf{e}_i$. The basic algebra laws of vectors are similar to those of scalars, which include commutativity, associativity, and distributivity:

$$
\begin{aligned}
&\mathbf{x}_1+\mathbf{x}_2 = \mathbf{x}_2+\mathbf{x}_1,\\
&\mathbf{x}_1+(\mathbf{x}_2+\mathbf{x}_3) = (\mathbf{x}_1+\mathbf{x}_2)+\mathbf{x}_3,\\
&\lambda_1(\lambda_2\mathbf{x}) = (\lambda_1\lambda_2)\mathbf{x} = \lambda_2(\lambda_1\mathbf{x}),\\
&(\lambda_1+\lambda_2)\mathbf{x} = \lambda_1\mathbf{x}+\lambda_2\mathbf{x},\\
&\lambda(\mathbf{x}_1+\mathbf{x}_2) = \lambda\mathbf{x}_1+\lambda\mathbf{x}_2.
\end{aligned}
$$

Denoting the Kronecker delta δ_{ij}, named after Leopold Kronecker (1823–1891), as 1, if $i = j$, or 0, if $i \neq j$. As a consequence, the following laws are valid:

$$
\mathbf{e}_i\cdot\mathbf{e}_j = \delta_{ij}, \quad \mathbf{x}_1\cdot\mathbf{x}_2 = \mathbf{x}_2\cdot\mathbf{x}_1 = x_{1i}x_{2i},
$$

$$\mathbf{x}_1 \cdot (\mathbf{x}_2 + \mathbf{x}_3) = \mathbf{x}_1 \cdot \mathbf{x}_2 + \mathbf{x}_1 \cdot \mathbf{x}_3,$$
$$\lambda(\mathbf{x}_1 \cdot \mathbf{x}_2) = (\lambda \mathbf{x}_1) \cdot \mathbf{x}_2 = \mathbf{x}_1 \cdot (\lambda \mathbf{x}_2) = (\mathbf{x}_1 \cdot \mathbf{x}_2)\lambda,$$
$$\mathbf{x}_1 \text{ is perpendicular to } \mathbf{x}_2, \text{ if } \mathbf{x}_1 \cdot \mathbf{x}_2 = 0, \quad \text{with } \mathbf{x}_1, \mathbf{x}_2 \neq \mathbf{0}.$$

Furthermore, we define the dot or scalar product of two vectors $\mathbf{x}_1$ and $\mathbf{x}_2$ as

$$\mathbf{x}_1 \cdot \mathbf{x}_2 = x_{1i} x_{2i} = |\mathbf{x}_1||\mathbf{x}_2| \cos\theta, \quad 0 \leq \theta \leq \pi,$$

where θ is the angle between vectors.

The angle θ between two nonzero vectors is defined as the ordinary interior angle $0 \leq \theta \leq \pi$ between these two vectors are positioned tail to tail. The physical significance of the dot product is the projection of one vector onto the other. It is obvious that the magnitude of a vector is in fact the square root of the dot product of the vector with itself.

A set of n vectors $\mathbf{x}_i$, $i = 1, \ldots, k \leq n$, is called linearly dependent if there exists a set of scalar coefficients $c_1, \ldots, c_k$, not all zero, such that $c_i \mathbf{x}_i = \mathbf{0}$. Otherwise, the set of vectors is called linearly independent. In fact, for an n-dimensional space, a set of linearly independent vectors consists of at most n vectors. Moreover, such a set of linearly independent vectors $\mathbf{x}_1, \ldots, \mathbf{x}_n \in \mathcal{S}$ is employed to span $\mathcal{S}$ and form a basis of $\mathcal{S}$ if every vector in $\mathcal{S}$ is a linear combination of $\mathbf{x}_1, \ldots, \mathbf{x}_n$. Succinctly, the dimension of a vector space denoted as $\dim(\mathcal{S})$, is the number of elements in a basis e.g., $\dim(\mathcal{R}^n) = n$. In addition, the corresponding coefficients are called the coordinates with respect to the basis. One example of the basis is the so-called standard basis with unit vectors $\mathbf{e}_i$. It is obvious that by adjoining a set of $k < n$ independent vectors suitable $n - k$ vectors, one can form a basis.

Now, we introduce a matrix, denoted in this book with boldface capital letters, as an array of numbers formed in rows and columns. For example, matrix $\mathbf{A} = [A_{ij}]$ with m rows and n columns can be expressed as

$$\mathbf{A} = [A_{ij}] = \begin{bmatrix} A_{11} & A_{12} & \ldots & A_{1n} \\ A_{21} & A_{22} & \ldots & A_{2n} \\ \vdots & \vdots & \ddots & \vdots \\ A_{m1} & A_{m2} & \ldots & A_{mn} \end{bmatrix}.$$

In fact, each $m \times n$ matrix $\mathbf{A}$ with m rows and n columns is associated with a linear mapping $\mathbf{A} : \mathcal{R}^n \to \mathcal{R}^m$. In addition, the transpose of this matrix, denoted as $\mathbf{A}^T = [A_{ji}]$, corresponds to a linear mapping $\mathbf{A}^T : \mathcal{R}^m \to \mathcal{R}^n$. Of course, an n-dimensional vector can be viewed as an $n \times 1$ matrix, and is traditionally denoted as a column, i.e., $\mathbf{x} =< x_1, \ldots, x_n >= [x_i] = [x_1, \ldots, x_n]^T$. Furthermore, let $\mathbf{B}$ be an $n \times l$ matrix, the composite map $\mathbf{AB}$ corresponds to an $m \times l$ matrix, $\mathbf{C} = [C_{ij}]$, with $C_{ij} = A_{ik}B_{kj}$. Defining $\mathbf{I}$ as an identity matrix (unit matrix), i.e., $I_{ij} = \delta_{ij}$, if there exists an $n \times m$ matrix $\mathbf{B}$ such that $\mathbf{BA} = \mathbf{I}_n$, where $\mathbf{I}_n$ is an $n \times n$ identity matrix, the matrix $\mathbf{B}$ is called the left inverse of the left invertible matrix $\mathbf{A}$. If there exists another $n \times m$ matrix $\mathbf{B}$ such that $\mathbf{AB} = \mathbf{I}_m$, where $\mathbf{I}_m$ is an $m \times m$ identity matrix, the matrix $\mathbf{B}$ is called the right inverse of the right invertible matrix $\mathbf{A}$. Of course, for a linear operator $\mathbf{A}$ on $\mathcal{R}^n$, if there exists an operator $\mathbf{B}$, such that $\mathbf{AB} = \mathbf{BA} = \mathbf{I}$, $\mathbf{A}$ is called invertible, i.e., both left and right invertible, hence, $\mathbf{B} = \mathbf{A}^{-1}$ and $\mathbf{A} = \mathbf{B}^{-1}$.

A non-empty subset $\mathcal{S} \subseteq \mathcal{R}^n$ is called a linear subspace (vector space, linear space), if $\mathcal{S}$ is closed under the operations of addition and scalar multiplication in $\mathcal{R}^n$. If $\mathcal{S}$ contains only $\mathbf{0}$, $\mathcal{S}$ is called the trivial subspace, and if $\mathcal{S} \neq \mathcal{R}^n$, it is called a proper subspace of $\mathcal{R}^n$. Moreover, if both $\mathcal{S}_1$ and $\mathcal{S}_2$ are subspaces of $\mathcal{R}^n$ and $\mathcal{S}_1 \subset \mathcal{S}_2$, $\mathcal{S}_1$ is a subspace of $\mathcal{S}_2$.

In $\mathcal{R}^3$, we also define the cross or vector product of two vectors $\mathbf{f}_1$ and $\mathbf{f}_2$ as

$$\mathbf{f}_1 \times \mathbf{f}_2 = |\mathbf{f}_1||\mathbf{f}_2| \sin\theta \mathbf{n}, \quad 0 \leq \theta \leq \pi,$$

where θ is the angle between vectors, $\mathbf{n}$ is a unit vector indicating the direction of the resulting vector $\mathbf{f}_3 = \mathbf{f}_1 \times \mathbf{f}_2$ which is perpendicular to the plane of $\mathbf{f}_1$ and $\mathbf{f}_2$, and $\mathbf{f}_1$, $\mathbf{f}_2$, and $\mathbf{f}_3$ form a right-handed system.

The physical significance of the magnitude of the cross-product corresponds to the area of the $\mathbf{f}_1$ and $\mathbf{f}_2$ parallelogram, and its direction is the direction of the unit normal vector $\mathbf{n}$. Moreover, if $\mathbf{f}_1$ and $\mathbf{f}_2$ are perpendicular, $\mathbf{f}_1$, $\mathbf{f}_2$, and $\mathbf{f}_3$ form a triad of orthogonal (perpendicular) vectors. Note that if $\mathbf{f}_1$ and $\mathbf{f}_2$ are non-collinear, or linearly independent, i.e., $\mathbf{f}_1 \times \mathbf{f}_2 \neq \mathbf{0}$, and tangent to the surface $\mathbf{s}$ at some point p, the unit normal vector $\mathbf{n}$ of that surface at the point p may be expressed as

$$\mathbf{n} = \frac{\mathbf{f}_1 \times \mathbf{f}_2}{|\mathbf{f}_1 \times \mathbf{f}_2|}.$$

Of course in general the unit normal vector of a surface varies with the location on that surface. At this point, we must unstipulate that in this book, we only concern ourselves with orientable surfaces, in which the normal vector with an opposite sign corresponds to the another side of the surface.

In a Cartesian coordinate system, we have

$$\mathbf{f}_1 \times \mathbf{f}_2 = \begin{vmatrix} \mathbf{e}_1 & \mathbf{e}_2 & \mathbf{e}_3 \\ f_{11} & f_{12} & f_{13} \\ f_{21} & f_{22} & f_{23} \end{vmatrix} = \varepsilon_{ijk} f_{1i} f_{2j} \mathbf{e}_k,$$

where the three-index permutation symbol ε_{ijk}, or alternating tensor, is 0, if two indices are the same; 1, if even permutation, i.e., in cyclic order of 1, 2, and 3; or –1, if odd permutation, i.e., in a cyclic order of 3, 2, and 1.

Moreover, it is easy to verify the ε–δ identity

$$\varepsilon_{ijk}\varepsilon_{irs} = \delta_{jr}\delta_{ks} - \delta_{js}\delta_{kr}.$$

As a consequence, the following laws are valid:

$$\begin{aligned}
&\mathbf{f}_1 \times \mathbf{f}_2 = -\mathbf{f}_2 \times \mathbf{f}_1, \\
&\mathbf{f}_1 \times (\mathbf{f}_2 + \mathbf{f}_3) = \mathbf{f}_1 \times \mathbf{f}_2 + \mathbf{f}_1 \times \mathbf{f}_3, \\
&a(\mathbf{f}_1 \times \mathbf{f}_2) = (a\mathbf{f}_1) \times \mathbf{f}_2 = \mathbf{f}_1 \times (a\mathbf{f}_2) = (\mathbf{f}_1 \times \mathbf{f}_2)a, \\
&\mathbf{e}_i \times \mathbf{e}_i = \mathbf{0}, \quad \mathbf{e}_1 \times \mathbf{e}_2 = \mathbf{e}_3, \quad \mathbf{e}_2 \times \mathbf{e}_3 = \mathbf{e}_1, \quad \mathbf{e}_3 \times \mathbf{e}_1 = \mathbf{e}_2; \\
&\mathbf{f}_1 \text{ is parallel to } \mathbf{f}_2, \text{ if } \mathbf{f}_2 \times \mathbf{f}_2 = \mathbf{0}, \quad \text{with } \mathbf{f}_1, \mathbf{f}_2 \neq \mathbf{0};
\end{aligned}$$

along with the combination of the dot and cross products:

$$\begin{aligned}
&\mathbf{f}_1 \cdot (\mathbf{f}_2 \times \mathbf{f}_3) = \mathbf{f}_2 \cdot (\mathbf{f}_3 \times \mathbf{f}_1) = \mathbf{f}_3 \cdot (\mathbf{f}_1 \times \mathbf{f}_2), \\
&\mathbf{f}_1 \times (\mathbf{f}_2 \times \mathbf{f}_3) = (\mathbf{f}_1 \cdot \mathbf{f}_3)\mathbf{f}_2 - (\mathbf{f}_1 \cdot \mathbf{f}_2)\mathbf{f}_3.
\end{aligned}$$

In particular, the scalar triple product $\mathbf{f}_1 \cdot (\mathbf{f}_2 \times \mathbf{f}_3)$ is the volume of the parallelepiped formed by $\mathbf{f}_1$, $\mathbf{f}_2$, and $\mathbf{f}_3$, and the physical significance of the vector triple product $\mathbf{f}_1 \times (\mathbf{f}_2 \times \mathbf{f}_3)$ is the vector which lies in the $\mathbf{f}_2$ and $\mathbf{f}_3$ plane and perpendicular to $\mathbf{f}_1$.

In practice, a scalar is called a tensor of order zero, a vector is a tensor of order one, and a tensor of order two can be expressed as a 3×3 matrix, also denoted in this book by a boldface uppercase letter or a dyadic product of two vectors, e.g., $\mathbf{F} = \mathbf{f}_1 \odot \mathbf{f}_2 = f_{1i} f_{2j} \mathbf{e}_i \mathbf{e}_j$, or $\mathbf{F} = \mathbf{f}_1 \mathbf{f}_2^T$, In this book, without specification, tensors are referred to as tensors of second order. Consider a linear mapping represented by an $m \times n$ matrix $\mathbf{A} : \mathcal{R}^n \rightarrow \mathcal{R}^m$, where $\mathcal{R}^n$ and $\mathcal{R}^m$ are n- and m-dimensional vector spaces, respectively. The kernel (null space) of $\mathbf{A}$ is defined as

$$\ker(\mathbf{A}) = \{\mathbf{x} \in \mathcal{R}^n | \quad \mathbf{Ax} = \mathbf{0}\},$$

and the image (column space) of $\mathbf{A}$ is defined as

$$\text{im}(\mathbf{A}) = \{\mathbf{y} \in \mathcal{R}^m | \quad \mathbf{Ax} = \mathbf{y} \text{ for some } \mathbf{x} \in \mathcal{R}^n\}.$$

Likewise, the transpose of the matrix $\mathbf{A}$ will denote a linear mapping $\mathbf{A}^T : \mathcal{R}^m \rightarrow \mathcal{R}^n$. The kernel (null space) of $\mathbf{A}^T$ or the left null space of $\mathbf{A}$ is defined as

$$\ker(\mathbf{A}^T) = \{\mathbf{y} \in \mathcal{R}^m | \quad \mathbf{A}^T\mathbf{y} = \mathbf{0}\},$$

and the image of $\mathbf{A}^T$ or the row space of $\mathbf{A}$ is defined as

$$\text{im}(\mathbf{A}^T) = \{\mathbf{x} \in \mathcal{R}^n | \quad \mathbf{A}^T\mathbf{y} = \mathbf{x} \text{ for some } \mathbf{y} \in \mathcal{R}^m\}.$$

Note that the rank of the matrix $\mathbf{A}$ is indeed the dimension of the image (column space) which is the same as the dimension of the image of its transpose $\mathbf{A}^T$ or the row space of $\mathbf{A}$. Of course, $\ker(\mathbf{A})$ and $\text{im}(\mathbf{A})$ are subspaces of $\mathcal{R}^n$ and $\mathcal{R}^m$, respectively, and

$$\dim(\text{im}(\mathbf{A})) + \dim(\ker(\mathbf{A})) = n.$$

Again, $\ker(\mathbf{A}^T)$ and $\text{im}(\mathbf{A}^T)$ are subspaces of $\mathcal{R}^m$ and $\mathcal{R}^n$, respectively, and

$$\dim(\text{im}(\mathbf{A}^T)) + \dim(\ker(\mathbf{A}^T)) = m.$$

It is now clear that the matrix $\mathbf{A}$ is characterized by its four fundamental subspaces, namely, the null space $\ker(\mathbf{A})$ and the row space $\text{im}(\mathbf{A}^T)$ within $\mathcal{R}^n$ and the left null space $\ker(\mathbf{A}^T)$ and the column space $\text{im}(\mathbf{A})$ within $\mathcal{R}^m$. Based on the definition, all the vectors within the null space must be perpendicular to all the vectors within the row space, while all the vectors within the left null space must be perpendicular to all the vectors within the column space. Therefore, within $\mathcal{R}^n$, the subspace $\ker(\mathbf{A})$ is orthogonal to the subspace $\text{im}(\mathbf{A}^T)$; while the subspace $\ker(\mathbf{A}^T)$ is orthogonal to the subspace $\text{im}(\mathbf{A})$. In fact the fundamental theorem of linear algebra or Fredholm alternative, named after Erik Ivar Fredholm (1866–1927), is stated in the context of the four fundamental subspaces, namely, row space and null space within $\mathcal{R}^n$ as well as column space and left null space within $\mathcal{R}^m$.

Theorem 1.1 (Fundamental Theorem of Linear Algebra). *For any matrix* $\mathbf{A}$ *and vector* $\mathbf{b}$, *one and only one of the following systems has a solution* (1) $\mathbf{Ax} = \mathbf{b}$ *or* (2) $\mathbf{A}^T\mathbf{y} = \mathbf{0}, \mathbf{y}^T\mathbf{b} \neq \mathbf{0}$. *The understanding of this alternative can be simply viewed as either the vector* $\mathbf{b}$ *is in the column space of* $\mathbf{A}$ *as in* (1) *or the vector* $\mathbf{b}$ *is not in the column space of* $\mathbf{A}$ *as in* (2).

Furthermore, $\mathbf{A}$ is a one-to-one mapping iff $\ker(\mathbf{A}) = \{\mathbf{0}\}$, and the kernel of a linear mapping $\mathbf{A} : \mathcal{R}^n \to \mathcal{R}^m$ corresponds to the solution space of the linear algebraic equation $\mathbf{Ax} = \mathbf{0}$ with $\mathbf{x} \in \mathcal{R}^n$. Obviously, solving the system $\mathbf{Ax} = \mathbf{0}$ is equivalent to finding a basis for $\ker(\mathbf{A})$. If the mapping $\mathbf{A} : \mathcal{R}^n \to \mathcal{R}^m$ is bijective, $\mathbf{A}$ is called an isomorphism and $\mathcal{R}^n$ and $\mathcal{R}^m$ are called isomorphic vector spaces with $n = m$. Hence, suppose $n = m$, we have the following equivalent logical statements:

$$\ker(\mathbf{A}) = \mathbf{0} \Leftrightarrow \text{im}(\mathbf{A}) = \mathcal{R}^m \Leftrightarrow \mathbf{A} \text{ is an isomorphism} \Leftrightarrow \mathbf{A} \text{ is invertible.}$$

A crucial fact of the four fundamental subspaces is that the dimension of the column space must be equal to the dimension of the row space, which is easy to establish through Gauss elimination or row reduction procedures, named after Johann Carl Friedrich Gauss (1777–1855). Thus, we define the rank of the matrix $\mathbf{A}$ as r with $\dim(\text{im}(\mathbf{A}^T)) = \dim(\text{im}(\mathbf{A})) = r$. Obviously, we should have $r \leq m$ and $r \leq n$. Moreover, if $r = m$, the matrix $\mathbf{A}$ has full row rank, and the right inverse of the matrix $\mathbf{A}$ is denoted as $\mathbf{A}^T(\mathbf{AA}^T)^{-1}$. Likewise, if $r = n$, the matrix $\mathbf{A}$ has full column rank, and the left inverse of the matrix $\mathbf{A}$ is denoted as $(\mathbf{A}^T\mathbf{A})^{-1}\mathbf{A}^T$.

Consider two sets of base vectors, $\mathbf{e}_i$ and $\hat{\mathbf{e}}_i \in \mathcal{R}^n$, with the corresponding coordinates x_i and $\hat{x}_i$. Thus the same vector $\mathbf{x}$ can also be expressed as $x_i\mathbf{e}_i$ or $\hat{x}_i\mathbf{e}_i$. Since base vector $\hat{\mathbf{e}}_i$ must be a linear combination of base vector $\mathbf{e}_i$, and we can define an $n \times n$ transformation matrix $\mathbf{T}$, called the transformation matrix for the basis, such that $\hat{\mathbf{e}}_i = T_{ij}\mathbf{e}_j$, with $T_{ij} = \hat{\mathbf{e}}_i \cdot \mathbf{e}_j$. Then any vector, including the base vectors, represented in the system $\mathbf{e}_i$ as

$\mathbf{f}$, can be expressed in the system $\hat{\mathbf{e}}_i$ as $\hat{\mathbf{f}}$, with the following associations:

$$\hat{f}_i = T_{ij} f_j \quad \text{or} \quad \hat{\mathbf{f}} = \mathbf{Tf}. \tag{1.1}$$

$\mathbf{T}$ is often called the coordinate transformation with respect to the rotation of axes. It is easy to verify that there exists another $n \times n$ matrix $\mathbf{S}$, named as the transformation matrix for the coordinates, such that the coordinate change from one base vector $\hat{\mathbf{e}}_i$ to another $\mathbf{e}_j$ is defined by $\mathbf{e}_i = S_{ij}\hat{\mathbf{e}}_j$. Thus, $S_{ij} = \mathbf{e}_i \cdot \hat{\mathbf{e}}_j$, namely, $\mathbf{T}^T = \mathbf{S}$, and furthermore, $S_{ik}T_{kj} = \delta_{ij}$, i.e., $\mathbf{ST} = \mathbf{I}$ and $\mathbf{T}^T = \mathbf{T}^{-1}$. Obviously, these invertible matrices themselves can be viewed as linear mappings.

Consider linear mapping $\mathbf{A} : \mathcal{R}^n \to \mathcal{R}^n$, corresponding to one basis $\mathbf{e}_i$, i.e., there is $\mathbf{y} = \mathbf{Ax} \in \mathcal{R}^n, \quad \forall \mathbf{x} \in \mathcal{R}^n$. Suppose the same linear mapping can be expressed as $\mathbf{B} : \mathcal{R}^n \to \mathcal{R}^n$ in another basis $\hat{\mathbf{e}}_i$ and $\hat{\mathbf{e}}_i = T_{ij}\mathbf{e}_j$, using the corresponding dyadic product of two vectors, e.g., $\mathbf{A} = A_{ij}\mathbf{e}_i\mathbf{e}_j$ and $\mathbf{B} = B_{ij}\hat{\mathbf{e}}_i\hat{\mathbf{e}}_j$, we then have

$$\begin{aligned} \mathbf{B} &= \mathbf{TAT}^T = \mathbf{TAT}^{-1}, \\ \mathbf{A} &= \mathbf{SBS}^T = \mathbf{SBS}^{-1}, \end{aligned} \tag{1.2}$$

where $\mathbf{A}$ and $\mathbf{B}$ are called similar matrices.

More specifically, in $\mathcal{R}^3$ we have sets of linearly independent vectors consisting of at most three vectors, and any vector can be expressed as a unique linear combination of a set of three linearly independent vectors, often called a set of base vectors, or basis. In a Cartesian coordinate system, $\mathbf{e}_i$ form a set of base vectors, and are orthonormal, i.e., a triad of mutually orthogonal unit vectors. Of course, in general, a set of base vectors does not have to be unit vectors or orthogonal to each other, and vectors in one basis must be a linear combination of the vectors of the other basis.

1.1.2 *Grad, Div, Curl, and Curvilinear Coordinates*

Consider a scalar function $f(x_1, x_2, x_3)$, abbreviated as $f(\mathbf{x})$, and a vector function $\mathbf{f}(x_1, x_2, x_3)$, abbreviated as $\mathbf{f}(\mathbf{x}) = f_i(\mathbf{x})\mathbf{e}_i$. The gradient of f is defined by

$$\text{grad } f = \nabla f = \frac{\partial f}{\partial x_i}\mathbf{e}_i, \tag{1.3}$$

while the divergence of **f** is defined by

$$\operatorname{div}\mathbf{f} = \boldsymbol{\nabla}\cdot\mathbf{f} = \frac{\partial f_1}{\partial x_1}\mathbf{e}_1 + \frac{\partial f_2}{\partial x_2}\mathbf{e}_2 + \frac{\partial f_3}{\partial x_3}\mathbf{e}_3. \tag{1.4}$$

The direction of the gradient of a scalar function f at some point is normal to the f-constant contour surface at that point and its magnitude is equal to the directional derivative in the normal direction **n**, i.e., $\frac{df}{dn} = \boldsymbol{\nabla} f\cdot\mathbf{n}$. Moreover, the physical significance of the divergence of a vector **f** at some point is the outflow per unit volume at that point. Thus, the Laplacian operator, named after Pierre Simon de Laplace (1749–1827), is given as

$$\Delta f = \nabla^2 f = \boldsymbol{\nabla}\cdot(\boldsymbol{\nabla} f) = \frac{\partial^2 f}{\partial x_1^2} + \frac{\partial^2 f}{\partial x_2^2} + \frac{\partial^2 f}{\partial x_3^2} = \frac{\partial^2 f}{\partial x_i \partial x_i},$$

and consequently, a bi-harmonic (bi-Laplacian) operator can be written as

$$\nabla^4 f = \nabla^2(\nabla^2) f = \frac{\partial^2}{\partial x_i \partial x_i}\frac{\partial^2 f}{\partial x_j \partial x_j}. \tag{1.5}$$

Furthermore, the curl of **f** is defined by

$$\operatorname{curl}\mathbf{f} = \boldsymbol{\nabla}\times\mathbf{f} = \begin{vmatrix} \mathbf{e}_1 & \mathbf{e}_2 & \mathbf{e}_3 \\ \dfrac{\partial}{\partial x_1} & \dfrac{\partial}{\partial x_2} & \dfrac{\partial}{\partial x_3} \\ f_1 & f_2 & f_3 \end{vmatrix} = \varepsilon_{ijk}\frac{\partial f_j}{\partial x_i}\mathbf{e}_k. \tag{1.6}$$

Finally, we have for grad, div, and curl:

$$\begin{aligned}
&\boldsymbol{\nabla}\times(\boldsymbol{\nabla} f) = \mathbf{0}, \quad \boldsymbol{\nabla}\cdot(\boldsymbol{\nabla}\times\mathbf{f}) = 0;\\
&\boldsymbol{\nabla}\times(\boldsymbol{\nabla}\times\mathbf{f}) = \boldsymbol{\nabla}(\boldsymbol{\nabla}\cdot\mathbf{f}) - \nabla^2\mathbf{f}, \quad \boldsymbol{\nabla}(f_1 + f_2) = \boldsymbol{\nabla} f_1 + \boldsymbol{\nabla} f_2;\\
&\boldsymbol{\nabla}\cdot(\mathbf{f}_1 + \mathbf{f}_2) = \boldsymbol{\nabla}\cdot\mathbf{f}_1 + \boldsymbol{\nabla}\cdot\mathbf{f}_2, \quad \boldsymbol{\nabla}\times(\mathbf{f}_1 + \mathbf{f}_2) = \boldsymbol{\nabla}\times\mathbf{f}_1 + \boldsymbol{\nabla}\times\mathbf{f}_2;\\
&\boldsymbol{\nabla}\cdot(a\mathbf{f}) = (\boldsymbol{\nabla} a)\cdot\mathbf{f} + (\boldsymbol{\nabla}\cdot\mathbf{f}), \quad \boldsymbol{\nabla}\times(a\mathbf{f}) = (\boldsymbol{\nabla} a)\times\mathbf{f} + a(\boldsymbol{\nabla}\times\mathbf{f});\\
&\boldsymbol{\nabla}\cdot(\mathbf{f}_1\times\mathbf{f}_2) = \mathbf{f}_2\cdot(\boldsymbol{\nabla}\times\mathbf{f}_1) - \mathbf{f}_1\cdot(\boldsymbol{\nabla}\times\mathbf{f}_2),\\
&\boldsymbol{\nabla}\times(\mathbf{f}_1\times\mathbf{f}_2) = (\mathbf{f}_2\cdot\boldsymbol{\nabla})\mathbf{f}_1 - \mathbf{f}_2(\boldsymbol{\nabla}\cdot\mathbf{f}_1) - (\mathbf{f}_1\cdot\boldsymbol{\nabla})\mathbf{f}_2 + \mathbf{f}_1(\boldsymbol{\nabla}\cdot\mathbf{f}_2),\\
&\boldsymbol{\nabla}(\mathbf{f}_1\cdot\mathbf{f}_2) = (\mathbf{f}_2\cdot\boldsymbol{\nabla})\mathbf{f}_1 + (\mathbf{f}_1\cdot\boldsymbol{\nabla})\mathbf{f}_2 + \mathbf{f}_2\times(\boldsymbol{\nabla}\times\mathbf{f}_1) + \mathbf{f}_1\times(\boldsymbol{\nabla}\times\mathbf{f}_2).
\end{aligned} \tag{1.7}$$

Let us now introduce the curvilinear coordinate systems. We first provide some definitions. Suppose that the rectangular coordinates $x_1, x_2,$

and x_3 are expressed in terms of new coordinates u_1, u_2, and u_3 with $x_i = x_i(u_1, u_2, u_3)$, or in abbreviated form $x_i = x_i(\mathbf{u})$, and that, conversely, these relations can be inverted to express u_1, u_2, and u_3 in terms of x_1, x_2, and x_3, i.e., $u_i = u_i(x_1, x_2, x_3)$, or in abbreviated form $u_i = u_i(\mathbf{x})$, with suitable restrictions. Then, in some regions, any point (x_1, x_2, x_3) corresponds to a point (u_1, u_2, u_3) in the new coordinate system. Moreover, we define the u_1-curve as the curve generated in space with u_2 and u_3 held constant, and u_1-surface as the surface with u_1 held constant. In general, the coordinate curves are not necessarily mutually perpendicular at every point, and such coordinates are called curvilinear coordinates.

One simple way of describing contravariant and covariant vectors is to introduce the following chain rules

$$du_i = \frac{\partial u_i}{\partial x_k} dx_k, \qquad \frac{\partial \phi}{\partial u_i} = \frac{\partial \phi}{\partial x_k} \frac{\partial x_k}{\partial u_i}, \tag{1.8}$$

where du_i and dx_i represent the differential elements of coordinates u_i and x_i and $\frac{\partial \phi}{\partial u_i}$ and $\frac{\partial \phi}{\partial x_i}$ stand for the gradient of an invariant function ϕ with respect to coordinates u_i and x_i.

According to the convention, we call dx_i as the components of a contravariant vector and $\frac{\partial \phi}{\partial x_i}$ as the components of a covariant vector. In general, a set of quantities t^i, associated with a point p, are said to be the components of a contravariant vector, if the corresponding quantities in the transformed coordinate system can be written as

$$\hat{t}^i = t^k \frac{\partial u_i}{\partial x_k}. \tag{1.9}$$

Analogously, a set of quantities t_i, associated with a point p, are said to be the components of a covariant vector, if the corresponding quantities in the transformed coordinate system can be written as

$$\hat{t}_i = t_k \frac{\partial x_k}{\partial u_i}. \tag{1.10}$$

Clearly, for the degenerate case in which the new coordinates u_i are linearly related to the original coordinates x_i, $\frac{\partial x_k}{\partial u_i}$ and $\frac{\partial u_i}{\partial x_k}$ are in fact the transformation matrices. Likewise, a set of quantities T^{ij} is said to be the components of a contravariant tensor of order two, if it transforms according to the following equation:

$$\hat{T}^{ij} = T^{mn} \frac{\partial u_i}{\partial x_m} \frac{\partial u_j}{\partial x_n}.$$

A set of quantities T_{ij} is said to be components of a covariant tensor of order two, if it transforms according to the following equation:

$$\hat{T}_{ij} = T_{mn} \frac{\partial x_m}{\partial u_i} \frac{\partial x_n}{\partial u_j}.$$

In addition, a mixed tensor of order two corresponds to the transformation

$$\hat{T}^i_j = T^m_n \frac{\partial u_i}{\partial x_m} \frac{\partial x_n}{\partial u_j}.$$

Suppose we have a set of contravariant vectors $\mathbf{g}^i$ and a set of covariant vectors $\mathbf{g}_i$ as two sets of base vectors, and in addition,

$$\mathbf{g}_i \cdot \mathbf{g}^j = \delta_i^j, \tag{1.11}$$

with the mixed Kronecker delta $\delta_i^j = \delta_{ij}$, which implies that the contravariant vectors form the dual basis, or reciprocal basis, of the covariant vectors.

Hence, using Eq. (1.11), we can derive

$$f_i = \mathbf{f} \cdot \mathbf{g}_i \quad \text{and} \quad f^i = \mathbf{f} \cdot \mathbf{g}^i,$$

where any vector $\mathbf{f}$ can be expressed as $f^i \mathbf{g}_i$ or $f_i \mathbf{g}^i$.

If we introduce another set of covariant vectors $\hat{\mathbf{g}}_i$ as another basis, analogous to the linear transformation in Eq. (1.1), we can certainly express $\hat{\mathbf{g}}_i$ as a linear combination of $\mathbf{g}_i$, and vice versa:

$$\begin{aligned} \hat{\mathbf{g}}_i &= A_{ij} \mathbf{g}_j, \\ \mathbf{g}_i &= \hat{A}_{ij} \hat{\mathbf{g}}_j. \end{aligned} \tag{1.12}$$

Note that base vectors are not necessarily unit vectors or mutually orthogonal to each other. Therefore, unlike the merely rotation of an orthonormal basis, in general, we have $\hat{\mathbf{A}} = \mathbf{A}^{-1} \neq \mathbf{A}^T$.

Therefore, it can be shown that if we express the vector $\mathbf{f}$ with the basis $\hat{\mathbf{g}}_i$, the corresponding contravariant components must be written as

$$\begin{aligned} \hat{f}^i &= \hat{A}_{ji} f^j, \\ f^i &= A_{ji} \hat{f}^j, \end{aligned} \tag{1.13}$$

which, compared to Eq. (1.12), uses a backward matrix in a forward transformation, or vice versa, and originates the name contravariant for the contravariant components f^i.

Of course, similar coordinate transformations in the form of Eq. (1.12) can be also operated on contravariant bases $\mathbf{g}_i$ and $\hat{\mathbf{g}}_i$. Hence, if we express the vector $\mathbf{f}$ with the basis $\hat{\mathbf{g}}^i$, we can easily obtain

$$\hat{f}_i = A_{ij} f_j,$$
$$f_i = \hat{A}_{ij} \hat{f}_j,$$

which, compared with Eq. (1.12), uses a forward matrix in a forward transformation, or vice versa, and originates the name *covariant.*

Moreover, if we denote $B_{ij} = \mathbf{g}_i \cdot \mathbf{g}_j$ and $B^{ij} = \mathbf{g}^i \cdot \mathbf{g}^j$, we have

$$\begin{aligned} f_i &= B_{ij} f^j, \\ f^i &= B^{ij} f_j, \end{aligned} \tag{1.14}$$

which are used in the process of the so-called raising or lowering indices.

Therefore, without much difficulty, we can show that B_{ij} is the linear transformation from contravariant components or vectors to covariant components or vectors, and B^{ij} is the linear transformation from covariant components or vectors to covariant components or vectors. Thus, the matrices B_{ij} and B^{ij} must be the inverse of each other, i.e., $[B_{ij}] = [B^{ij}]^{-1}$. In fact, B^{ij} and B_{ij} are called the contravariant and covariant components of a unit tensor $\mathbf{I}$, i.e., $I_{ij} = \delta_{ij}$. A typical example is the so-called metric tensor illustrated with the following chain rules:

$$\begin{aligned} dx_k dx_k &= \frac{\partial x_k}{\partial u_i}\frac{\partial x_k}{\partial u_j} du_i du_j = G_{ij} du_i du_j, \\ du_k du_k &= \frac{\partial u_k}{\partial x_i}\frac{\partial u_k}{\partial x_j} dx_i dx_j = G^{ij} dx_i dx_j. \end{aligned} \tag{1.15}$$

From the definitions of tensors, we can associate one vector, through a tensor, with another vector,

$$\mathbf{a} = \mathbf{T}\mathbf{b},$$

and more specifically, analogous to Eq. (1.14), we can have

$$a^i = T^{ij} b_j, \quad a_i = T_{ij} b^j, \quad a_i = T_i^{\ j} b_j.$$

Using the same coordinate transformation in Eq. (1.12), we can easily derive

$$\hat{T}^{ij} = \hat{A}_{mi} T^{mn} \hat{A}_{nj},$$
$$\hat{T}_{ij} = A_{im} T_{mn} A_{jn},$$
$$\hat{T}_i^{\ j} = A_{im} T_m^{\ n} \hat{A}_{nj}.$$

Here, we introduce both superscript and subscript to delineate contravariant and covariant vectors. In practice, we often prefer the subscript and introduce the superscript only if the subscript is adopted for another purpose. In the curvilinear system, the tangential vector to the u_i-curve at a point p is defined as

$$\mathbf{u}_i = \frac{\partial \mathbf{r}}{\partial u_i} = \frac{\partial \mathbf{r}}{\partial s_i}\frac{\partial s_i}{\partial u_i} = h_i \mathbf{e}_u^i, \quad \text{(without summation)}, \tag{1.16}$$

where s_i is the arc length along the u_i-curve, $\mathbf{e}_u^i$ denotes the unit vector $\frac{\partial \mathbf{r}}{\partial s_i}$, and h_i represents the length of $\mathbf{u}_i$.

It is clear from the definition of the unit vector $\mathbf{e}_u^i$ that we have

$$\mathbf{e}_u^i \cdot \mathbf{e}_u^j = \delta_{ij}, \tag{1.17}$$

implying the coordinate curves are mutually orthogonal at every point.

Equation (1.16) can also be converted to the scalar differential form

$$ds_i = h_i du_i, \quad \text{(without summation)}.$$

Therefore, the arc length along any curve can be represented as

$$ds^2 = h_i^2 du_i^2,$$

from which we can easily identify that $h_i^2 = G_{ii}$ as defined in Eq. (1.15).

Since, from the chain rule, we have $\mathbf{u}_i = \frac{\partial \mathbf{r}}{\partial x_k}\frac{\partial x_k}{\partial u_i}$, according to Eq. (1.10), $\mathbf{u}_i$ are covariant vectors and can be written as $\mathbf{g}_i$. Conversely,

$$\mathbf{e}_i = \frac{\partial \mathbf{r}}{\partial x_i} = \frac{\partial \mathbf{r}}{\partial u_k}\frac{\partial u_k}{\partial x_i} = \frac{\partial u_k}{\partial x_i}\mathbf{g}_k, \tag{1.18}$$

and $\mathbf{e}_i$ is in fact $\hat{\mathbf{g}}_i$ defined with the transformation $A_{ij} = \frac{\partial u_i}{\partial x_j}$ and $\hat{A}_{ij} = \frac{\partial x_i}{\partial u_j}$.

In the curvilinear coordinate system, we define the normal vector to the u_i-surface at a point p as

$$\mathbf{n}_i = \frac{\partial u_i}{\partial x_j}\mathbf{e}_j, \tag{1.19}$$

which can be viewed as contravariant vectors and written as $\mathbf{g}^i$.

Then, according to Eq. (1.9), we have,

$$\hat{\mathbf{g}}^i = \mathbf{g}^k \frac{\partial x_i}{\partial u_k} = \frac{\partial u_k}{\partial x_j}\mathbf{e}_j \frac{\partial x_i}{\partial u_k} = \mathbf{e}_i. \tag{1.20}$$

Equation (1.20) in fact suggests that in the Cartesian system there is no distinction between contravariant and covariant vectors, namely, $\mathbf{e}_i = \mathbf{e}^i$.

On the u_1-surface, the differential element of surface area $d\mathbf{s}_1$ is given by the vector product $\mathbf{u}_2 du_2 \times \mathbf{u}_3 du_3$ following the assumed right-hand rule, and hence with $\mathbf{e}_u^1 = \mathbf{e}_u^2 \times \mathbf{e}_u^3$, we obtain

$$d\mathbf{s}_1 = h_2 h_3 du_2 du_3 \mathbf{n}_1.$$

Analogously, we have for the differential element of volume $d\Omega$,

$$d\Omega = (\mathbf{u}_1 du_1 \times \mathbf{u}_2 du_2) \cdot \mathbf{u}_3 du_3 = h_1 h_2 h_3 du_1 du_2 du_3.$$

In general, we refer to curvilinear coordinate systems as orthogonal curvilinear coordinate systems, with typical examples such as the cylindrical coordinate system and the spherical coordinate system. In these cases, the unit vectors corresponding to contravariant and covariant vectors form the identical triad of mutually orthogonal unit vectors. However, we must point out that in the definition of nonlinear strain and stress components, the contravariant and covariant base vectors are no longer orthogonal to each other.

We now proceed to determine the expressions for the gradient, divergence, curl, and the Laplacian operator ∇^2 in the curvilinear coordinate system. To determine the gradient of a scalar function f, we begin with the relation

$$df = \nabla f \cdot d\mathbf{r}. \tag{1.21}$$

In terms of the coordinates u_1, u_2, and u_3, we have

$$df = \frac{\partial f}{\partial u_i} du_i, \quad d\mathbf{r} = \mathbf{u}_i du_i, \quad \nabla f = \lambda_i \mathbf{e}_u^i,$$

where λ_i is expressed as

$$\lambda_i = \frac{1}{h_i} \frac{\partial f}{\partial u_i} \quad \text{(without summation)}.$$

Thus, the gradient of a scalar function f in orthogonal curvilinear coordinate systems can be written as

$$\nabla f = \frac{\mathbf{e}_u^1}{h_1} \frac{\partial f}{\partial u_1} + \frac{\mathbf{e}_u^2}{h_2} \frac{\partial f}{\partial u_2} + \frac{\mathbf{e}_u^3}{h_3} \frac{\partial f}{\partial u_3}. \tag{1.22}$$

Consequently, we obtain

$$\nabla u_i = \frac{\mathbf{e}_u^i}{h_i}, \quad \text{(without summation)},$$

and following Eq. (1.7), we have

$$\nabla \times \frac{\mathbf{e}_u^i}{h_i} = \mathbf{0}, \quad \text{(without summation).} \tag{1.23}$$

Furthermore, according to the mutual orthogonality in Eq. (1.17), we also obtain

$$\frac{\mathbf{e}_u^1}{h_2 h_3} = \frac{\mathbf{e}_u^2}{h_2} \times \frac{\mathbf{e}_u^3}{h_3} = (\nabla u_2) \times (\nabla u_3), \tag{1.24}$$

and following Eq. (1.7), we have

$$\nabla \cdot \frac{\mathbf{e}_u^1}{h_2 h_3} = \nabla \cdot \frac{\mathbf{e}_u^2}{h_3 h_1} = \nabla \cdot \frac{\mathbf{e}_u^3}{h_1 h_2} = 0. \tag{1.25}$$

Therefore, for a vector $\mathbf{f} = f_i \mathbf{e}_u^i$, using Eqs. (1.7), (1.22), and (1.25), we have

$$\nabla \cdot (f_1 \mathbf{e}_u^1) = \frac{\mathbf{e}_u^1}{h_2 h_3} \cdot \nabla (h_2 h_3 f_1) = \frac{1}{h_1 h_2 h_3} \frac{\partial}{\partial u_1} (h_2 h_3 f_1). \tag{1.26}$$

Hence, treating the other terms in a similar fashion, we obtain

$$\nabla \cdot \mathbf{f} = \frac{1}{h_1 h_2 h_3} \left[\frac{\partial}{\partial u_1} (h_2 h_3 f_1) + \frac{\partial}{\partial u_2} (h_3 h_1 f_2) + \frac{\partial}{\partial u_3} (h_1 h_2 f_3) \right]. \tag{1.27}$$

Likewise, combining Eqs. (1.22) and (1.27), the expression ∇^2, or $\nabla \cdot \nabla$ can be written as

$$\nabla^2 = \frac{1}{h_1 h_2 h_3} \left[\frac{\partial}{\partial u_1} \left(\frac{h_2 h_3}{h_1} \frac{\partial}{\partial u_1} \right) + \frac{\partial}{\partial u_2} \left(\frac{h_3 h_1}{h_2} \frac{\partial}{\partial u_2} \right) + \frac{\partial}{\partial u_3} \left(\frac{h_1 h_2}{h_3} \frac{\partial}{\partial u_3} \right) \right]. \tag{1.28}$$

Furthermore, using Eqs. (1.7), (1.22), and (1.23), we also obtain

$$\nabla \times (f_1 \mathbf{e}_u^1) = \nabla (h_1 f_1) \times \frac{\mathbf{e}_u^1}{h_1} = -\frac{\mathbf{e}_u^3}{h_1 h_2} \frac{\partial}{\partial u_2} (h_1 f_1) + \frac{\mathbf{e}_u^2}{h_3 h_1} \frac{\partial}{\partial u_3} (h_1 f_1). \tag{1.29}$$

Again, treating the other terms in a similar fashion, we have

$$\nabla \times \mathbf{f} = \frac{1}{h_1 h_2 h_3} \begin{vmatrix} h_1 \mathbf{e}_u^1 & h_2 \mathbf{e}_u^2 & h_3 \mathbf{e}_u^3 \\ \dfrac{\partial}{\partial u_1} & \dfrac{\partial}{\partial u_2} & \dfrac{\partial}{\partial u_3} \\ h_1 f_1 & h_2 f_2 & h_3 f_3 \end{vmatrix}. \tag{1.30}$$

1.1.3 *Linear spaces and complex variables*

Complex analysis was mainly founded in the 19th century by Augustin-Louis Cauchy (1789–1857), Bernhard Riemann (1826–1866), Karl Weierstrass (1815–1897), and Johann Carl Friedrich Gauss (1777–1855). The concepts of complex variables are important in a number of areas, such as linear algebra, Fourier transform, Laplace transform, and contour integral. In this section, we briefly introduce some fundamentals. The concepts of contour integral and conformal mapping along with Laurent series have been summarized in Wang (2022) along with their respective applications.

A complex variable z is denoted as $z = x + iy$, where $i = \sqrt{-1}$ is the imaginary unit, and x and y are real numbers, denoted as the real part of z, i.e., $x = \mathrm{Re}(z)$, and the imaginary part of z, i.e., $y = \mathrm{Im}(z)$, respectively. Geometrically, each complex number z corresponds to the point (x, y) in a Cartesian plane, a so-called complex plane, with x and y as abscissa and ordinate, respectively. Therefore, the distance between two complex numbers z_1 and z_2 is defined as the distance between two points (x_1, y_1) and (x_2, y_2) in the complex plane, i.e.,

$$|z_1 - z_2| = \sqrt{(x_1 - x_2)^2 + (y_1 - y_2)^2}. \tag{1.31}$$

Naturally, the absolute value (modulus) of a complex number z is defined as

$$|z| = \sqrt{x^2 + y^2}. \tag{1.32}$$

The complex conjugate of a complex number corresponds to a complex number with the same real part and the imaginary part with an opposite sign, denoted as $\bar{z} = x - iy$. Furthermore, two complex numbers are the same if and only if both imaginary and real parts are identical. Thus, the complex number is zero if and only if the corresponding real and imaginary parts are equal. The complex number z can also be written in the polar coordinate frame as

$$z = r\cos\theta + ir\sin\theta, \tag{1.33}$$

with $x = r\cos\theta$ and $y = r\sin\theta$, where r is called the modulus of z (amplitude), and θ is the argument of z (polar angle), denoted as $\arg z$.

It is obvious that θ has infinitely many values with its 2π periodicity, and it can be expressed as

$$\theta = \theta_p + 2n\pi, \quad n \in \mathcal{Z}, \tag{1.34}$$

where θ_p is the principal value of θ with $r\cos\theta_p = x$, $r\sin\theta_p = y$, and $r = \sqrt{x^2+y^2}$; and of course, any interval of length 2π, e.g., $[0,2\pi)$ or $[-\pi,\pi)$ can be used for θ_p.

The simplest function of a complex variable is the integral power function $f(z) = z^n$, with $n \in \mathcal{N}$. Analogous to the power series of a real variable, a complex variable function can be expressed as a complex power series about $z = 0$ or $z = z_o$

$$f(z) = \sum_{n=0}^{N} a_n z^n \quad \text{or} \quad f(z) = \sum_{n=0}^{N} a_n (z - z_o)^n, \tag{1.35}$$

with $a_n \in \mathcal{C}$ and $n \in \mathcal{N}$, and if $|z - z_o| \lim_{n\to+\infty} |\frac{a_{n+1}}{a_n}| < 1$ with $z_o \in \mathcal{C}$, the infinite series $\sum_{i=0}^{+\infty} a_n (z - z_o)^n$ converges.

Hence, by adopting the work of Karl Weierstrass, the exponential function of a complex variable $f(z) = e^z$ can be expressed as

$$e^z = \sum_{n=0}^{+\infty} \frac{z^n}{n!}. \tag{1.36}$$

As a consequence, $e^{z_1+z_2} = e^{z_1} e^{z_2}$ and if $n \in \mathcal{N}$, $(e^z)^n = e^{nz}$. Furthermore, similar to the trigonometric function of a real variable, we define trigonometric (circular) functions of a complex variable as

$$\begin{aligned} \sin z &= \frac{e^{iz} - e^{-iz}}{2i} = \sum_{n=0}^{+\infty} (-1)^n \frac{z^{2n+1}}{(2n+1)!}, \\ \cos z &= \frac{e^{iz} + e^{-iz}}{2} = \sum_{n=0}^{+\infty} (-1)^n \frac{z^{2n}}{(2n)!}. \end{aligned} \tag{1.37}$$

We then have Euler's formula, named after Leonhard Euler (1707–1783):

$$e^{iz} = \cos z + i\sin z \quad \text{and} \quad z = r(\cos\theta + i\sin\theta) = re^{i\theta}. \tag{1.38}$$

Furthermore, from Eq. (1.38), we can deduce de Moivre's theorem, named after Abraham de Moivre (1667–1754):

$$(\cos\theta + i\sin\theta)^n = \cos n\theta + i\sin n\theta \quad \text{and} \quad z^n = r^n(\cos n\theta + i\sin n\theta). \tag{1.39}$$

Analogous to Eq. (1.37), hyperbolic functions of a complex variable are defined as

$$\begin{aligned}\sinh z &= \frac{e^z - e^{-z}}{2} = \sum_{n=0}^{+\infty} \frac{z^{2n+1}}{(2n+1)!},\\ \cosh z &= \frac{e^z + e^{-z}}{2} = \sum_{n=0}^{+\infty} \frac{z^{2n}}{(2n)!}.\end{aligned} \tag{1.40}$$

By comparing Eq. (1.40) with Eq. (1.37), we can relate hyperbolic functions with trigonometric functions:

$$\begin{aligned}\sinh iz &= i\sin z, \quad \sin iz = i\sinh z,\\ \cosh iz &= \cos z, \quad \cos iz = \cosh z;\end{aligned} \tag{1.41}$$

and eventually derive the following formulas for trigonometric and hyperbolic functions of a complex variable z:

$$\begin{aligned}e^z &= e^{x+iy} = e^x \cos y + ie^x \sin y,\\ \sin z &= \sin(x+iy) = \sin x \cosh y + i\cos x \sinh y,\\ \cos z &= \cos(x+iy) = \cos x \cosh y - i\sin x \sinh y,\\ \sinh z &= \sinh(x+iy) = \sinh x \cos y + i\cosh x \sin y,\\ \cosh z &= \cosh(x+iy) = \cosh x \cos y + i\sinh x \sin y.\end{aligned} \tag{1.42}$$

Theorem 1.2 (Fundamental Theorem of Algebra). *The algebraic equation*

$$a_0 + a_1 x + a_2 x^2 + a_3 x^3 + \cdots + a_n x^n = 0, \quad a_n \neq 0, \tag{1.43}$$

with complex coefficients a_i, $n \in \bar{\mathcal{N}}$, *and* $n > 1$, *always has a complex number as a solution.*

Although Theorem 1.2 indicates that the set of complex numbers is algebraically closed, we must note that for other purposes, mathematicians have developed systems of hypercomplex, such as quaternions and Cayley numbers.

1.1.4 *Eigenvalues and eigenvectors*

One of the most important aspects in linear algebra is the study of eigenvalues and eigenvectors, which is directly related with dynamics, stability analysis, and numerical methods. Let $\mathbf{A}$ be an $n \times n$ matrix. Denote $\mathbf{A}_{ij}$ as the $(n-1) \times (n-1)$ matrix obtained by deleting the ith row and jth column. The determinant of the matrix $\mathbf{A}_{ij}$ is called a minor and is denoted as $\det(\mathbf{A}_{ij})$, or $|\mathbf{A}_{ij}|$, and if the determinant of a 1×1 matrix $[a]$ is defined as a, the determinant of the $n \times n$ matrix is then defined inductively as $\det(\mathbf{A}) = |\mathbf{A}| = \sum_{j=1}^{n}(-1)^{i+j}a_{ij}\det(\mathbf{A}_{ij})$, where i is a fixed integer with $1 \leq i \leq n$. The basic properties of the determinant function are as follows:

$$\begin{aligned} &\det(\mathbf{I}) = 1, \quad \det(\mathbf{AB}) = \det(\mathbf{A})\det(\mathbf{B}), \\ &\det(\mathbf{A}) \neq 0 \Leftrightarrow \mathbf{A} \text{ invertible } \Leftrightarrow \ker(\mathbf{A}) = \{\mathbf{0}\}. \end{aligned} \tag{1.44}$$

Furthermore, if the kernel of the operator $\mathbf{A} - \alpha\mathbf{I} : \mathcal{S} \subseteq \mathcal{R}^n \to \mathcal{S}$ is nontrivial, it is called the α-eigenspace of $\mathbf{A}$, which consists of all eigenvectors belonging to α, the so-called eigenvalue of $\mathbf{A}$. The solution of the eigenvalue is equivalent to the solution of

$$\det(\mathbf{A} - \alpha\mathbf{I}) = 0, \tag{1.45}$$

the result of which is a polynomial $p(\alpha)$, often called the characteristic polynomial of $\mathbf{A}$.

Thus, let $\lambda_1, \ldots, \lambda_r$ be the distinct roots of $p(\lambda)$ and the integer $n_k \geq 1$, with $k = 1, \ldots, r$, stands for the multiplicity of λ_k with $\sum_{k=1}^{r} n_k = \dim(\mathcal{S})$. We can express the characteristic polynomial as

$$p(\alpha) = \prod_{k=1}^{r}(\alpha - \lambda_k)^{n_k}. \tag{1.46}$$

Similar to the eigenspace

$$\ker(\mathbf{A} - \lambda_k\mathbf{I}) \subset \mathcal{S}, \tag{1.47}$$

the generalized eigenspace of $\mathbf{A}$ associated with λ_k is a subspace of $\mathcal{S}$, denoted as

$$\ker(\mathbf{A} - \lambda_k\mathbf{I})^{n_k} \subset \mathcal{S}. \tag{1.48}$$

If we change the basis of the vector space from $\mathbf{e}_i$ to $\hat{\mathbf{e}}_i$, with $\hat{\mathbf{e}}_i = \mathbf{T}\mathbf{e}_i$, the linear operator $\mathbf{A}$ is transformed as $\mathbf{B} = (\mathbf{T}^T)^{-1}\mathbf{A}(\mathbf{T}^T)$ and consequently, the characteristic polynomial of $\mathbf{B}$ can be written as

$$\det(\mathbf{B} - \alpha\mathbf{I}) = \det(\mathbf{T}^T)^{-1}\det(\mathbf{A} - \alpha\mathbf{I})\det(\mathbf{T}^T) = \det(\mathbf{A} - \alpha\mathbf{I}) = 0, \tag{1.49}$$

which means that the characteristic polynomial $p(\alpha)$ is independent of the basis, i.e., invariant.

Another invariant is the trace of a matrix $\mathbf{A} = [A_{ij}]$, denoted as $\text{tr}(\mathbf{A}) = \sum_{i=1}^{n} A_{ii}$, which has the following properties

$$\text{tr}(\mathbf{AB}) = \text{tr}(\mathbf{A})\text{tr}(\mathbf{B}) \quad \text{and} \quad \text{tr}(\mathbf{A}+\mathbf{B}) = \text{tr}(\mathbf{A}) + \text{tr}(\mathbf{B}). \tag{1.50}$$

Moreover, supposing the vector space $\mathcal{S}$ has a set of base vectors $\hat{\mathbf{e}}_i$ as the eigenvectors of $\mathbf{A}$, then $\mathbf{A}$ is considered diagonalizable and the corresponding eigenvalues are called simple, i.e., there exists an invertible matrix $\mathbf{S}$ such that

$$\mathbf{SAS}^{-1} = \text{diag}\{\lambda_1, \ldots, \lambda_n\} = \begin{bmatrix} \lambda_1 & 0 & \ldots & 0 \\ 0 & \lambda_2 & \ldots & 0 \\ \vdots & \vdots & \ddots & \vdots \\ 0 & 0 & \ldots & \lambda_n \end{bmatrix}. \tag{1.51}$$

So far, we have discussed linear operators in real spaces. However, the elementary properties can be directly extended to complex spaces, which consist of complex elements. This is of particular importance when dealing with FSI systems. Just like the real Cartesian space, the complex Cartesian space $\mathcal{C}^n$ has a set of n-tuples complex number, denoted as $(z_1, z_2, \ldots, z_n)$. In particular, a linear operator $\mathbf{A}$ in a complex space is diagonalizable if the characteristic polynomial has distinct roots. To proceed further, we have to introduce the concept of the direct sum of subspaces. Consider a set of subspaces $\mathcal{S}_1, \mathcal{S}_2, \ldots, \mathcal{S}_n \subset \mathcal{S}$. If every vector $\mathbf{x} \in \mathcal{S}$ can be expressed uniquely with the vectors $\mathbf{x}_i \in \mathcal{S}_i$ as $[\mathbf{x}_1^T \quad \ldots \quad \mathbf{x}_n^T]^T$, then the vector space $\mathcal{S}$ is called a direct sum of $\mathcal{S}_i$ and denoted as

$$\mathcal{S} = \mathcal{S}_1 \oplus \cdots \oplus \mathcal{S}_n = \bigoplus_{i=1}^{n} \mathcal{S}_i. \tag{1.52}$$

Moreover, if each $\mathcal{S}_i$ is invariant under a linear operator $\mathbf{A}_i$, i.e., $\mathbf{A}_i : \mathcal{S}_i \to \mathcal{S}_i$, then the linear operator $\mathbf{A} : \mathcal{S} \to \mathcal{S}$ is the direct sum of $\mathbf{A}_i$, denoted as

$$\mathbf{A} = \mathbf{A}_1 \oplus \cdots \oplus \mathbf{A}_n = \bigoplus_{i=1}^{n} \mathbf{A}_i, \tag{1.53}$$

where each $\mathcal{S}_i$ is invariant under $\mathbf{A}$, i.e., $\mathbf{A}(\mathcal{S}_i) \subset \mathcal{S}_i$ or $\mathbf{Ax} = \mathbf{A}_i\mathbf{x}_i$ with $\mathbf{x} = [\mathbf{0}^T \quad \ldots \quad \mathbf{x}_i^T \quad \ldots \quad \mathbf{0}^T]^T$, denoted as $\mathbf{x} \in \mathcal{S}_i$.

Obviously, if the basis of $\mathbf{A}$ is the union of the basis elements of the $\mathcal{S}_i$, we have

$$\mathbf{A} = \text{diag}\{\mathbf{A}_1, \ldots, \mathbf{A}_n\} = \begin{bmatrix} \mathbf{A}_1 & \mathbf{0} & \ldots & \mathbf{0} \\ \mathbf{0} & \mathbf{A}_2 & \ldots & \mathbf{0} \\ \vdots & \vdots & \ddots & \vdots \\ \mathbf{0} & \mathbf{0} & \ldots & \mathbf{A}_n \end{bmatrix}. \tag{1.54}$$

It can be also verified that

$$\det(\mathbf{A}) = \det(\mathbf{A}_1) \cdot \cdots \cdot \det(\mathbf{A}_n) = \prod_{i=1}^{n} \det(\mathbf{A}_i). \tag{1.55}$$

Theorem 1.3 (Eigenspace of Real Matrix). *If $\mathbf{A} : \mathcal{S} \to \mathcal{S}$ is a real operator, then the set of its eigenvalues is preserved under complex conjugation, i.e., if $\zeta = \xi + i\eta$ is an eigenvalue, so is its complex conjugate $\bar{\zeta} = \xi - i\eta$. Therefore, we can write the eigenvalues of $\mathbf{A}$ as: (r all real eigenvalues) $\lambda_1, \ldots, \lambda_r$ and ($2s$ all complex eigenvalues) $\zeta_1, \bar{\zeta}_1, \ldots, \zeta_s, \bar{\zeta}_s$. Moreover, supposing $\mathbf{A}$ has distinct eigenvalues, then both $\mathcal{S}$ and $\mathbf{A}$ have a direct sum decomposition:*

$$\mathcal{S} = \mathcal{S}_1 \oplus \mathcal{S}_2 \quad \textit{and} \quad \mathbf{A} = \mathbf{A}_1 \oplus \mathbf{A}_2, \tag{1.56}$$

with

$$\mathbf{A}_1 : \mathcal{S}_1 \to \mathcal{S}_1 \quad \textit{and} \quad \mathbf{A}_2 : \mathcal{S}_2 \to \mathcal{S}_2, \tag{1.57}$$

where $\mathbf{A}_1$ has real eigenvalues and $\mathbf{A}_2$ has complex eigenvalues.

Consider a linear mapping $\mathbf{A} : \mathcal{S} \to \mathcal{S}$ be an operator on a real vector space with distinct complex eigenvalues ζ_1, $\bar{\zeta}_1$, ..., ζ_s, $\bar{\zeta}_s$. Then, as a direct consequence of Theorem 1.3, there is an invariant direct sum decomposition for $\mathcal{S}$ and a corresponding direct sum decomposition for $\mathcal{A}$, such that

$$\mathcal{S} = \bigoplus_{i=1}^{r} \mathcal{S}_i \quad \text{and} \quad \mathbf{A} = \bigoplus_{i=1}^{r} \mathbf{A}_i, \tag{1.58}$$

where $\mathcal{S}_i$ and $\mathbf{A}_i$ are two-dimensional and $\mathbf{A}_i$ has eigenvalues ζ_i and $\bar{\zeta}_i$.

Therefore, consider a linear operator $\mathbf{A}$ on a two-dimensional vector space $\mathcal{S} \subset \mathcal{R}^2$ with nonreal eigenvalues ζ and $\bar{\zeta}$, i.e., $\zeta = \xi + i\eta$ and $\eta \neq 0$.

It can be shown that we have two eigenvectors $\varphi = \mathbf{\Phi} + i\psi$ and $\bar{\varphi} = \mathbf{\Phi} - i\psi$ corresponding to ζ and $\bar{\zeta}$, respectively. Hence,

$$\begin{aligned} \mathbf{A}\varphi &= \mathbf{A}(\mathbf{\Phi} + i\psi) = \zeta\varphi = (\xi + i\eta)(\mathbf{\Phi} + i\psi), \\ \mathbf{A}\bar{\varphi} &= \mathbf{A}(\mathbf{\Phi} - i\psi) = \bar{\zeta}\bar{\varphi} = (\xi - i\eta)(\mathbf{\Phi} - i\psi), \end{aligned} \tag{1.59}$$

we have

$$\begin{aligned} \mathbf{A}\mathbf{\Phi} &= \xi\mathbf{\Phi} - \eta\psi, \\ \mathbf{A}\psi &= \eta\mathbf{\Phi} + \xi\psi. \end{aligned} \tag{1.60}$$

Note that for complex vectors and matrices, the transpose or the Hermitian, named after Charles Hermite (1822–1901), corresponds to a transpose for real vectors and matrices along with a complex conjugate operation, denoted as $(\bar{\mathbf{A}})^T$ or $\mathbf{A}^H$, for a typical matrix $\mathbf{A}$. If $\mathbf{\Phi}$ and ψ are normalized such that $\mathbf{\Phi}^T\mathbf{\Phi} = \psi^T\psi = 1$ and $\mathbf{\Phi}^T\psi = 0$, we then have

$$\begin{aligned} \bar{\varphi}^T\mathbf{A}\varphi &= \mathbf{\Phi}^T\mathbf{A}\mathbf{\Phi} + \psi^T\mathbf{A}\psi - i\psi^T\mathbf{A}\mathbf{\Phi} + i\mathbf{\Phi}^T\mathbf{A}\psi = 2\xi + 2i\eta, \\ \varphi^T\mathbf{A}\varphi &= \mathbf{\Phi}^T\mathbf{A}\mathbf{\Phi} - \psi^T\mathbf{A}\psi + i\psi^T\mathbf{A}\mathbf{\Phi} + i\mathbf{\Phi}^T\mathbf{A}\psi = 0. \end{aligned} \tag{1.61}$$

From Eq. (1.61), we can easily derive

$$\begin{aligned} &\mathbf{\Phi}^T\mathbf{A}\mathbf{\Phi} = \xi, \quad \psi^T\mathbf{A}\psi = \xi, \\ &\psi^T\mathbf{A}\mathbf{\Phi} = -\eta, \quad \mathbf{\Phi}^T\mathbf{A}\psi = \eta. \end{aligned} \tag{1.62}$$

It is obvious that $\mathbf{\Phi}$ and ψ are independent vectors, and if we assign $\mathbf{S}^{-1} = [\psi \quad \mathbf{\Phi}]$, we have

$$\mathbf{B} = \mathbf{S}\mathbf{A}\mathbf{S}^{-1} = \begin{bmatrix} \xi & -\eta \\ \eta & \xi \end{bmatrix}. \tag{1.63}$$

Compared with the linear transformation (1.2), the matrix $\mathbf{T} = (\mathbf{S}^{-1})^T$ corresponds to the transformation matrix from the standard basis $\mathbf{e}_1$ and $\mathbf{e}_2$ to the basis ψ and $\mathbf{\Phi}$. Equation (1.63) demonstrates why the operator on a complex space having distinct complex eigenvalues is called semi-simple, while the operator on a real space having distinct real eigenvalues is called simple.

As a natural extension of Taylor's expansion for scalar functions, e.g., Eq. (1.36), exponentials of operators can also be expanded in the same form, i.e.,

$$e^{\mathbf{A}} = \sum_{k=0}^{+\infty} \frac{\mathbf{A}^k}{k!}. \tag{1.64}$$

Moreover, if $\mathbf{AB} = \mathbf{BA}$, we have $e^{\mathbf{A}+\mathbf{B}} = e^{\mathbf{A}}e^{\mathbf{B}}$; if $\mathbf{A} = \begin{bmatrix} a & -b \\ b & a \end{bmatrix}$, we have $e^{\mathbf{A}} = e^{a}\begin{bmatrix} \cos b & -\sin b \\ \sin b & \cos b \end{bmatrix}$; and if $\mathbf{B} = \mathbf{SAS}^{-1}$, then $e^{\mathbf{B}} = \mathbf{S}e^{\mathbf{A}}\mathbf{S}^{-1}$.

Theorem 1.4 (Decomposition). *Let* $\mathbf{A}$ *be an operator on* $\mathcal{S}$, *where* $\mathcal{S}$ *is a complex vector space, or else* $\mathcal{S}$ *is real and* $\mathbf{A}$ *has real eigenvalues. Then* $\mathcal{S}$ *is the direct sum of the generalized eigenspaces of* $\mathbf{A}$. *The dimension of each generalized eigenspace equals the multiplicity of the corresponding eigenvalue.* $\mathbf{A}$ *can be uniquely decomposed into* $\mathbf{N}$ *and* $\mathbf{D}$, *where* $\mathbf{N}$ *is a so-called nilpotent matrix, i.e., there exists a natural number m such that* $\mathbf{N}^m = \mathbf{0}$, *and* $\mathbf{D}$ *is diagonalizable.*

Suppose for an $n_i \times n_i$ matrix $\mathbf{A}_i : \mathcal{S}_i \to \mathcal{S}_i$, there is only one eigenvalue λ_i with a multiplicity of $n_i = \dim(\mathcal{S}_i)$. Based on Theorem 1.4, if we consider $\mathbf{A}_i = \mathbf{N}_i + \mathbf{D}_i$, with $\mathbf{N}_i = \mathbf{A}_i - \lambda_i \mathbf{I}_i$ and $\mathbf{D}_i = \lambda_i \mathbf{I}_i$, where $\mathbf{I}_i$ is an $n_i \times n_i$ unity matrix, since $\mathbf{N}_i\mathbf{D}_i = \mathbf{D}_i\mathbf{N}_i$, we have $e^{\mathbf{A}_i} = e^{\mathbf{N}_i}e^{\mathbf{D}_i}$. Furthermore, based on the fact that $\dim(\mathbf{N}_i^n) = \dim(\mathcal{S}_i)$, i.e., $\ker(\mathbf{N}_i^n) = \mathcal{S}_i$, we know $\mathbf{N}_i$ is nilpotent and $\mathbf{N}_i^{n_i} = \mathbf{0}$, hence,

$$e^{\mathbf{A}_i} = e^{\lambda_i} \sum_{k=0}^{n_i - 1} \frac{\mathbf{N}_i^k}{k!}.$$

This in fact changes the infinite series of the exponential operator $e^{\mathbf{A}}$ to a finite series.

More importantly, if we decompose an $n \times n$ matrix $\mathbf{A}$ into $\mathbf{A} = \mathbf{A}_1 \oplus \cdots \oplus \mathbf{A}_r$, where each matrix $\mathbf{A}_i$ has only one eigenvalue with multiplicity $n_i = \dim(\mathbf{A}_i) \geq 1$ and

$$\sum_{i=1}^{r} n_i = \dim(\mathcal{S}).$$

Apply the previous results, with $\mathbf{D}_i = \lambda_i \mathbf{I}_i$ and $\mathbf{N}_i = \mathbf{A}_i - \mathbf{D}_i$, $\mathbf{N}_i$ is a nilpotent matrix of order n_i, and if we assign

$$\mathbf{D} = \mathbf{D}_1 \oplus \cdots \oplus \mathbf{D}_r \quad \text{and} \quad \mathbf{N} = \mathbf{N}_1 \oplus \cdots \oplus \mathbf{N}_r, \tag{1.65}$$

it is obvious that $\mathbf{D}$ is diagonalizable, $\mathbf{N}$ is nilpotent of $m = \max(n_1, \ldots, n_r)$, and $\mathbf{DN} = \mathbf{ND}$.

Theorem 1.5 (Cayley–Hamilton). *Consider* $\mathbf{A}$ *be an operator on* $\mathcal{S}$ *(real or complex). If the characteristic polynomial is denoted as*

$$p(\alpha) = \sum_{k=0}^{n} a_k \alpha^k,$$

then $p(\mathbf{A}) = \mathbf{0}$, *i.e.,*

$$\sum_{k=0}^{n} a_k \mathbf{A}^k \mathbf{x} = \mathbf{0}, \quad \forall \mathbf{x} \in \mathcal{S}. \tag{1.66}$$

This theorem is named after Arthur Cayley (1821–1895) and William R. Hamilton (1805–1865).

Introducing a matrix

$$\mathbf{N} = \begin{bmatrix} 0 & 1 & 0 & \dots & 0 & 0 \\ 0 & 0 & 1 & \dots & 0 & 0 \\ \vdots & \vdots & \ddots & \ddots & \vdots & \vdots \\ 0 & 0 & 0 & \dots & 0 & 1 \\ 0 & 0 & 0 & \dots & 0 & 0 \end{bmatrix}, \tag{1.67}$$

since $\mathbf{N}\mathbf{e}_1 = \mathbf{0}$ and $\mathbf{N}\mathbf{e}_i = \mathbf{e}_{i-1}$ with $2 \leq i \leq n$, we have $\mathbf{N}^n \mathbf{e}_k = 0$ for $1 \leq k \leq n$, i.e., $\mathbf{N}^n = \mathbf{0}$.

Thus, $\mathbf{N}$ is nilpotent of order n and is called the elementary nilpotent matrix with $\dim(\ker(\mathbf{N})) = 1$ and $\dim(\mathbf{N}) = n-1$. In addition, we also call the matrix $\lambda\mathbf{I} + \mathbf{N}$ the elementary Jordan matrix, denoted as

$$\mathbf{J} = \begin{bmatrix} \lambda & 1 & 0 & \dots & 0 & 0 \\ 0 & \lambda & 1 & \dots & 0 & 0 \\ \vdots & \vdots & \ddots & \ddots & \vdots & \vdots \\ 0 & 0 & 0 & \dots & \lambda & 1 \\ 0 & 0 & 0 & \dots & 0 & \lambda \end{bmatrix}. \tag{1.68}$$

Note that in this book, we use the upper nilpotent matrix and upper Jordan matrix. The properties are the same for lower nilpotent matrix and lower Jordan matrix and they are similar matrices.

Theorem 1.6 (Jordan Form). *Let* $\mathbf{A}$ *be an operator on* $\mathcal{S}$, *where* $\mathcal{S}$ *is a complex vector space, or else* $\mathcal{S}$ *is real and* $\mathbf{A}$ *has real eigenvalues.*

Then $\mathbf{A}$ *can be transposed into* $\mathbf{N}+\lambda\mathbf{I}$ *with* $\mathbf{N}$ *nilpotent. Moreover,* $\mathbf{N}$ *can be uniquely decomposed into* $\mathbf{N}=\text{diag}\{\mathbf{N}_1,\ldots,\mathbf{N}_r\}$*, where* $\mathbf{N}_i$ *is the elementary nilpotent matrix whose size is a non-increasing function of* i *with* $1\leq i\leq r$*, with* $\sum_{i=1}^{r} n_i = dim(\mathbf{A})$ *and* $r = dim(ker(\mathbf{A}))$.

This theorem is named after Marie Ennemond Camille Jordan (1838–1922).

Finally, consider a linear operator $\mathbf{A}$, which can be decomposed into a diagonal matrix and a nilpotent matrix, i.e., $\mathbf{A}=\lambda\mathbf{I}+\mathbf{N}$, the corresponding exponential operator can be expressed as

$$e^{t\mathbf{A}} = e^{t\lambda}\sum_{k=0}^{n-1}\frac{t^k\mathbf{N}^k}{k!} = e^{t\lambda}\begin{bmatrix} 1 & t & \frac{t^2}{2!} & \cdots & \frac{t^{n-2}}{(n-2)!} & \frac{t^{n-1}}{(n-1)!} \\ 0 & 1 & t & \cdots & \frac{t^{n-3}}{(n-3)!} & \frac{t^{n-2}}{(n-2)!} \\ \vdots & \vdots & \ddots & \ddots & \vdots & \vdots \\ 0 & 0 & 0 & \cdots & 1 & t \\ 0 & 0 & 0 & \cdots & 0 & 1 \end{bmatrix}. \tag{1.69}$$

1.1.5 *Singular value decomposition*

Consider an $m\times m$ matrix $\mathbf{A}$. Because of the definition of the determinant, the characteristic polynomial of the matrix $\mathbf{A}$ is the same as that of its transpose. Suppose the eigenvector of the matrix $\mathbf{A}$ corresponding to the eigenvalue λ_i is $\mathbf{u}_i$. Since λ_i must be one of the eigenvalues of the matrix $\mathbf{A}^T$ and the corresponding eigenvector is $\mathbf{v}_i$. In practice, we often call $\mathbf{u}_i$ the right eigenvector and $\mathbf{v}_i$ the left eigenvector. It is clear that the right eigenvectors span the row space whereas the left eigenvectors make the column space. Now, supposing there exist two distinct eigenvalues λ_i and λ_j, we have

$$\mathbf{v}_j^T\mathbf{A}\mathbf{u}_i = \lambda_i\mathbf{v}_j^T\mathbf{u}_i \quad \text{and} \quad \mathbf{u}_i^T\mathbf{A}^T\mathbf{v}_j = \lambda_j\mathbf{u}_i^T\mathbf{v}_j, \tag{1.70}$$

which implies that $\mathbf{u}_i$ is orthogonal to $\mathbf{v}_j$ if $\lambda_i\neq\lambda_j$.

Consider an $m\times n$ matrix $\mathbf{A}$. We can easily construct two symmetric matrices $\mathbf{A}^T\mathbf{A}$ and $\mathbf{A}\mathbf{A}^T$. Suppose the rank of the matrix $\mathbf{A}$ is r, there will be r nonzero eigenvalues of these two symmetric matrices. In fact, these r

eigenvalues are positive. However, although it can be shown that the positive eigenvalues of the $n \times n$ matrix $\mathbf{A}^T\mathbf{A}$ are the same as those of the $m \times m$ matrix $\mathbf{AA}^T$, the number of zero eigenvalues will be different. For the symmetric matrix $\mathbf{A}^T\mathbf{A}$, the dimension of the null space will be $n-r$, which means there are $n-r$ null space vectors. It can be easily shown that these $n-r$ null space vectors of the matrix $\mathbf{A}^T\mathbf{A}$ are also the null space vectors of the matrix $\mathbf{A}$. Likewise, there will be $m-r$ null space vectors of the matrix $\mathbf{AA}^T$ and they are the same as the left null space vector of the matrix $\mathbf{A}$.

Consider the four fundamental spaces of the matrix. It is easy to confirm that the orthogonal subspace of the null space must be the row space while the orthogonal subspace of the left null space is the column space. As a consequence, the eigenvectors of the matrix $\mathbf{A}^T\mathbf{A}$ denoted as $\mathbf{v}_i$ span the row space of the matrix $\mathbf{A}$ and the eigenvectors of the matrix $\mathbf{AA}^T$ represented with $\mathbf{u}_i$ construct the column space of the matrix $\mathbf{A}$. In fact, $\mathbf{u}_i$ and $\mathbf{v}_i$ are similar to the right and left eigenvectors of the square matrix.

Denote a matrix $\mathbf{U}$, the first r columns of which are the eigenvectors of the symmetric matrix $\mathbf{A}^T\mathbf{A}$, and the last $m-r$ columns are constructed with the left null space vectors. In addition, introduce a matrix $\mathbf{V}$, the first r columns of which are the eigenvectors of the symmetric matrix $\mathbf{AA}^T$, and the last $n-r$ columns are made of the null space vectors. The singular value decomposition (SVD) of the matrix $\mathbf{A}$ is then depicted as

$$\mathbf{A} = \mathbf{U}\Sigma\mathbf{V}^T, \tag{1.71}$$

where the first r diagonal entities of the $m \times n$ matrix Σ are the square root of the nonzero eigenvalues of the matrix $\mathbf{A}^T\mathbf{A}$ or $\mathbf{AA}^T$ or the singular values of the matrix $\mathbf{A}$.

This SVD could be considered the true makeup of any matrix $\mathbf{A}$ and specifically illustrates the four fundamental subspaces and the eigenspaces of the row and column spaces. This decomposition will be abundantly used in the model reduction procedures as well as the principle component analysis.

1.2 Kinematics and Conservation Laws

Kinematical descriptions play an important role in the study of FSI systems. In this section, we present the generic kinematic descriptions in

Cartesian, cylindrical, and spherical coordinate systems. Moreover, in order to handle some FSI systems with moving boundaries, we also need to adopt a so-called arbitrary Lagrangian–Eulerian (ALE) for a gradual transition of mesh motions between Eulerian and Lagrangian descriptions which can ensure the mesh regularities within the entire computational domain.

1.2.1 *Mass points, rigid bodies, and continua*

To study the mechanical behavior of objects in nature, we introduce kinematics to mathematically describe the geometrical locations of material particles or mesh points. Kinematical considerations center on the behavior in time of displacements, velocities, and accelerations, or the parametric family of configurations, i.e., mappings and deformations. By introducing a real number system t to represent time, the habitual three-dimensional representation of space and time can be viewed as a four-dimensional continuum.

Two types of kinematical descriptions are commonly used, namely, the Lagrangian description and the Eulerian description. In the Lagrangian description, named after Joseph–Louis Lagrange (1736–1813), state variables are associated with material particles identified with their initial positions. In the Eulerian description, named after Leonhard Euler (1707–1783), state variables are described at spatial locations without regard to specific material particles. Therefore, it is more convenient to use the Eulerian description with velocities as the primitive variables in fluid mechanics, and the Lagrangian description with displacements as the primitive variable in solid mechanics. Furthermore, to represent the motion of solids in computational mechanics, we commonly use the Lagrangian mesh, the motion of which coincides with the material particle motion. Likewise, an Eulerian mesh, fixed in space, is used to represent the motion of fluids. In numerical approaches for FSI problems, if we are to follow the moving boundaries, in order to avoid excessive mesh distortions we often need to arbitrarily select proper mesh motions for continuum media, in which the mesh position for the description of the state variables is neither fixed in space nor attached with material particles.

A point mass is a single mass point whose spatial dimension is negligibly small in comparison with the distances involved in the problem under consideration. A rigid body is an idealization of a continuum medium of which the internal deformation is neglected in comparison to the motion of the body as a whole. Naturally, the dynamical behavior of a point mass is governed by the dynamical motion of that point, often along a trajectory or curve. In this case, the Lagrangian description must be used. Since a rigid body can also be considered as a distribution of mass rigidly fixed to a coordinate frame, rotations of such a rigid frame must also be considered in the description of the dynamical behavior of the rigid body.

Let us start with the parametric representation of a curve,

$$\mathbf{x}(s) = x_i(s)\mathbf{e}_i,$$

with the real parameter s.

Thus, a simple closed curve corresponds to a close curve such that $s_1 \neq s_2$ implies that $\mathbf{x}(s_1) \neq \mathbf{x}(s_2)$. If the parameter s stands for the arc length, i.e., $ds = \sqrt{dx_i dx_i}$, the unit tangent vector can be written as

$$\mathbf{t} = \frac{dx_i(s)}{ds}\mathbf{e}_i. \tag{1.72}$$

Moreover, differentiating $\mathbf{t} \cdot \mathbf{t} = 1$ with respect to s, we have

$$\frac{d\mathbf{t}}{ds} = \kappa\mathbf{n} = \frac{1}{\rho}\mathbf{n}, \tag{1.73}$$

where $\mathbf{n}$ is the unit normal vector, κ is the curvature, and ρ is the radius of curvature.

Denoting the unit vector $\mathbf{m}$ normal to the osculating plane of $\mathbf{t}$ and $\mathbf{n}$ as the so-called binormal vector, we then form a triad of orthogonal unit vectors ($\mathbf{n}$, $\mathbf{t}$, $\mathbf{m}$), as shown in Fig. 1.1. From the right-hand rule we have $\mathbf{m} = \mathbf{n} \times \mathbf{t}$ and $\mathbf{n} = \mathbf{t} \times \mathbf{m}$. Hence, using Eq. (1.7), we can easily verify

$$\frac{d\mathbf{m}}{ds} = \nu\mathbf{n}, \tag{1.74}$$

$$\frac{d\mathbf{n}}{ds} = -\nu\mathbf{m} - \kappa\mathbf{t}, \tag{1.75}$$

where ν is called the torsion of the curve.

Equations (1.73)–(1.75) are often called the Serret–Frenet formula, named after Joseph Alfred Serret (1819–1885) and Fredari–Jean Frenet (1816–1900). Two of the most widely used orthogonal curvilinear coordinate systems are cylindrical and spherical coordinate systems, as shown in Fig. 1.2.

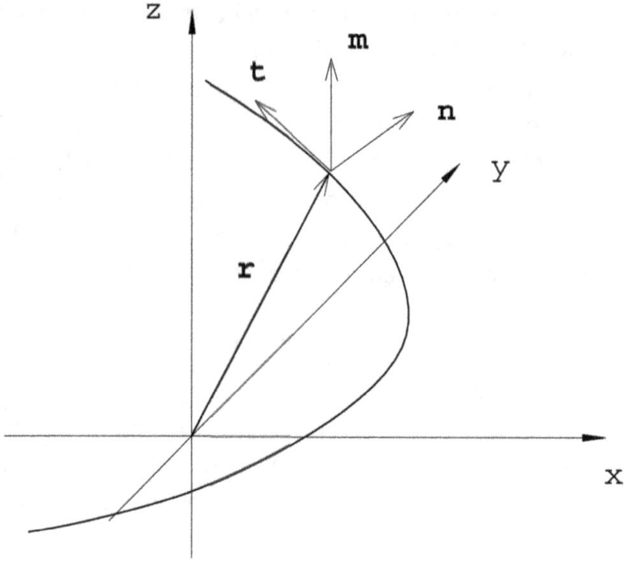

Fig. 1.1. Normal, tangent, and binormal unit vectors.

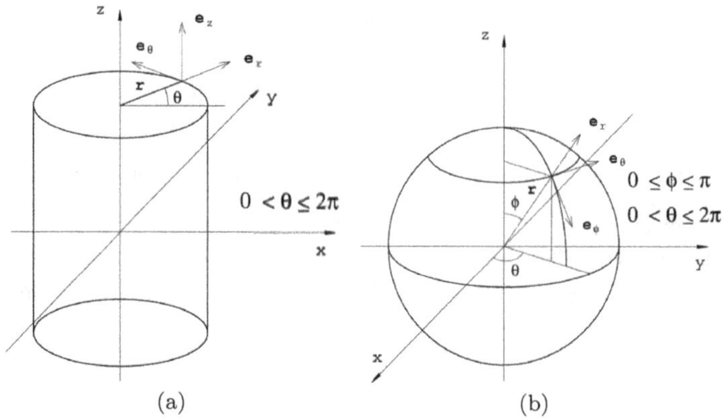

Fig. 1.2. Cylindrical and spherical coordinate systems.

The common denotations for these two coordinate frames are cylindrical system (r, θ, z): $x = r\cos\theta$, $y = r\sin\theta$, $z = z$ with $r \geq 0$, $0 \leq \theta < 2\pi$, and $z \in \mathcal{R}$; and spherical system (r, ϕ, θ): $x = r\sin\phi\cos\theta$, $y = r\sin\phi\sin\theta$, and $z = r\cos\phi$, with $r \geq 0$, $0 \leq \phi \leq \pi$, and $0 \leq \theta < 2\pi$.

Hence, we can express the unit vectors in the cylindrical coordinate system as

$$\begin{aligned}\mathbf{e}_r &= \cos\theta\mathbf{e}_1 + \sin\theta\mathbf{e}_2,\\ \mathbf{e}_\theta &= -\sin\theta\mathbf{e}_1 + \cos\theta\mathbf{e}_2,\\ \mathbf{e}_z &= \mathbf{e}_3,\end{aligned} \tag{1.76}$$

and consequently,

$$\begin{aligned}&\frac{\partial\mathbf{e}_r}{\partial r} = \mathbf{0}, \quad \frac{\partial\mathbf{e}_r}{\partial\theta} = \mathbf{e}_\theta, \quad \frac{\partial\mathbf{e}_r}{\partial z} = \mathbf{0};\\ &\frac{\partial\mathbf{e}_\theta}{\partial r} = \mathbf{0}, \quad \frac{\partial\mathbf{e}_\theta}{\partial\theta} = -\mathbf{e}_r, \quad \frac{\partial\mathbf{e}_\theta}{\partial z} = \mathbf{0};\\ &\frac{\partial\mathbf{e}_z}{\partial r} = \mathbf{0}, \quad \frac{\partial\mathbf{e}_z}{\partial\theta} = \mathbf{0}, \quad \frac{\partial\mathbf{e}_z}{\partial z} = \mathbf{0}.\end{aligned} \tag{1.77}$$

Similarly, the spherical coordinate system can be represented as

$$\begin{aligned}\mathbf{e}_r &= \sin\phi\cos\theta\mathbf{e}_1 + \sin\phi\sin\theta\mathbf{e}_2 + \cos\phi\mathbf{e}_3,\\ \mathbf{e}_\phi &= \cos\phi\cos\theta\mathbf{e}_1 + \cos\phi\sin\theta\mathbf{e}_2 - \sin\phi\mathbf{e}_3,\\ \mathbf{e}_\theta &= -\sin\theta\mathbf{e}_1 + \cos\theta\mathbf{e}_2,\end{aligned}$$

and consequently,

$$\begin{aligned}&\frac{\partial\mathbf{e}_r}{\partial r} = \mathbf{0}, \quad \frac{\partial\mathbf{e}_r}{\partial\phi} = \mathbf{e}_\phi, \quad \frac{\partial\mathbf{e}_r}{\partial\theta} = \sin\phi\mathbf{e}_\theta;\\ &\frac{\partial\mathbf{e}_\phi}{\partial r} = \mathbf{0}, \quad \frac{\partial\mathbf{e}_\phi}{\partial\phi} = -\mathbf{e}_r, \quad \frac{\partial\mathbf{e}_\phi}{\partial\theta} = \cos\phi\mathbf{e}_\theta;\\ &\frac{\partial\mathbf{e}_\theta}{\partial r} = \mathbf{0}, \quad \frac{\partial\mathbf{e}_\theta}{\partial\phi} = \mathbf{0}, \quad \frac{\partial\mathbf{e}_\theta}{\partial\theta} = -\sin\phi\mathbf{e}_r - \cos\phi\mathbf{e}_\phi.\end{aligned}$$

In essence, for any vector $\mathbf{a}$ of fixed length, i.e., $\frac{d}{dt}(\mathbf{a}\cdot\mathbf{a}) = 2\frac{d\mathbf{a}}{dt}\cdot\mathbf{a} = 0$, we can introduce an angular velocity $\boldsymbol{\omega}$ from the expression

$$\frac{d\mathbf{a}}{dt} = \boldsymbol{\omega}\times\mathbf{a}. \tag{1.78}$$

1.2.2 *Arbitrary Lagrangian–Eulerian descriptions*

In the modeling of deformable solids, if the deformation is not excessively large, so-called Lagrangian descriptions are introduced. More specifically, if the reference configuration is the initial configuration, the corresponding Lagrangian description is often called the total Lagrangian description; whereas if the reference configuration is the last time step configuration, the corresponding Lagrangian description is often called the updated Lagrangian description. There are of course pros and cons for both Lagrangian descriptions as presented in detail in Bathe (1996). In fluid mechanics, stationary meshes or grids are often introduced to describe the flow fields, as such Eulerian descriptions are preferred along with the needed convective terms. The arbitrary Lagrangian–Eulerian (ALE) is employed with a gradual transition of mesh motions between Eulerian and Lagrangian descriptions for the maintenance of mesh regularities within the entire computational domain. Thus, if $\mathcal{B}$ is a rigid body, $\boldsymbol{\omega}$ must represent the three rotations in space, and it is easy to verify that if $\mathcal{B}$ undergoes an angular velocity $\dot{\theta}(t)$ and is expressed in cylindrical coordinate system, we have $\boldsymbol{\omega} = \dot{\theta}\mathbf{e}_z$ and

$$\frac{d\mathbf{e}_r}{dt} = \boldsymbol{\omega} \times \mathbf{e}_r = \dot{\theta}\mathbf{e}_\theta.$$

In fact, if we use Eq. (1.77) and chain differentiation, we have

$$\frac{d\mathbf{e}_r}{dt} = \frac{\partial \mathbf{e}_r}{\partial r}\frac{dr}{dt} + \frac{\partial \mathbf{e}_r}{\partial \theta}\frac{d\theta}{dt} + \frac{\partial \mathbf{e}_r}{\partial z}\frac{dz}{dt} = \dot{\theta}\frac{\partial \mathbf{e}_r}{\partial \theta} = \dot{\theta}\mathbf{e}_\theta.$$

Now, if we present the curve as

$$\mathbf{r} = r\mathbf{e}_r + z\mathbf{e}_z, \tag{1.79}$$

the velocity and acceleration are represented as

$$\begin{aligned}
\mathbf{v} &= \dot{r}\mathbf{e}_r + r\dot{\theta}\mathbf{e}_\theta + \dot{z}\mathbf{e}_z,\\
\mathbf{a} &= (\ddot{r} - r\dot{\theta}^2)\mathbf{e}_r + (r\ddot{\theta} + 2\dot{r}\dot{\theta})\mathbf{e}_\theta + \ddot{z}\mathbf{e}_z.
\end{aligned}$$

Likewise, if we employ spherical coordinate system, we obtain similar expressions:

$$\begin{aligned}
\mathbf{r} &= r\mathbf{e}_r,\\
\mathbf{v} &= \dot{r}\mathbf{e}_r + r\dot{\phi}\mathbf{e}_\phi + r\dot{\theta}\sin\phi\,\mathbf{e}_\theta,\\
\mathbf{a} &= (\ddot{r} - r\dot{\phi}^2 - r\dot{\theta}^2\sin^2\phi)\mathbf{e}_r + (r\ddot{\phi} + 2\dot{r}\dot{\phi} - r\dot{\theta}^2\sin\phi\cos\phi)\mathbf{e}_\phi\\
&\quad + (r\ddot{\theta}\sin\phi + 2\dot{r}\dot{\phi}\sin\phi + 2r\dot{\theta}\dot{\phi}\cos\phi)\mathbf{e}_\theta.
\end{aligned}$$

Let a body $\mathcal{B}$ occupy a region $\mathcal{R}_x$ in the Euclidean space $\mathcal{R}^3$ at time t_o. The motion of any material particle of $\mathcal{B}$ is defined by a mapping or function $\mathbf{y} : \mathcal{R}_x \times t \rightarrow \mathcal{R}^3$. The region occupied by the body $\mathcal{B}$ at time t or the image of $\mathcal{R}_x$ at time t is denoted by $\mathcal{R}_y$. Thus, the position at time t of a material particle which at time t_o occupied the position $\mathbf{x}$ or the image of $\mathbf{x} \in \mathcal{R}_x$ at time t is denoted by

$$\mathbf{y} = \mathbf{y}(\mathbf{x}, t). \tag{1.80}$$

We then define another domain, the referential domain $\mathcal{R}_z$, e.g., the finite element mesh domain, and another mapping or function $\hat{\mathbf{y}} : \mathcal{R}_z \times t \rightarrow \mathcal{R}^3$ whose range is also $\mathcal{R}_y$. This mapping $\hat{\mathbf{y}}$ defines the motion of the mesh. The motion of any mesh point or the image of $\mathbf{z} \in \mathcal{R}_z$ at time t is defined by

$$\mathbf{y} = \hat{\mathbf{y}}(\mathbf{z}, t).$$

The choice of $\mathcal{R}_z \neq \mathcal{R}_x$ only serves to underscore the difference between the mesh motion and the material motion. In practice, the mesh configuration has to coincide with the material configuration, i.e., $\mathcal{R}_z = \mathcal{R}_x$. Nevertheless, although in general, $\hat{\mathbf{y}} \neq \mathbf{y}$, it is worth mentioning that if $\mathcal{R}_z = \mathcal{R}_x$ and $\hat{\mathbf{y}} = \mathbf{y}$ for all t the mesh-referential description is then identical to the Lagrangian description and the mesh motion is identical to the material motion. Likewise, if $\mathcal{R}_z = \mathcal{R}_y$ and $\hat{\mathbf{y}} = \mathbf{1}$ (identity mapping) the mesh-referential description specializes the Eulerian description. The mappings $\mathbf{y}$ and $\hat{\mathbf{y}}$ induce another mapping $\tilde{\mathbf{y}} : \mathcal{R}_x \times t \rightarrow \mathcal{R}_z$ and we can write

$$\mathbf{z} = \tilde{\mathbf{y}}(\mathbf{x}, t) = \hat{\mathbf{y}}^{-1}(\mathbf{y}(\mathbf{x}, t), t). \tag{1.81}$$

A material particle which occupied $\mathbf{x}$ at time t_o is in general, at different times t_1 and t_2, represented by different point $\mathbf{z}_1$ and $\mathbf{z}_2$ in the referential domain. The mapping $\tilde{\mathbf{y}}$ defines such a relation. Consider a physical quantity f (e.g., density, particle velocity components). To express this quantity in the Eulerian description, we write

$$f = f(\mathbf{y}, t), \tag{1.82}$$

while in the Lagrangian description, we write

$$f = f(\mathbf{y}(\mathbf{x}, t), t) = \tilde{f}(\mathbf{x}, t). \tag{1.83}$$

We then introduce the "mesh-referential description", in which we associate the value of f with $\mathbf{z} \in \mathcal{R}_z$ via the mesh motion $\hat{\mathbf{y}}(\mathbf{z}, t)$, i.e.,

$$f = f(\hat{\mathbf{y}}(\mathbf{z}, t), t) = \hat{f}(\mathbf{z}, t), \tag{1.84}$$

which means that we attach the variable f to mesh points.

In the Eulerian description, the time rate of change is the rate observed at a fixed point in space. The so-called spatial time derivative is defined as

$$f' = \frac{\partial}{\partial t} f(\mathbf{y}, t), \tag{1.85}$$

whereas the time rate of change of f with respect to a fixed material particle, the so-called material time derivative, is defined as

$$\dot{f} = \frac{\partial}{\partial t} \tilde{f}(\mathbf{x}, t). \tag{1.86}$$

Finally, the time rate of change of f with respect to a given reference point $\mathbf{z} \in \mathcal{R}_z$ (a given mesh point), the so-called mesh-referential time derivative, is defined as

$$\check{f} = f^* = \frac{\partial}{\partial t} \hat{f}(\mathbf{z}, t). \tag{1.87}$$

The spatial time derivative and the material time derivative are related by

$$\dot{f} = \frac{\partial f(\mathbf{y}, t)}{\partial t} + \frac{\partial f(\mathbf{y}, t)}{\partial \mathbf{y}} \frac{\partial \mathbf{y}}{\partial t} \bigg|_{\mathbf{y} = \mathbf{y}(\mathbf{x}, t)} = f' + \mathbf{v} \cdot \nabla f. \tag{1.88}$$

The convective term $\mathbf{v} \cdot \nabla f$ arises from the fact that a fixed point in space $\mathbf{y}$ does not always represent the same material particle. Likewise, using the mesh-referential time derivative and the spatial time derivative, we obtain

$$\check{f} = f^* = \frac{\partial f(\mathbf{y}, t)}{\partial t} + \frac{\partial f(\mathbf{y}, t)}{\partial \mathbf{y}} \cdot \frac{\partial \mathbf{y}}{\partial t} \bigg|_{\mathbf{y} = \hat{\mathbf{y}}(\mathbf{z}, t)} = f' + \hat{\mathbf{v}} \cdot \nabla f, \tag{1.89}$$

where $\hat{\mathbf{v}} = \check{\mathbf{y}} = \frac{\partial \hat{\mathbf{y}}}{\partial t}$ is the velocity of a mesh point.

Note that if we denote the mesh displacement as $\hat{\mathbf{u}} = \hat{\mathbf{u}}(\mathbf{z}, t) = \hat{\mathbf{y}}(\mathbf{z}, t) - \hat{\mathbf{y}}(\mathbf{z}, t_o)$, the mesh velocity can be also represented as $\hat{\mathbf{u}}^*$. We also note that if the mesh motion coincides with the material particle motion, i.e., $\hat{\mathbf{v}} = \mathbf{v}$, the mesh-referential time derivative is the material derivative, also denoted as $\frac{d}{dt}$. On the other hand, if the mesh is fixed in space, i.e., $\hat{\mathbf{v}} = \mathbf{0}$, then the mesh-referential time derivative and the spatial time derivative are identical.

1.2.3 *Transport theorems and variational principles*

We begin with a brief introduction to Green's theorem, named after George Green (1793–1841), Stokes' theorem, named after George Gabriel Stokes (1819–1903), and Gauss' theorem, named after Johann Carl Friedrich Gauss (1777–1855). These theorems are commonly used in the derivation of transport theorems and variational principles. In this book, a curve is called a closed curve, if the end points coincide; if not, it is called an arc. A curve is simple if it does not intersect itself, and a closed curve is called a simple closed curve if it does not intersect itself. A curve is smooth if it poses a tangent vector that varies continuously along the length, and is piecewise smooth if it is comprised of finite number of smooth segments.

Let $\mathbf{f}$ be a continuous vector field defined on a smooth curve C given by a vector function $\mathbf{r}(t)$ with $t \in [a,b]$ or $\mathbf{r}(s)$ with $s \in [0,L]$ where L is the total arc length defined as $L = \int_C ds$. Then the line integral of $\mathbf{f}$ along C is

$$\int_C \mathbf{f} \cdot d\mathbf{r} = \int_0^L \mathbf{f} \cdot \mathbf{t} ds = \int_a^b \mathbf{f}(\mathbf{r}(t)) \cdot \mathbf{r}'(t) dt, \tag{1.90}$$

with the unit tangent vector $\mathbf{t}$ defined as $d\mathbf{r}/ds$.

One important result of the line contour integral is for a differentiable function $f(\mathbf{r})$, whose gradient ∇f, often called a conservative vector field, is continuous along C. Hence, employing the Fundamental Theorem of Calculus, i.e.,

$$\int_a^b f'(x) dx = f(b) - f(a), \tag{1.91}$$

we have

$$\int_C \nabla f \cdot d\mathbf{r} = \int_C df = f(\mathbf{r}(b)) - f(\mathbf{r}(a)). \tag{1.92}$$

Equation (1.92) clearly shows that the contour integral for the conservative vector field is path independent.

Theorem 1.7 (Green's theorem). *Consider a continuous differentiable or C^1 vector field $\mathbf{f} = f_1 \mathbf{e}_1 + f_2 \mathbf{e}_2$ defined in a two-dimensional region S bounded by a piecewise smooth simple closed curve C, positively oriented namely counterclockwise. Then*

$$\int_C \mathbf{f} \cdot d\mathbf{r} = \int_C (f_1 dx + f_2 dy) = \int_S \left(\frac{\partial f_2}{\partial x_1} - \frac{\partial f_1}{\partial x_2} \right) dS. \tag{1.93}$$

Consider a smooth surface S represented by $\mathbf{r}(u,v)$ with two parameters u and v and $(u,v) \in D$. The surface area is defined as

$$A = \int_D ||\mathbf{r}_u \times \mathbf{r}_v|| dudv, \tag{1.94}$$

with $\mathbf{r}_u = \frac{\mathbf{r}}{u}$ and $\mathbf{r}_v = \frac{\mathbf{r}}{v}$.

The surface integral of a continuous function $f(\mathbf{r})$ can be expressed as

$$\int_S f(\mathbf{r}) dS = \int_D f(\mathbf{r}(u,v)) ||\mathbf{r}_u \times \mathbf{r}_v|| dudv. \tag{1.95}$$

For a continuous vector field $\mathbf{f}(\mathbf{r})$ defined on an oriented surface S with the unit normal vector $\mathbf{n}$ often defined as

$$\mathbf{n} = \frac{\mathbf{r}_u \times \mathbf{r}_v}{||\mathbf{r}_u \times \mathbf{r}_v||}, \tag{1.96}$$

the surface integral is expressed as

$$\int_S \mathbf{f}(\mathbf{r}) \cdot \mathbf{n} dS = \int_D \mathbf{f}(\mathbf{r}(u,v)) \cdot (\mathbf{r}_u \times \mathbf{r}_v) dudv. \tag{1.97}$$

Theorem 1.8 (Stokes' theorem). *Let S be a simple surface enclosed by the boundary curve C given with positive, namely counterclockwise, orientation. Let $\mathbf{f}$ be a C^1 vector field defined in S. Let $\mathbf{t}$ be the unit tangential vector along the curve C chosen so that the vector $\mathbf{n} \times \mathbf{t}$ points into S from V. Then*

$$\int_S \nabla \times \mathbf{f} \cdot \mathbf{n} dS = \int_C \mathbf{f} \cdot \mathbf{t} ds. \tag{1.98}$$

In fact, Green's theorem is the simple two-dimensional representation of Stokes' theorem. The geometric divergence of a vector function at a point p is a scalar function of position and represents the outflow per unit volume at the point p. In fact, the other definition of the divergence can be written as

$$\text{div}\,\mathbf{f} = \lim_{\Delta V \to 0} \frac{1}{\Delta V} \int_S \mathbf{f} \cdot \mathbf{n} dS. \tag{1.99}$$

Of course, the divergence is uniquely defined with no dependence on any particular coordinate system. The detailed physical account is discussed in Schey (1992). We can have the divergence theorem for n-dimensional systems as proven in Munkres (1991).

Theorem 1.9 (Divergence or Gauss' theorem). *Let V be a simple solid region enclosed by the boundary surface S given with positive, namely outward, orientation. Let* **f** *be a continuous differentiable vector field and* **n** *be the unit normal vector on the boundary surface S that points outwards from V. Then*

$$\int_V \boldsymbol{\nabla}\cdot\mathbf{f}dV = \int_S \mathbf{f}\cdot\mathbf{n}dS. \tag{1.100}$$

Thus, it can be easily established that

$$\int_V \nabla^2 f dV = \int_S \boldsymbol{\nabla} f\cdot\mathbf{n}dS, \tag{1.101}$$

$$\int_V \boldsymbol{\nabla}\times\mathbf{f}dV = \int_S \mathbf{n}\times\mathbf{f}dS. \tag{1.102}$$

In particular, employing the permutation symbol, we can easily prove that Eq. (1.102) can be rewritten in the following form:

$$-\int_V \boldsymbol{\nabla}\times(\mathbf{r}f)dV = \int_S (\mathbf{r}\times\mathbf{n})fdS. \tag{1.103}$$

Consequently, from the divergence theorem it is easy to derive two forms of Green's theorem.

The first form of Green's theorem:

$$\int_S \mathbf{n}\cdot(f_1\boldsymbol{\nabla}f_2)dS = \int_V \left[f_1\nabla^2 f_2 + (\boldsymbol{\nabla}f_1)\cdot(\boldsymbol{\nabla}f_2)\right]dV. \tag{1.104}$$

The second form of Green's theorem:

$$\int_S \mathbf{n}\cdot(f_1\boldsymbol{\nabla}f_2 - f_2\boldsymbol{\nabla}f_1)\,dS = \int_V \left(f_1\nabla^2 f_2 - f_2\nabla^2 f_1\right)dV. \tag{1.105}$$

To simplify the notations, we replace the original volumetric integral sign $\iiint$ with $\int_V$, and the surface integral sign $\iint$ with $\int_S$, where V and S denote the volume and surface, respectively.

To express the fundamental laws of mechanics, i.e., the laws of mass, linear (angular) momentum, and energy conservation in integral form, we define a material volume as a volume which always contains the same material particles. In other words, a material volume $\tilde{V}(t)$ at some time

t is bounded by a closed surface $\tilde{S}(t)$ in which every point moves with the material velocity $\tilde{\mathbf{v}}$. We define a control volume as a volume arbitrarily chosen for the convenience of the observer. Unlike the material volume which is governed by the motions of the same material particles, a proper definition of a control volume $V(t)$ with the bounding surface $S(t)$ must include the velocity $\mathbf{v}(\mathbf{x},t)$ for all time t.

Let $f(\mathbf{x},t)$ be a continuous property per unit volume, and consider integral $I(t) = \int_{V(t)} f(\mathbf{x},t)d\Omega$ with $\mathbf{v}(\mathbf{x},t)$. Thus, the derivative with respect to t is expressed as

$$\frac{dI(t)}{dt} = \lim_{\Delta t \to 0} \frac{[I(t+\Delta t) - I(t)]}{\Delta t},$$

and according to the Taylor's expansion, named after Brook Taylor (1685–1731), we have

$$f(\mathbf{x},t+\Delta t) = f(\mathbf{x},t) + \frac{\partial}{\partial t} f(\mathbf{x},t)\Delta t + O(\Delta t^2). \tag{1.106}$$

Moreover,

$$\frac{d}{dt}I(t) = \int_{V(t)} \frac{\partial f(\mathbf{x},t)}{\partial t} dV + \lim_{\Delta t \to 0} \frac{\left[\int_{V(t+\Delta t)} f dV - \int_{V(t)} f dV\right]}{\Delta t}. \tag{1.107}$$

From the geometric arguments, analogous to the Leibniz rule for one-dimensional integral, named after Gottfried Wilhelm von Leibniz (1646–1716), we obtain the following variation of the integral:

$$\int_{V(t+\Delta t)} f(\mathbf{x},t)dV = \int_{V(t)} f dV + \int_{S(t)} f \mathbf{v} \cdot \mathbf{n} \Delta t dS. \tag{1.108}$$

Hence, derived from Eq. (1.107), the first Reynolds transport theorem (kinematic transport theorem), named after Osborne Reynolds (1842–1912), can be expressed as

$$\frac{d}{dt} \int_{V(t)} f(\mathbf{x},t)dV = \int_{V(t)} \left[\frac{\partial f}{\partial t} + \boldsymbol{\nabla} \cdot (f\mathbf{v})\right] dV. \tag{1.109}$$

Of course, considering the material volume and the control volume coinciding at time t, namely at time t, $V(t) = \tilde{V}(t)$ and $S(t) = \tilde{S}(t)$, the first kinematic transport theorem can then be written as

$$\frac{d}{dt}\int_{\tilde{V}(t)} f(\mathbf{x},t)dV = \int_{V(t)} \frac{\partial f}{\partial t}dV + \int_{S(t)} f\tilde{\mathbf{v}}\cdot\mathbf{n}dS, \tag{1.110}$$

or

$$\frac{d}{dt}\int_{\tilde{V}(t)} f(\mathbf{x},t)dV = \frac{d}{dt}\int_{V(t)} fdV + \int_{\tilde{S}(t)} f(\tilde{\mathbf{v}}-\mathbf{v})\cdot\mathbf{n}dS, \tag{1.111}$$

with the unit normal vector $\mathbf{n}$ pointing outward from the volume.

Note that in Eq. (1.111), the time derivative on the right-hand side follows the time evolution of the control volume V instead of the material volume $\tilde{V}$. Equation (1.110) does not use the control volume velocity $\mathbf{v}(\mathbf{x},t)$ and is convenient for steady problems. Equation (1.111) only requires the relative material velocity $\tilde{\mathbf{v}}(\mathbf{x},t)-\mathbf{v}(\mathbf{x},t)$ and is particularly useful for unsteady problems such as rocket buster analysis. If $f(\mathbf{x},t)$ stands for the density $\rho(\mathbf{x},t)$ and $\tilde{\mathbf{v}}(\mathbf{x},t)=\mathbf{v}(\mathbf{x},t)$, i.e., the selected control volume coincides with the material volume at all time, according to the law of mass conservation, we have

$$\int_{\Omega}\left[\frac{\partial\rho}{\partial t}+\nabla\cdot(\rho\mathbf{v})\right]dV = 0. \tag{1.112}$$

Since Eq. (1.112) holds for arbitrary material volumes, the law of mass conservation can then be written as

$$\frac{\partial\rho}{\partial t}+\nabla\cdot(\rho\mathbf{v}) = 0. \tag{1.113}$$

Moreover, since Eq. (1.113) holds for any material volume, we can use Eq. (1.7) to get

$$\frac{d\rho}{dt}+\rho\nabla\cdot\mathbf{v} = 0. \tag{1.114}$$

In general, ρ is a function of the position $\mathbf{x}$, pressure p, temperature T, etc. For incompressible materials, i.e., $d\rho/dt=0$, we simply have

$$\nabla\cdot\mathbf{v} = 0. \tag{1.115}$$

If in the first kinematic transport theorem, we replace f in Eq. (1.109) with ρf, we get

$$\frac{d}{dt}\int_{V(t)} \rho f dV = \int_{V(t)}\left[\frac{\partial}{\partial t}(\rho f)+\nabla\cdot(\rho f\mathbf{v})\right]dV. \tag{1.116}$$

Similarly, with the chain rule and in Eq. (1.7), we obtain

$$\frac{\partial}{\partial t}(\rho f)+\nabla\cdot(\rho f\mathbf{v})=\rho\frac{df}{dt}+f\left[\frac{\partial\rho}{\partial t}+\nabla\cdot(\rho\mathbf{v})\right]. \tag{1.117}$$

Following the law of mass conservation, we obtain the *second Reynolds transport theorem*,

$$\frac{d}{dt}\int_{V(t)}\rho f dV=\int_{V(t)}\rho\frac{df}{dt}dV. \tag{1.118}$$

In the ALE finite element formulation needed for the study of FSI systems, the mesh velocity $\hat{\mathbf{v}}$ is specified by some updating rules to the material velocity $\mathbf{v}$. Fundamentally, there are two ways of relating the mesh and material particle motions. The first is used for problems with closed domains or complete prescribed boundaries. In this ALE description, the mesh velocity can be assigned arbitrarily as long as the mesh boundary coincides with the material boundary. The second is for problems with unknown material boundaries, such as large amplitude free surfaces and FSI interfaces. In these cases, the mesh configurations cannot be assigned arbitrarily, due to the fact that the mesh boundary has to move with the material boundary. Their positions in general must be solved for any time t.

Obviously, in both cases, choices for the mesh velocities are not unique. The kinematic relations developed in the study of continuum mechanics can be used with mesh motions by reinterpreting the particle motion as the motion of a mesh point. Define mesh deformation gradient as $\hat{\mathbf{D}}=\partial\mathbf{y}/\partial\mathbf{z}$, mesh Jacobian determinant as $\hat{J}=\det(\hat{\mathbf{D}})$, mesh velocity gradient as $\hat{\mathbf{L}}=\partial\hat{\mathbf{v}}/\partial\mathbf{y}$, spatial gradient operator as $\nabla=\partial/\partial\mathbf{y}$, and referential gradient operator as $\hat{\nabla}=\partial/\partial\mathbf{z}$. Relationships between the referential and spatial derivatives both in time and space for a scalar f and a vector $\mathbf{a}$ can be derived as follows:

$$\begin{aligned}
\check{f}&=f'+\nabla f\cdot\hat{\mathbf{v}}, &\quad \check{\mathbf{a}}&=\mathbf{a}'+(\nabla\mathbf{a})\hat{\mathbf{v}};\\
\hat{\nabla}f&=\hat{\mathbf{D}}^T\nabla f, & \hat{\nabla}\mathbf{a}&=(\nabla\mathbf{a})\hat{\mathbf{D}};\\
\hat{\mathbf{L}}&=\hat{\mathbf{D}}^*\hat{\mathbf{D}}^{-1}, & \check{J}^m&=J^m tr(\hat{\mathbf{L}})=J^m\nabla\cdot\hat{\mathbf{v}};\\
(\nabla f)^*&=-\hat{\mathbf{L}}^T\nabla f+\nabla\check{f}, & (\nabla\mathbf{a})^*&=-(\nabla\mathbf{a})\hat{\mathbf{L}}+\nabla\check{\mathbf{a}}.
\end{aligned} \tag{1.119}$$

Consider a subregion $V\subset\mathcal{R}_y$ and a surface $S\subset\mathcal{R}_y$ that represent the same mesh region and the same mesh surface at all time, using

$$(\nabla\cdot\mathbf{a})^*=-(\nabla\mathbf{a})\cdot\hat{\mathbf{L}}+\nabla\cdot\check{\mathbf{a}}, \tag{1.120}$$

we have

$$\left(\int_V f dV\right)^* = \int_V \left[\check{f} + f tr(\hat{\mathbf{L}})\right] dV,$$
$$\left(\int_S \mathbf{a}\cdot\mathbf{n}dS\right)^* = \int_S \left[\check{\mathbf{a}} + \mathbf{a}tr(\hat{\mathbf{L}}) - \hat{\mathbf{L}}\mathbf{a}\right]\cdot\mathbf{n}dS, \tag{1.121}$$

where $\mathbf{n}$ is the surface normal vector corresponding to S.

The above relations will be elaborated upon and used extensively in the linearization of nonlinear governing equations in Chapter 4. Of course, the proof of the above relations require the use of Eq. (1.7) and the concepts illustrated in this chapter.

We call a particular collection of continua in question a system. A closed system means that the system does not exchange matter with its surroundings. An isolated system means that no interactions between the system and its surroundings occur. In addition, any process within the system which is thermally insulated is called adiabatic. Therefore, the conservation of mass simply states that the mass of the same system stays constant, which has been adequately explained during the discussion of the second Reynolds transport theorem. Newton's law of motion (the law of linear momentum conservation) equates the rate of change of linear momentum in the material volume to the total external force exerted on the volume, of which the ith component can be expressed as $\int_V f_i^b dV + \int_S f_i^s dS$, where f_i^b and f_i^s stand for the ith component of the body force and the surface traction vectors, respectively. If $f(\mathbf{x},t)$ in Eq. (1.118) represents the ith velocity component, with the second Reynolds transport theorem, we then have

$$\int_V f_i^b dV + \int_S f_i^s dS = \frac{d}{dt}\int_V \rho v_i dV = \int_V \rho\left(\frac{\partial v_i}{\partial t} + \mathbf{v}\cdot\nabla v_i\right) dV. \tag{1.122}$$

Denoting the stress tensor as σ_{ij}, the surface traction can be expressed as $f_i^s = \sigma_{ij}n_j$. Therefore, employing Gauss' theorem and the second Reynolds transport theorem, we have

$$\sigma_{ij,j} + f_i^b = \rho\frac{dv_i}{dt}. \tag{1.123}$$

In addition to the conservation of mass and linear momentum, we must also discuss the conservation of energy. To properly present the concepts, we start with the laws of thermodynamics. The first law of thermodynamics (the law of energy conservation) states that the rate of increase of the total energy in the material volume U, which includes the potential, kinetic, and internal energy associated with the thermal state of the material, to the sum of the net external heat input Q and the net external work of the volume W, namely, the net thermal energy input to the system and the net work done by the system to the external environment.

$$dQ = dU - dW. \tag{1.124}$$

Naturally, the first law of thermodynamics implies that if a thermally insulated system can be taken from one state to another by alternative paths, the work done on the system W has the same value for every adiabatic path. The second law of thermodynamics for a homogeneous system states that there exist two single-valued functions of state, namely, the absolute temperature T and the entropy S. The commonly used unit for the temperature T is Kelvin, named after William Thomson Kelvin (1824–1907). With this unit, the temperature is always a positive function.

In general, the energy of the system consists of the kinetic energy, the potential energy, e.g., the gravitational energy, and the internal energy. The key aspect of energy conservation is to understand the concept of internal energy. For example, the internal energy of a gas is a function of its state. It has been shown in various experiments that only two of the three states, or measurable properties, namely, pressure p, density ρ, and temperature T, are independent. In practice, we often define the specific quantities with respect to the unit mass of gas. For example, the specific volume v is equivalent to V/m or $1/\rho$ where V represent the volume occupied by the gas and m the mass of the gas. For a perfect gas where molecules are far enough apart so that intermolecular forces are negligible and occupied volume is much larger than the molecular volume itself, we have

$$pV = nR_uT, \quad \text{or} \quad pV = mRT, \quad \text{or} \quad p = \rho RT, \quad \text{or} \quad pv = RT, \tag{1.125}$$

where R is a constant for the gas and can be expressed as $R_u/\bar{M}$ with the universal gas constant R_u and the molar mass of the gas molecule.

Moreover, the increment of the entropy S can be further decomposed as

$$dS = \frac{dQ}{T} + dS_{\text{gen}}, \tag{1.126}$$

where dQ/T denotes the part of the increase due to interaction with the surroundings and dS_{gen} denotes the part of the increase due to changes taking place inside the system.

Note that dS_{gen} is always non-negative. In particular, if it is zero, the process is said to be reversible. However, if dQ/T is zero, the process is said to be adiabatic. Therefore, it is the combination of reversible and adiabatic conditions which is equivalent to the case of isentropic, or $dS = 0$. If dS_{gen} is positive, the process is said to be irreversible. In general, the internal energy per unit mass, or specific energy of gas, u can be expressed as

$$du = \left.\frac{\partial u}{\partial v}\right|_T dv + \left.\frac{\partial u}{\partial T}\right|_v dT. \tag{1.127}$$

Now, if heat is added to the gas and the temperature rise is observed according to the first law of thermodynamics, we have

$$dq = pdv + \left.\frac{\partial u}{\partial T}\right|_v dT + \left.\frac{\partial u}{\partial v}\right|_T dv, \tag{1.128}$$

where q stands for the heat input per unit mass of gas.

According to Joule's experiments, the internal energy of a perfect gas is contained entirely within the molecule itself, as energy of translation and rotation, and the intermolecular forces are not important. Therefore, for a perfect gas, we have

$$\left.\frac{\partial u}{\partial v}\right|_T = 0. \tag{1.129}$$

Employing Eq. (1.129), directly from Eq. (1.128), we define the specific heat at constant volume C_v as

$$C_v = \left.\frac{\partial q}{\partial T}\right|_v = \left.\frac{\partial u}{\partial T}\right|_v. \tag{1.130}$$

Hence, integrating Eq. (1.130), assuming that the internal energy vanishes at zero temperature, which is in accordance with the third law of thermodynamics, we derive

$$du = C_v dT. \tag{1.131}$$

If Eq. (1.131) holds, the internal energy of the gas is simply proportional to the temperature. Moreover, applying Eq. (1.129), the first law of thermodynamics can be expressed as

$$dq = d(pv) - vdp + C_v dT. \tag{1.132}$$

In engineering practice, instead of the internal energy per unit mass u, we introduce the enthalpy per unit mass, or specific enthalpy, h as

$$h = pv + u. \tag{1.133}$$

Hence, from Eq. (1.133), employing the first law of thermodynamics, we have

$$dh = vdp + pdv + du = vdp + dq, \tag{1.134}$$

which implies that the change of enthalpy is identical to the heat added to the gas at a constant pressure.

Using Eq. (1.125), from Eq. (1.132), we can express the specific heat at constant pressure C_p as

$$dh = d(pv) + C_v dT = C_p dT, \tag{1.135}$$

with

$$C_p = \left.\frac{\partial q}{\partial T}\right|_p = \frac{d(pv)}{dT} + C_v = R + C_v. \tag{1.136}$$

Furthermore, using the definition for the entropy S, in reversible process, the entropy per unit mass s can be written as

$$ds = C_p \frac{dT}{T} - R\frac{dp}{p}, \tag{1.137}$$

or rather

$$s - s_o = C_p \ln \frac{T}{T_o} - R \ln \frac{p}{p_o}, \tag{1.138}$$

where s_o, T_o, and p_o represent the initial entropy per unit mass (a quantity related to the so-called relative pressure and relative specific volume), temperature, and pressure.

Defining the ratio of the specific heat at constant volume and the specific heat at constant pressure or the adiabatic exponent γ as

$$\gamma = \frac{C_p}{C_v},$$

we have $C_v = R/(\gamma - 1)$.

Hence, we can easily derive from Eqs. (1.125), (1.136), and (1.137), for isentropic process, i.e., the entropy remains a constant,

$$pv^{\gamma} = C, \tag{1.139}$$

where C is a constant.

According to the basic principle in classical kinetic theory, the internal energy of each degree of freedom of the individual molecules of a gas can be expressed as $KT/2$, where K represents the Boltzmann constant. Denoting the number of molecules of a gas per unit mass as n, we have $R = nK$. Thus, if there are N degrees of freedom, the internal energy per unit mass can be calculated as

$$u = \frac{NRT}{2}. \tag{1.140}$$

Combining Eqs. (1.131) and (1.140), we have

$$C_p = \frac{NR}{2} + R = \left(\frac{N}{2} + 1\right) R,$$

and consequently,

$$\gamma = \frac{N+2}{N}.$$

For a monatomic gas which has three degrees of freedom in translation, we have $\gamma = 1.67$; for a diatomic gas which has five degrees of freedom (three translations and two rotations), we have $\gamma = 1.4$. For air, which is a mixture of diatomic molecules such as Nitrogen (78.06%), Oxygen (21%), and Argon (0.94%), we also have $\gamma = 1.4$.

Internal energy provides a thermal dynamic quantification of the molecular motion in microscopic scale, whereas kinetic energy provides a macroscopic representation of a motion of continuum. In Eq. (1.128), although the molecule of gas is obviously not stationary, the continuum is stationary. Therefore, there is no kinetic energy term other than the internal energy.

In an ideal gas, like the specific internal energy e, the specific enthalpy h is evidently a function of the temperature. Furthermore, for isentropic ideal gas, the specific enthalpy can also be expressed as

$$h = \frac{\gamma RT}{\gamma - 1} = \frac{\gamma pv}{\gamma - 1}. \tag{1.141}$$

Based on the physical fact the heat is transferred from high temperature to low temperature, in engineering practice we introduce the heat flux vector $\mathbf{h}$ calculated as $-k\nabla T$, where k is the so-called heat conductivity. Consequently, we have the rate of heat change expressed as

$$\frac{dQ}{dt} = \int_V \rho \frac{dq}{dt} dV = -\int_S h_i n_i dS, \tag{1.142}$$

where $\mathbf{n}$ is the surface normal unit vector normally pointing outwards and q is the specific heat transfer to the system.

Obviously, the rate of work done on the system with the material volume V from the surrounding can be expressed as $\int_V f_i^b v_i dV + \int_S f_i^s v_i dS$, where the kinetic energy is simply $\int_V \rho v_i v_i/2dV$. Moreover, the internal energy is defined as $\int_V \rho u dV$, where u is the internal energy per unit mass. By using the second Reynolds transport theorem, we can easily derive

$$\int_V f_i^b v_i dV + \int_S f_i^s v_i dS = \int_V \rho \left(\frac{du}{dt} + v_i \frac{dv_i}{dt} \right) dV - \int_V \rho \frac{dq}{dt} dV.$$

Applying Eq. (1.123), for small deformations we can easily derive that

$$\rho \frac{du}{dt} = \rho \frac{dq}{dt} + \sigma_{ij} \frac{d\varepsilon_{ij}}{dt},$$

or employ Eq. (1.142),

$$\rho \frac{du}{dt} = -\frac{\partial h_i}{\partial x_i} + \sigma_{ij} \frac{d\varepsilon_{ij}}{dt}.$$

Consider a general three-dimensional body V with the surface area $S = \partial V$, as shown in Fig. 1.3. Assume $\mathbf{u}_p$ as the prescribed displacement on the Dirichlet boundary $S_u \subset S$, named after Johann Peter Gustav Lejeune Dirichlet (1805–1859) and $\mathbf{f}^{S_f}$ as the prescribed surface traction on the Neumann boundary $S_f \subset S$, named after Carl Neumann (1832–1925). In addition, $S_u \cup S_f = S$ and $S_u \cap S_f = \varnothing$. For simplicity, we represent stress and strain tensors with the corresponding vector forms:

$$\boldsymbol{\sigma} = < \sigma_{11} \quad \sigma_{22} \quad \sigma_{33} \quad \sigma_{12} \quad \sigma_{23} \quad \sigma_{31} >,$$
$$\boldsymbol{\varepsilon} = < \varepsilon_{11} \quad \varepsilon_{22} \quad \varepsilon_{33} \quad \gamma_{12} \quad \gamma_{23} \quad \gamma_{31} >,$$

with $\gamma_{ij} = 2\varepsilon_{ij}$, for $i \neq j$.

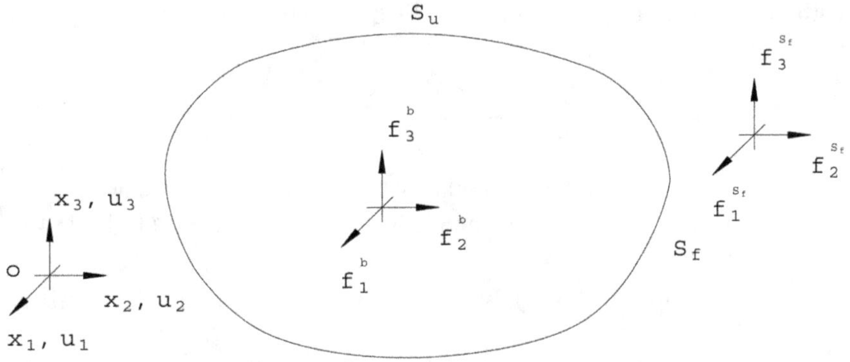

Fig. 1.3. A typical three-dimensional continuous body.

If the material is linear elastic, the stress-strain relation can be denoted as

$$\boldsymbol{\sigma} = \mathbf{C}\boldsymbol{\varepsilon}, \tag{1.143}$$

or

$$\boldsymbol{\sigma} = \mathbf{C}\boldsymbol{\varepsilon} + \boldsymbol{\sigma}^I, \tag{1.144}$$

if there exists a stress vector $\boldsymbol{\sigma}^I$ representing the initial stress such as the thermal stress.

On the other hand, the stress vector $\boldsymbol{\sigma}$ and the strain vector $\boldsymbol{\varepsilon}$ in Eq. (1.143) can be rewritten as tensorial quantities expressed by the following 3×3 matrices:

$$\boldsymbol{\sigma} = \begin{bmatrix} \sigma_{11} & \sigma_{12} & \sigma_{13} \\ \sigma_{21} & \sigma_{22} & \sigma_{23} \\ \sigma_{31} & \sigma_{32} & \sigma_{33} \end{bmatrix} \quad \text{and}$$

$$\boldsymbol{\varepsilon} = \begin{bmatrix} \varepsilon_{11} & \varepsilon_{12} & \varepsilon_{13} \\ \varepsilon_{21} & \varepsilon_{22} & \varepsilon_{23} \\ \varepsilon_{31} & \varepsilon_{32} & \varepsilon_{33} \end{bmatrix}.$$

Thus, the relation between stress and strain tensors can be represented as

$$\sigma_{ij} = C_{ijrs}\varepsilon_{rs} = [\lambda\delta_{ij}\delta_{rs} + \mu(\delta_{ir}\delta_{js} + \delta_{is}\delta_{jr})]\varepsilon_{rs} = \lambda\delta_{ij}\varepsilon_{rr} + \mu\varepsilon_{ij} + \mu\varepsilon_{ji}, \tag{1.145}$$

with the Lamé constants, named after Gabriel Lamé (1795–1870), λ and μ defined as

$$\lambda = \frac{E(1-\nu)}{(1+\nu)(1-2\nu)} \quad \text{and} \quad \mu = \frac{E}{2(1+\nu)},$$

where E and ν are Young's modulus and Poisson's ratio, named after Thomas Young (1773–1829) and Simé–Denis Poisson (1781–1840), respectively.

In addition, in Eq. (1.143), we can write the constant material matrix as

$$\mathbf{C} = \lambda \begin{bmatrix} 1 & \frac{\nu}{(1-\nu)} & \frac{\nu}{(1-\nu)} & 0 & 0 & 0 \\ \frac{\nu}{(1-\nu)} & 1 & \frac{\nu}{(1-\nu)} & 0 & 0 & 0 \\ \frac{\nu}{(1-\nu)} & \frac{\nu}{(1-\nu)} & 1 & 0 & 0 & 0 \\ 0 & 0 & 0 & G/\lambda & 0 & 0 \\ 0 & 0 & 0 & 0 & G/\lambda & 0 \\ 0 & 0 & 0 & 0 & 0 & G/\lambda \end{bmatrix}.$$

Moreover, with the consideration of linear thermal expansion coefficient α, for the normal stresses, we often employ the following

$$\varepsilon_{11} = \frac{1}{E}(\sigma_{11} - \nu\sigma_{22} - \nu\sigma_{33}) + \alpha T,$$

$$\varepsilon_{22} = \frac{1}{E}(-\nu\sigma_{11} + \sigma_{22} - \nu\sigma_{33}) + \alpha T,$$

$$\varepsilon_{33} = \frac{1}{E}(-\nu\sigma_{11} - \nu\sigma_{22} + \sigma_{33}) + \alpha T.$$

Note that the thermal expansion has no effect on the shear stress components. If we assign the volumetric strain as $\varepsilon_v = \varepsilon_{rr}$, and the deviatoric strain as

$$\varepsilon'_{ij} = \varepsilon_{ij} - \varepsilon_v \frac{\delta_{ij}}{3}, \tag{1.146}$$

Eq. (1.145) can be expressed as

$$\sigma_{ij} = \left(\lambda + \frac{2\mu}{3}\right) \varepsilon_v \delta_{ij} + 2\mu \varepsilon'_{ij}. \tag{1.147}$$

Furthermore, if we denote the pressure as

$$p = -\frac{\sigma_{rr}}{3} = -\kappa \varepsilon_v, \tag{1.148}$$

and the deviatoric stress as

$$\mathbf{s} = \boldsymbol{\sigma} + p\delta_{ij}, \tag{1.149}$$

with the bulk modulus $\kappa = \lambda + \frac{2\mu}{3} = \frac{E}{3(1-2\nu)}$, Eq. (1.145) can also be rewritten as

$$\sigma_{ij} = -p\delta_{ij} + 2G\varepsilon'_{ij},$$

with $G = \mu$ as the shear modulus.

For small displacements, the linear strain-displacement relation (so-called compatibility condition) can be expressed as

$$\boldsymbol{\varepsilon} = \boldsymbol{\partial}_\varepsilon \mathbf{u},$$

with

$$\boldsymbol{\partial}_\varepsilon = \begin{bmatrix} \frac{\partial}{\partial x_1} & 0 & 0 \\ 0 & \frac{\partial}{\partial x_2} & 0 \\ 0 & 0 & \frac{\partial}{\partial x_3} \\ \frac{\partial}{\partial x_2} & \frac{\partial}{\partial x_1} & 0 \\ 0 & \frac{\partial}{\partial x_3} & \frac{\partial}{\partial x_2} \\ \frac{\partial}{\partial x_3} & 0 & \frac{\partial}{\partial x_1} \end{bmatrix}.$$

The principle of virtual displacements (the principle of virtual work) states that the equilibrium of the body V requires that for any compatible small virtual displacements $\bar{\mathbf{u}}$, i.e., $\bar{\mathbf{u}} = \mathbf{0}$ on S_u and $\bar{\boldsymbol{\varepsilon}} = \boldsymbol{\partial}_\varepsilon \bar{\mathbf{u}}$, the total internal virtual work is equal to the total external virtual work:

$$\int_V \bar{\boldsymbol{\varepsilon}}^T \boldsymbol{\sigma} dV = \int_V \bar{\mathbf{u}}^T \mathbf{f}^b dV + \int_{S_f} \bar{\mathbf{u}}^{S_f^T} \mathbf{f}^{S_f} dS,$$

where $\bar{\boldsymbol{\varepsilon}}$ represents the virtual strain vector associated with the virtual displacement $\bar{\mathbf{u}}$.

Here, we introduce the variational indicator (so-called energy form):

$$\Pi_o = \int_V \frac{1}{2}\boldsymbol{\varepsilon}^T \mathbf{C}\boldsymbol{\varepsilon} dV - \int_V \mathbf{u}^T \mathbf{f}^b dV - \int_{S_f} \mathbf{u}^{S_f^T} \mathbf{f}^{S_f} dS, \tag{1.150}$$

with the displacement boundary condition $\mathbf{u}^{S_u} = \mathbf{u}_p$ on S_u.

Invoking the stationarity of Π_o, i.e., evaluating $\delta\Pi_o = 0$ with respect to the displacement $\mathbf{u}$ which also appears in the strain, we obtain

$$\int_V \delta\boldsymbol{\varepsilon}^T \mathbf{C}\boldsymbol{\varepsilon} dV - \int_V \delta\mathbf{u}^T \mathbf{f}^b dV - \int_{S_f} \delta\mathbf{u}^{S_f^T} \mathbf{f}^{S_f} dS = 0. \tag{1.151}$$

It is obvious that $\delta\mathbf{u}^{S_u} = \delta\mathbf{u}_p = \mathbf{0}$ on S_u. By using Eq. (1.143), we can easily identify that $\bar{\boldsymbol{\varepsilon}} = \delta\boldsymbol{\varepsilon}$, $\bar{\mathbf{u}} = \delta\mathbf{u}$, and $\bar{\mathbf{u}}^{S_f} = \delta\mathbf{u}^{S_f}$. In fact, the principle of virtual displacements corresponding to the minimum of the energy form in Eq. (1.150) and is often called the minimum energy principle. Of course, we assume that the displacement boundary condition and the compatibility condition are satisfied. To arrive at more general forms of variational principles, we first relax the compatibility condition and introduce the following variation indicator:

$$\Pi_1 = \Pi_o - \int_V \boldsymbol{\lambda}_\varepsilon^T (\boldsymbol{\varepsilon} - \boldsymbol{\partial}_\varepsilon \mathbf{u}) dV, \tag{1.152}$$

where $\boldsymbol{\lambda}_\varepsilon$ is a Lagrangian multiplier.

Invoking the stationarity of Π_1, i.e., evaluating $\delta\Pi_1 = 0$ with respect to $\mathbf{u}$, $\boldsymbol{\varepsilon}$, and $\boldsymbol{\lambda}_\varepsilon$, we can easily verify that $\boldsymbol{\lambda}_\varepsilon = \boldsymbol{\sigma} = \mathbf{C}\boldsymbol{\varepsilon}$ and recover the compatibility condition. Since $\boldsymbol{\lambda}_\varepsilon^T(\boldsymbol{\partial}_\varepsilon \delta\mathbf{u}) = \sigma_{ij}\delta u_{i,j} = (\sigma_{ij}\delta u_j)_{,j} - \sigma_{ij,j}\delta u_i$, using the divergence theorem, we also recover the force equilibrium equation (1.123).

The variational principle associated with Π_1 is often called Hellinger–Reissner principle, named after Ernst Hellinger (1883–1950) and Eric Reissner (1913–1996). Furthermore, to relax the compatibility condition as well as the prescribed displacements on S_u, we introduce another variational indicator

$$\Pi_2 = \Pi_o - \int_V \boldsymbol{\lambda}_\varepsilon^T (\boldsymbol{\varepsilon} - \boldsymbol{\partial}_\varepsilon \mathbf{u}) dV - \int_{S_u} \boldsymbol{\lambda}_u^T (\mathbf{u}^{S_u} - \mathbf{u}_p) dS, \tag{1.153}$$

where $\boldsymbol{\lambda}_\varepsilon$ and $\boldsymbol{\lambda}_u$ are Lagrangian multipliers.

Invoking the stationarity of Π_2, i.e., evaluating $\delta\Pi_2 = 0$ with respect to $\mathbf{u}$, $\boldsymbol{\varepsilon}$, $\boldsymbol{\lambda}_\varepsilon$, and $\boldsymbol{\lambda}_u$, we can easily identify $\boldsymbol{\lambda}_\varepsilon = \boldsymbol{\sigma}$ and $\boldsymbol{\lambda}_u = \mathbf{f}^{S_u}$ and recover the equilibrium condition, the compatibility condition, and the displacement boundary condition, along with $f_i^{S_u} = \sigma_{ij}n_j$ on S_u and $f_i^{S_f} = \sigma_{ij}n_j$ on S_f. The variational principle associated with Π_2 is often called Hu–Washizu principle, named after Hu Haichang (1928–2011) and Kyuichiro Washizu (1921–1981).

The fifteen equations governing six stress components, six strain components, and three displacements can be directly derived from Hu–Washizu principle as well as the geometrical (essential, Dirichlet) boundary conditions, named after Johann Peter Gustav Lejeune Dirichlet (1805–1859), and force (natural, Neumann) boundary conditions, named after Carl Gottfried Neumann (1832–1925). In this section, we only provide a foundation of continuum mechanics approaches. The detailed discussion on geometrical nonlinearity with large displacements and deformations and material nonlinearity will be presented in the discussions of the immersed methods and their applications.

1.3 Some Mathematical Tools

In this section, we briefly review some mathematical tools pertaining to immersed methods, for example, the Fourier transform in periodic spatial domain in the Immersed Boundary Method.

1.3.1 *Fourier series, Fourier transform, and Laplace transform*

One of the most important concepts in engineering mathematics is the use of Fourier Series and Transforms, named after Jean Baptiste Joseph Fourier (1768–1830). Based on the following integrals, for m and $n \in \mathcal{Z}$,

$$\frac{2}{T}\int_0^T \cos\frac{2\pi mt}{T}\cos\frac{2\pi nt}{T}dt = \begin{cases} 1 & m = n, \\ 0 & m \neq n; \end{cases}$$
$$\frac{2}{T}\int_0^T \sin\frac{2\pi mt}{T}\sin\frac{2\pi nt}{T}dt = \begin{cases} 1 & m = n, \\ 0 & m \neq n; \end{cases} \tag{1.154}$$

and

$$\int_0^T \sin\frac{2\pi mt}{T}\cos\frac{2\pi nt}{T}dt = \int_0^T \sin\frac{2\pi mt}{T}dt = \int_0^T \cos\frac{2\pi nt}{T}dt = 0, \tag{1.155}$$

we can express any T-periodic function $f(t)$ with $t \in (-\infty, +\infty)$ by a Fourier series:

$$f(t) = a_o + \sum_{n=1}^{+\infty} \left(a_n \cos \frac{2\pi nt}{T} + b_n \sin \frac{2\pi nt}{T} \right), \tag{1.156}$$

with

$$\begin{aligned} a_o &= \frac{1}{T} \int_0^T f(t) dt, \\ a_n &= \frac{2}{T} \int_0^T f(t) \cos \frac{2\pi nt}{T} dt, \\ b_n &= \frac{2}{T} \int_0^T f(t) \sin \frac{2\pi nt}{T} dt. \end{aligned}$$

Furthermore, using the following integral with respect to a series of even functions, for m and $n \in \mathcal{Z}$,

$$\int_0^{T/2} \cos \frac{2\pi mt}{T} \cos \frac{2\pi nt}{T} dt = \begin{cases} T/2 & m = n = 0, \\ T/4 & m = n \neq 0, \\ 0 & m \neq n; \end{cases} \tag{1.157}$$

any even T-periodic function $f_e(t)$ with $t \in (-\infty, +\infty)$ can be expressed as

$$f_e(t) = A_o + \sum_{n=1}^{+\infty} A_n \cos \frac{2\pi nt}{T}, \tag{1.158}$$

with

$$\begin{aligned} A_o &= \frac{2}{T} \int_0^{T/2} f_e(t) dt, \\ A_n &= \frac{4}{T} \int_0^{T/2} f_e(t) \cos \frac{2\pi nt}{T} dt. \end{aligned}$$

Analogously, using the following integral with respect to a series of odd functions, for m and $n \in \mathcal{Z}$,

$$\int_0^{T/2} \sin \frac{2\pi mt}{T} \sin \frac{2\pi nt}{T} dt = \begin{cases} T/4 & m = n, \\ 0 & m \neq n; \end{cases} \tag{1.159}$$

any odd T-periodic function $f_o(t)$ with $t \in (-\infty, +\infty)$ can be expressed as

$$f_o(t) = \sum_{n=1}^{+\infty} B_n \sin \frac{2\pi nt}{T}, \tag{1.160}$$

with

$$B_n = \frac{4}{T}\int_0^{T/2} f_o(t)\sin\frac{2\pi nt}{T}dt.$$

In addition, it is easy to verify that for a T-periodic function $f(t)$, we have

$$\begin{aligned}\int_{T/2}^{T} f(t)\sin\frac{2\pi nt}{T}dt &= \int_0^{T/2} -f(-t)\sin\frac{2\pi nt}{T}dt,\\ \int_{T/2}^{T} f(t)\cos\frac{2\pi nt}{T}dt &= \int_0^{T/2} f(-t)\cos\frac{2\pi nt}{T}dt.\end{aligned} \tag{1.161}$$

Using Eq. (1.161) and the definitions of odd and even functions, we can show that odd functions have sine expansions whereas even functions have cosine expansions. Moreover, the coefficients of the sine and cosine expansions only involve half period integrations. Since any T-periodic function $f(t)$ with $t \in (-\infty, +\infty)$ can be written as a combination of a T-periodic odd function $f_o(t) = [f(t) - f(-t)]/2$ and a T-periodic even function $f_e(t) = [f(t) + f(-t)]/2$, using Eq. (1.161), the Fourier series of $f(t)$ written in the form of (1.156) can be viewed as the combination of forms (1.158) and (1.160) with

$$a_o = \frac{2}{T}\int_0^{T/2}\frac{[f(t)+f(-t)]}{2}dt = \frac{1}{T}\int_0^{T} f(t)dt = A_o,$$

$$a_n = \frac{4}{T}\int_0^{T/2}\frac{[f(t)+f(-t)]}{2}\cos\frac{2\pi nt}{T}dt = \frac{2}{T}\int_0^{T} f(t)\cos\frac{2\pi nt}{T}dt = A_n,$$

$$b_n = \frac{4}{T}\int_0^{T/2}\frac{[f(t)-f(-t)]}{2}\sin\frac{2\pi nt}{T}dt = \frac{2}{T}\int_0^{T} f(t)\sin\frac{2\pi nt}{T}dt = B_n.$$

Furthermore, employing the complex variable representations in Eq. (1.42), we can rewrite Eq. (1.156) in the following complex exponential form:

$$\begin{aligned} f(t) &= a_o + \sum_{n=1}^{+\infty}\left(a_n\frac{e^{i2\pi nt/T} + e^{-i2\pi nt/T}}{2} + b_n\frac{e^{i2\pi nt/T} - e^{-i2\pi nt/T}}{2i}\right)\\ &= a_o + \sum_{n=1}^{+\infty}\left(\frac{a_n - ib_n}{2}e^{i2\pi nt/T} + \frac{a_n + ib_n}{2}e^{-i2\pi nt/T}\right) = \sum_{k=-\infty}^{+\infty} c_k e^{i2\pi kt/T},\end{aligned}$$

with $k \in \mathcal{Z}$ and

$$c_k = \frac{1}{T}\int_0^{T} f(t)e^{-i2\pi kt/T}dt.$$

At this point, we are ready to discuss the convergence issue of this Fourier series expansion. By using integration by parts, we also have

$$c_k = \frac{i}{2\pi k}\left[f(T) - f(0)\right] + \frac{T}{(2\pi k)^2}\left[f'(T) - f'(0)\right] - \frac{T}{(2\pi k)^2}\int_0^T f''(t)e^{-i2\pi kt/T}dt,$$

from which we can obtain the rate of convergence of the Fourier coefficients as $k \to \pm\infty$.

For instance, if $f(T) \neq f(0)$, the Fourier coefficients c_k converge as $O(\frac{1}{k})$, and if $f(T) = f(0)$ and $f'(T) \neq f'(0)$, the Fourier coefficients c_k converge as $O(\frac{1}{k^2})$. Furthermore, any function $f(t)$ with $t \in (-\infty, +\infty)$ can be considered as a T-periodic function with $T \to +\infty$. It can be proven (refer to Hildebrand (1976)) that for the even function $f_e(t)$, we have

$$f_e(t) = \int_0^{+\infty} A(\omega)\cos\omega t d\omega, \qquad A(\omega) = \frac{2}{\pi}\int_0^{+\infty} f_e(t)\cos\omega t dt; \tag{1.162}$$

whereas for the odd function $f_o(t)$,

$$f_o(t) = \int_0^{+\infty} B(\omega)\sin\omega t d\omega, \qquad B(\omega) = \frac{2}{\pi}\int_0^{+\infty} f_o(t)\sin\omega t dt. \tag{1.163}$$

Again, since any function can be viewed as a combination of an odd function with an even function, analogous to the Fourier series expansion, based on Eqs. (1.163) and (1.162), we derive the following Fourier transform:

$$g(t) = \frac{1}{2\pi}\int_{-\infty}^{+\infty} G(\omega)e^{i\omega t}d\omega, \qquad G(\omega) = \int_{-\infty}^{+\infty} g(t)e^{-i\omega t}dt. \tag{1.164}$$

In some cases, the Fourier transform is also written as

$$g(t) = \int_{-\infty}^{+\infty} G(f)e^{i2\pi ft}df, \qquad G(f) = \int_{-\infty}^{+\infty} g(t)e^{-i2\pi ft}dt. \tag{1.165}$$

The important prerequisite for the existence of the Fourier series or transform is that $\int_{-\infty}^{+\infty}|f(t)|dt$ exists, which means c_k or $F(f) \to 0$ as k or $f \to \pm\infty$.

An interesting aspect of the Fourier series is the Gibbs' phenomenon, named after Josiah Willard Gibbs (1839–1903). Consider the δ function or Dirac function, named after Paul Adrien Maurice Dirac (1902–1984):

$$f(t) = \delta(t) = \begin{cases} 1 & t = 0, \\ 0 & \text{elsewhere}; \end{cases} \tag{1.166}$$

with its Fourier transform $F(f) = 1$.

If we truncate $F(f) = 1$ to a symmetric square pulse, we get the approximation for $\delta(t)$ as

$$\tilde{f}(t) = \int_{-k}^{k} F(f) e^{i2\pi ft} df = \frac{\sin kt}{\pi t}, \tag{1.167}$$

a so-called sin*c* function, where $t = \frac{n\pi}{k}$, $n \in \mathcal{Z}$ corresponds to $\tilde{f}(t) = 0$.

Now, suppose we have a Heeaviside function (or step function), named after Oliver Heaviside (1850–1925):

$$f(t) = H(t) = \int_{-\infty}^{t} \delta(s) ds. \tag{1.168}$$

Its approximate Fourier transform can be expressed as

$$\tilde{F}(t) = \int_{-\infty}^{t} \frac{\sin ks}{\pi s} ds = \frac{1}{2} + \int_{0}^{t} \frac{\sin ks}{\pi s} ds, \tag{1.169}$$

with

$$\int_{-\infty}^{0} \frac{\sin x}{x} dx = \int_{0}^{+\infty} \frac{\sin x}{x} dx = \frac{1}{2}\int_{-\infty}^{+\infty} \frac{\sin x}{x} dx = \frac{\pi}{2}. \tag{1.170}$$

Equation (1.170) can be easily proven using a complex contour integral $\oint_C \frac{e^{iz}}{z} dz$ as presented in detail in Wang (2022, 2008) (Fig. 1.4.).

The important fact is that the maximum of $\tilde{F}(f) = \tilde{F}(f)\big|_{t=\pi/k} = \frac{1}{2} + \frac{1}{\pi}\int_0^{\pi} \frac{\sin t}{t} dt$, which corresponds to $\frac{d}{dt}\int_0^t \frac{\sin ks}{\pi s} ds = \frac{\sin kt}{\pi t} = 0$ and is independent of k. Thus, if we have a jump (or discontinuity) of $f(t)$ at t_o, we can always define a new value $f(t_o) = [f_+(t_o) + f_-(t_o)]/2$, where $f_+(t_o) = \lim_{h\to 0} f(t_o + h)$ and $f_-(t_o) = \lim_{h\to 0} f(t_o - h)$, with $h > 0$. Nevertheless, since the overshoot is independent of k, i.e., the number of the

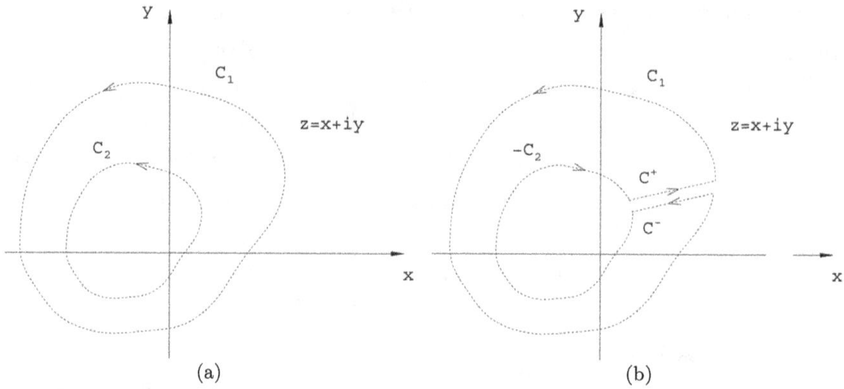

Fig. 1.4. Path independence of complex contour integrals.

Fourier series taken, the overshoot will always exist. This is called Gibbs' phenomenon, named after Josiah Willard Gibbs (1839–1903).

Let's prove the following important Convolution Theorem. Denote a pair of functions $u(t)$ and $v(t)$ with the corresponding Fourier transforms $U(f)$ and $V(f)$. For the following new function which is defined as the convolution $c(t)$

$$c(t) = \int_{-\infty}^{+\infty} u(s)v(t-s)ds.$$

We will be able to demonstrate that the Fourier transform of $c(t)$ as $C(f)$ is in fact the multiplication of two spectrum functions $U(f)$ and $V(f)$. Using Eq. (1.165), we have

$$\begin{aligned} C(f) &= \int_{-\infty}^{+\infty}\int_{-\infty}^{+\infty} u(s)v(t-s)e^{-i2\pi ft}dtds \\ &= \int_{-\infty}^{+\infty}\int_{-\infty}^{+\infty} u(s)e^{-i2\pi fs}dsv(t-s)e^{-i2\pi(t-s)}d(t-s) \\ &= U(f)V(f). \end{aligned} \tag{1.171}$$

Note that if we express the Fourier transform of $c(t)$, $u(t)$, and $v(t)$ with respect to the natural frequency ω instead of frequency f, following Eq. (1.165), it is easy to verify that we need the extra coefficient 2π,

$$C(\omega) = 2\pi U(\omega)V(\omega).$$

Before we expand the discussion from the discrete infinite series to the continuous transformation, let's first discuss the very important discrete sampling and the corresponding discrete Fourier transform (DFT) and the popular fast Fourier transform (FFT). Assume the time period is T and the sampling time step is Δt. Denote N as the number of discrete points (or sampling points) within a period T, thus $N = \frac{T}{\Delta t}$ as well as $\Delta f = \frac{1}{N\Delta t} = \frac{1}{T}$ with Δf as the increment of the frequency and Δt as the increment of the time. The sampling function and its spectrum are illustrated in Figs. 1.5 and 1.6.

Note that for T-periodic signals in time domain, the signal can be viewed as a convolution of an impulse train $\delta(t - nT)$ with the function within a period as illustrated in Fig. 1.7. The detailed proof of the important convolution relationships in functional and spectrum domains are presented in Eq. (1.171). The frequency domain of such an impulse train is in fact a multiple of $\frac{1}{T}$ with another impulse train $\delta(f - n\frac{1}{T})$. This pair in fact is reflected in the Fourier series expressed in complex domain. Due to the

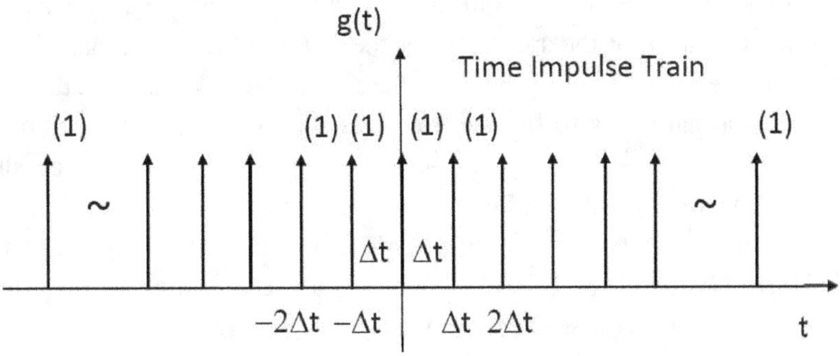

Fig. 1.5. Sampling train in the temporal or functional domain.

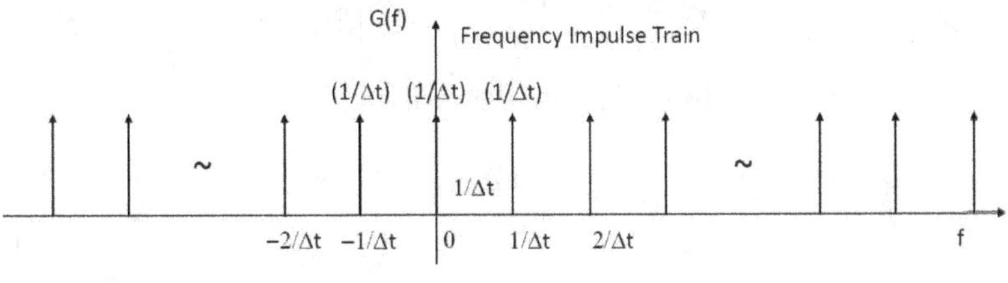

Fig. 1.6. Sampling train in the spectrum or frequency domain.

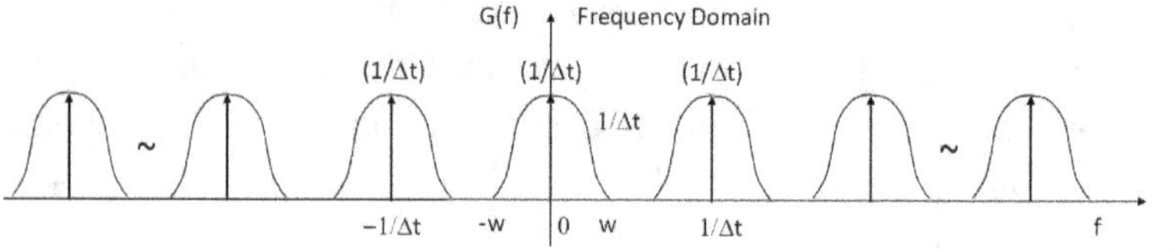

Fig. 1.7. Spectrum of the discretized function in the frequency domain.

discrete nature of the time domain signal, with an impulse chain

$$\sum_{-\infty}^{+\infty} \delta(t - n\Delta t),$$

the frequency domain will also be a convolution of the spectrum for the T-period function about the impulse chain

$$\sum_{-\infty}^{+\infty} \delta(f - m\Delta f).$$

Furthermore, we must avoid the so-called aliasing which requires the compact nature of the spectrum for the T-period function, namely, the bandwidth denoted within $[-w, w]$, where w is the half bandwidth in the frequency domain, has to be able to cover the entire spectrum or rather $2w \le \frac{1}{\Delta t}$ or $\Delta t \le \frac{1}{2w}$ or $\Delta t \le \frac{1}{2f_{\max}}$ or $\Delta t \le \frac{\pi}{\omega_{\max}}$. In signal processing, the sampling rate f_s, also defined as the reciprocal of the sampling time step, namely, $\frac{1}{\Delta t}$, is two times the so-called Nyquist frequency f_n, named after Harry Nyquist (1889–1976). Thus, the avoid aliasing, we must have $f_{\max} < f_n = \frac{f_s}{2}$. Of course, here we have $\Delta f = \frac{2w}{N}$. With $2w \le \frac{1}{\Delta t}$, we can also use $\Delta f = \frac{1}{N\Delta t} = \frac{1}{T}$. Therefore, we have the Fourier series expressed as

$$x(t) = \int_{-w}^{w} X(f) e^{i2\pi f t} df = \int_{0}^{1/\Delta t} X[m] \delta(f - m\Delta f) e^{i2\pi f t} df,$$

$$X(f) = \frac{1}{T} \int_{0}^{T} x(t) e^{-i2\pi f t} dt = \frac{1}{T} \int_{0}^{T} x[n] \delta(t - n\Delta t) e^{-i2\pi f t} dt;$$

and consequently the discrete Fourier transform as follows

$$\begin{aligned} x[n] &= \int_{0}^{1/\Delta t} X[m] \delta(f - m\Delta f) e^{i2\pi f t} df = \sum_{m=0}^{N-1} X[m] e^{i2\pi \frac{nm}{N}}; \\ X[m] &= \frac{1}{N} \int_{0}^{T} x[n] \delta(t - n\Delta t) e^{-i2\pi f t} dt = \frac{1}{N} \sum_{n=0}^{N-1} x[n] e^{-i2\pi \frac{nm}{N}}. \end{aligned} \tag{1.172}$$

DFT will be the basis for FFT utilizing the properties of divisions of polar angles and the symmetric nature with respect to both x and y axes, namely, all angles within the second, third, and fourth quadrants can be expressed as their counter parts in the first quadrant. Naturally, for FFT to work efficiently, we normally assign the number of samples N to be the power of 2 such that $N = 2^p$ with $p = \log_2 N$.

We are now ready to introduce the Laplace transform, named after Pierre–Simon Laplace (1749–1827) and often denoted as L of a real variable function $f(t)$,

$$F(s) = \int_0^{+\infty} e^{-st} f(t)dt, \tag{1.173}$$

with the inverse Laplace transform,

$$f(t) = \lim_{b\to+\infty} \frac{1}{2\pi i} \int_{a-ib}^{a+ib} e^{st} F(s)ds, \tag{1.174}$$

where $f(t)$ is of exponential order, i.e., there exist real constants M, t_m, and c, such that $|f(t)| \leq Me^{ct_m}$, $\forall t > t_m$; a is a real constant, and s is a complex number with $\mathrm{Re}(s) = a > c$.

Consider the following real variable function:

$$x(t) = \begin{cases} e^{-at} f(t) & t \geq 0, \\ 0 & t < 0. \end{cases} \tag{1.175}$$

Since $f(t)$ is of exponential order and $a > c$, $\int_{-\infty}^{+\infty} |f(t)|dt$ exists, by using the Fourier transform of Eq. (1.164), we have

$$\begin{aligned} X(\omega) &= \int_0^{+\infty} f(t)e^{-at}e^{-i\omega t}dt, \\ f(t)e^{-at} &= \frac{1}{2\pi} \int_{-\infty}^{+\infty} X(\omega)e^{i\omega t}d\omega. \end{aligned} \tag{1.176}$$

Thus, if we consider $s = a + iy$, with $a > c$, i.e., $ds = idy$, the inverse Laplace transform becomes

$$\begin{aligned} \lim_{b\to+\infty} \frac{1}{2\pi i} \int_{a-ib}^{a+ib} e^{st} F(s)ds &= \lim_{b\to+\infty} \frac{1}{2\pi i} \int_{a-ib}^{a+ib} \int_0^{+\infty} e^{st-su} f(u)duds \\ &= \lim_{b\to+\infty} \frac{1}{2\pi} e^{at} \int_{-b}^{b} e^{iyt}dy \int_0^{+\infty} e^{-iyu} \left[e^{-au} f(u)\right] du, \end{aligned}$$

and using the Fourier transform of Eq. (1.176),

$$\lim_{b\to+\infty}\frac{1}{2\pi i}\int_{a-ib}^{a+ib} e^{st}F(s)ds = \frac{1}{2\pi}e^{at}2\pi f(t)e^{-at} = f(t),$$

which proves that the inverse Laplace transform is indeed valid.

Moreover, let $f(t)$ have a period $T > 0$, i.e., $f(t) = f(t+T)$. It can be proven that

$$\mathcal{L}[f(t)] = F(s) = \frac{\int_0^T e^{-st}f(t)dt}{1-e^{-sT}}.$$

The evaluation of the inverse Laplace transform often requires the use of contour integrals to be discussed in Section 1.3.2.

1.3.2 *Contour integral and conformal mapping*

To better understand the inverse transformations, we must introduce the concepts of contour integral and conformal mapping in the complex plane. These concepts are of tremendous value in the solutions of problems of heat transfer, potential theory, fluid mechanics, elasticity, and many science and engineering disciplines.

The key ingredient of the complex contour integral and conformal mapping involves the differentiability (analyticity) of a complex function under consideration. Although, parallel to real variable theories, the limit and continuity of a complex variable function are defined the same as those of a real variable function, the differentiability is quite different. In fact, it is clear from the definition of the derivative of $f(z)$ at z, $f'(z) = \lim_{\Delta z\to 0}\frac{f(z+\Delta z)-f(z)}{\Delta z}$, $f'(z)$ can have different values based on different paths of Δz.

Theorem 1.10 (Cauchy–Riemann). *Suppose $f(z) = \phi(x,y) + i\psi(x,y)$ is defined in a domain S. The necessary condition of $f(z)$ being differentiable at z (analytic, regular, or holomophic) is that the following Cauchy–Riemann equations are satisfied:*

$$\frac{\partial\phi}{\partial x} = \frac{\partial\psi}{\partial y} \quad \textit{and} \quad \frac{\partial\phi}{\partial y} = -\frac{\partial\psi}{\partial x}; \tag{1.177}$$

and the sufficient condition corresponds to an additional requirement of both ϕ and ψ being C^1 at z in S.

Equation (1.177), so-called Cauchy–Riemann equations, named after August Louis Cauchy (1789–1857) and Georg Friedrich Bernhard Riemann (1826–1866), can also be expressed in the polar coordinate system as

$$\frac{\partial \phi}{\partial r} = \frac{1}{r}\frac{\partial \psi}{\partial \theta} \quad \text{and} \quad \frac{1}{r}\frac{\partial \phi}{\partial \theta} = -\frac{\partial \psi}{\partial r}. \tag{1.178}$$

In fact, assuming $f(z)$ is analytic, i.e., $f'(z)$ is uniquely defined, we have for the Cartesian coordinate system,

$$\frac{\partial f(x,y)}{\partial x} = f'(z)\frac{\partial (x+iy)}{\partial x} \quad \text{and} \quad \frac{\partial f(x,y)}{\partial y} = f'(z)\frac{\partial (x+iy)}{\partial y}; \tag{1.179}$$

and for the polar coordinate system

$$\frac{\partial f(r,\theta)}{\partial r} = f'(z)\frac{\partial (re^{\theta})}{\partial r} \quad \text{and} \quad \frac{\partial f(r,\theta)}{\partial \theta} = f'(z)\frac{\partial (re^{\theta})}{\partial \theta}. \tag{1.180}$$

It is then clear that by matching the real and imaginary parts of the equations in (1.179), we can easily confirm the Cauchy–Riemann equation in the Cartesian coordinate system (1.177). Likewise, from the equations in (1.180), the same relationship in the polar coordinate system can be derived as (1.178).

Obviously, if both ϕ and ψ are C^1 and $f(z)$ is analytic in $\mathcal{S}$, then both ϕ and ψ satisfy Laplace's equation, i.e., $\nabla^2\phi = \nabla^2\psi = 0$ in $\mathcal{S}$, and are called harmonic functions, or harmonic conjugate functions. Note that the complex conjugate function $f(z) = \bar{z}$ is not analytic. To make multiple-valued functions single-valued, we set up an artificial barrier, a so-called branch cut, which often starts from a branch point to the infinity or between branch points. Hence, each branch of a multiple-valued function is represented by a complex sheet and the whole complex plane can be viewed as those sheets superimposed on each other and joined together at the branch cut. Thus, the Riemann surface of a function, named after Bernhard Riemann (1826–1866), is a collection of those sheets. Obviously, $\ln z$ has infinitely many sheets within its Riemann surface, whereas the complex function

$$z^{1/3} = r^{1/3}e^{i\theta/3} = r^{1/3}e^{i\theta_p/3+i2\pi m/3}, \tag{1.181}$$

with $m \in \mathcal{Z}$, has three sheets in its Riemann surface corresponding to its three branches of solutions, i.e., $r^{1/3}e^{i\theta_p/3+i2\pi k}$, $r^{1/3}e^{i\theta_p/3+i2\pi(3k+1)/3}$, and $r^{1/3}e^{i\theta_p/3+i2\pi(3k+2)/3}$ with $k \in \mathcal{Z}$.

Furthermore, if $f(z)$ is analytic in a region, so are $f'(z)$, $f''(z)$, ..., $f^{(n)}(z)$ with $n \in \bar{\mathcal{N}}$ and $n \geq 3$ in the same region, i.e., all higher derivatives exist in that region. A point at which $f(z)$ is not analytic is a singular point or singularity of $f(z)$. More specifically, if we can find a circle $|z - z_o| = \delta > 0$ which encloses no singular point other than z_o, the point z_o is called an isolated singularity; if we can find $n \in \bar{\mathcal{N}}$ such that $\lim_{z \to z_o}(z - z_o)^n f(z)$ exists, then the singularity point z_o is called an nth order pole (the first-order pole is often called a simple pole). The singularity z_o is also called a removable singularity if $\lim_{z \to z_o} f(z)$ exists. A singularity of a multiple-valued function is also called a branch point. Finally, a singularity which is not a pole, a removable singularity, or a branch point is called an essential singularity.

We define the complex logarithmic function by mimicking the real variable formula as

$$f(z) = \ln z = \ln(r) + i\theta, \quad \text{with } r = |z|, \tag{1.182}$$

where the argument θ has infinitely many angles differing by integral multiplies of 2π which renders $\ln z$ multiple-valued.

We begin the discussion on the complex contour integral by defining a simply-connected domain as a domain in which any simple closed curve can be shrunk to a point without leaving the domain. The opposite of a simply-connected domain is a multiply-connected domain, which can be divided into a simply-connected domain with some simple closed continuous curves called Jordan curves, named after Marie Ennemond Camille Jordan (1838–1922), enclosing infinitesimally small area.

Analogous to the integral $\int_a^b f(x)dx$ for a real variable function, the complex integral is defined as $\int_C f(z)dz$, where the path C can generally be a piecewise smooth simple curve (an arc or a closed curve). In fact, if $f(z)$ is analytic in a simply-connected domain and z_o is any point within the domain, $\int_{z_o}^z f(\zeta)d\zeta = F(z) - F(z_o)$ is also analytic in the same domain with $F'(z) = f(z)$. According to the convention, the region of integration must lie to the left of curve C. Hence, if C is a closed curve, with the contour integral expressed as $\oint_C f(z)dz$, C is in the counterclockwise direction.

Theorem 1.11 (Cauchy's theorem). *If $f(z)$ is analytic in a simply-connected domain enclosed by a piecewise smooth simple closed curve*

C, then

$$\oint_C f(z)dz = 0. \qquad (1.183)$$

Thus, if $f(z)$ is analytic in a simply-connected domain, then $\int_C f(z)dz$ is independent of path C from point p_1 to point p_2 within the domain, i.e., $\int_{C_1} f(z)dz = \int_{C_2} f(z)dz$. Furthermore, $\oint_C f(z)dz$ is also independent of closed path C within the domain enclosing a point, which means if $f(z)$ is analytic on and between C_1 and C_2, as shown in Fig. 1.4, $\oint_{C_1} f(z)dz = \oint_{C_2} f(z)dz$. This can be easily proven by using a Jordan curve to cut the multiply-connected domain between C_1 and C_2 into a simply-connected domain and applying Cauchy's theorem.

Theorem 1.12 (Cauchy integral theorem). *Let $f(z)$ be analytic in a simply-connected domain enclosed by a piecewise counterclockwise smooth simple closed curve C and z_o be any point within the domain, then*

$$\oint_C \frac{f(z)}{z - z_o} dz = 2\pi i f(z_o). \qquad (1.184)$$

Although the Taylor series for an analytic function $f(z)$ in a domain $\mathcal{D}$ is the same as for a real variable function, i.e., $f(z) = \sum_{n=0}^{+\infty} \frac{f^{(n)}(a)}{n!}(z - a)^n$, with the disk $|z - a|$ lying entirely within $\mathcal{D}$, a more general series expansion in complex analysis is the Laurent series, named after Pierre Alphonse Laurent (1813–1854),

$$f(z) = \sum_{n=-\infty}^{+\infty} c_n (z - z_o)^n, \qquad (1.185)$$

employing Cauchy integral the coefficient c_n can be easily derived as $c_n = \frac{1}{2\pi i} \oint_C \frac{f(\zeta)}{(\zeta - z_o)^{n+1}} d\zeta$, with any piecewise smooth simple closed counterclockwise path C in a domain enclosed by two concentric circles centered at $z = z_o$.

Theorem 1.13 (Generalized Cauchy integral theorem). *Let $f(z)$ be analytic in a simply-connected domain enclosed by a piecewise smooth simple closed curve C oriented counterclockwise and let z_o be any point within the domain. Then*

$$\oint_C \frac{f(z)}{(z - z_o)^{n+1}} dz = \frac{2\pi i}{n!} f^{(n)}(z_o), \quad \text{with } n \in \bar{\mathcal{N}}. \qquad (1.186)$$

In addition, as an extension of Cauchy integral theorem and Generalized Cauchy integral theorem, we can easily prove the following Residue theorem, by employing the Laurent series.

Theorem 1.14 (Residue theorem). *Let $f(z)$ be an analytic function inside the domain enclosed by a piecewise smooth simple closed curve C oriented counterclockwise, and $f(z)$ is meromorphic, i.e., analytic everywhere except at a finite number of n_jth poles z_j, with $j = 1,\ldots,k$, where k is the number of poles and n_j is the order of the jth pole. Then*

$$\oint_C f(z)dz = 2\pi i \sum_{j=1}^{k} c_{-1}^{(j)},$$

with

$$c_{-1}^{(j)} = \frac{1}{(n_j-1)!} \lim_{z\to z_j} \frac{d^{n_j-1}}{dz^{n_j-1}} \left[(z-z_j)^{n_j} f(z)\right].$$

Cauchy's theorem is used to prove an important theorem named after Joseph Liouville (1809–1882).

Theorem 1.15 (Liouville theorem). *If $f(z)$ is analytic in the entire complex plane and $f(z)$ is bounded, then $f(z)$ must be a constant.*

For further illustrations, we use the following few examples to show that the complex contour integral is of great importance in evaluating the inverse Fourier and Laplace transforms.

Example: Evaluate the inverse Fourier transform $f(x) = \frac{1}{2\pi}\int_{-\infty}^{+\infty} \frac{e^{ix\omega}}{\omega^2+1} d\omega$.

For $x \geq 0$, consider a complex contour integral $\oint_C \frac{e^{ix\zeta}}{\zeta^2+1} d\zeta$ with $C = C_+ + C_\infty$ as shown in Fig. 1.8. According to the Residue theorem, we have

$$\oint_C \frac{e^{ix\zeta}}{\zeta^2+1} d\zeta = 2\pi i \frac{e^{ixi}}{i+i} = \pi e^{-x}.$$

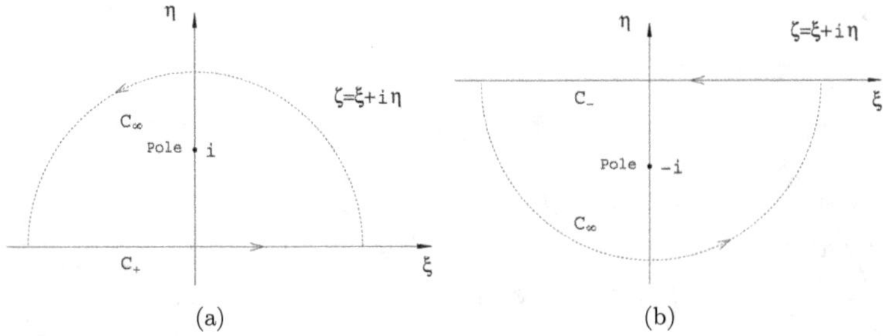

Fig. 1.8. Examples of complex contours.

Moreover, we also have $\int_{C_\infty} \frac{e^{ix\zeta}}{\zeta^2+1} d\zeta = 0$ as $R \to +\infty$. Therefore,

$$\frac{1}{2\pi}\int_{-\infty}^{+\infty} \frac{e^{ix\omega}}{\omega^2+1} d\omega = \frac{1}{2\pi}\int_{C_+} \frac{e^{ix\zeta}}{\zeta^2+1} d\zeta = \frac{1}{2\pi}\oint_{C} \frac{e^{ix\zeta}}{\zeta^2+1} d\zeta = \frac{1}{2}e^{-x}.$$

For $x < 0$, analogously, if we select the contour with $C = C_- + C_\infty$ as shown in Fig. 1.8 and use the Residue theorem, we have

$$\oint_C \frac{e^{ix\zeta}}{\zeta^2+1} d\zeta = 2\pi i \frac{e^{ix(-i)}}{-i-i} = -\pi e^x,$$

and obtain

$$\frac{1}{2\pi}\int_{-\infty}^{+\infty} \frac{e^{ix\omega}}{\omega^2+1} d\omega = -\frac{1}{2\pi}\int_{C_+} \frac{e^{ix\zeta}}{\zeta^2+1} d\zeta = -\frac{1}{2\pi}\oint_{C} \frac{e^{ix\zeta}}{\zeta^2+1} d\zeta = \frac{1}{2}e^{x}.$$

Therefore, combining both cases, we have

$$\frac{1}{2\pi}\int_{-\infty}^{+\infty} \frac{e^{ix\omega}}{\omega^2+1} d\omega = \frac{1}{2}e^{-|x|}.$$

Consider a complex variable function $\zeta = f(z)$ with $\zeta = \xi + i\eta$ and $z = x + iy$ as a mapping from a complex z plane to a complex ζ plane. To ensure a one-to-one correspondence between the domain $\mathcal{D}$ in the z plane and the domain $\bar{\mathcal{D}}$ in the ζ plane, we assume that $\zeta = \zeta(x,y)$ and $\eta = \eta(x,y)$ are single-valued, and so are the inverse functions $x = x(\xi,\eta)$ and $y = y(\xi,\eta)$. Note, however, from the Cauchy–Riemann equations, we have $\xi_x = \eta_y$ and $\xi_y = -\eta_x$. Naturally, we must then have the Jacobian expressed as

$$\frac{\partial(\xi,\eta)}{\partial(x,y)} = \begin{vmatrix} \xi_x & \xi_y \\ \eta_x & \eta_y \end{vmatrix} = \xi_x \eta_y - \xi_y \eta_x = \xi_x^2 + \xi_y^2 = |f'(z)|^2 \neq 0.$$

Theorem 1.16. *Consider an analytic function $\zeta = f(z) = \xi(x,y) + i\eta(x,y)$ in a domain $\mathcal{D}$ with $|f'(z)| \neq 0$. Assuming the mapping $\zeta = f(z)$ from the domain $\mathcal{D}$ in the z plane to the domain $\bar{\mathcal{D}}$ in the ζ plane is bijective, if $\phi(x,y)$ is harmonic in $\mathcal{D}$, then $\phi(x(\xi,\eta), y(\xi,\eta)) = \hat{\phi}(\xi,\eta)$ is harmonic in $\hat{\mathcal{D}}$.*

Consider two smooth curves C_1 and C_2 intersecting at z_o in the complex z plane. Assume a mapping $\zeta = f(z)$ with $f'(z_o) \neq 0$, and the corresponding two smooth curves $\hat{C}_1$ and $\hat{C}_2$ intersecting at $\zeta_o = f(z_o)$. If we parameterize C_1 and C_2 as $z_1(s) = x_1(s) + iy_1(s)$ and $z_2(s) = x_2(s) + iy_2(s)$ with the arc length s, the corresponding parameterized forms of $\hat{C}_1$ and $\hat{C}_2$ are also denoted as $\zeta_1(s) = \xi_1(s) + i\eta_1(s)$ and $\zeta_2(s) = \xi_2(s) + i\eta_2(s)$, and the angle between the two curves $\hat{C}_1$ and $\hat{C}_2$ at ζ_o can be expressed as

$$\begin{aligned} \arg\dot{\zeta}_2(0) - \arg\dot{\zeta}_1(0) &= \arg f'(z_o) + \arg\dot{z}_2(0) - \arg f'(z_o) - \arg\dot{z}_1(0) \\ &= \arg\dot{z}_2(0) - \arg\dot{z}_1(0), \end{aligned}$$

which is equal to the angle between the two curves C_1 and C_2 at z_o in the complex z plane.

Hence, the mapping $\zeta = f(z)$ is called the conformal mapping. In fact, any simply-connected domain $D \subset \mathcal{R}^2$ can be mapped one-to-one and conformally onto any other simply-connected domain $D' \subset \mathcal{R}^2$. In this section, we illustrated four typical examples of conformal mappings in Figs. 1.9–1.12. In Fig. 1.9, an upper plane is mapped into the interior of a unit circle through a bilinear transformation. If z_1, z_2, z_3, and z_4 are distinct, under the general bilinear transformation

$$\zeta = \frac{c_1 z + c_2}{c_3 z + c_4}, \tag{1.187}$$

where c_i are complex constants with $c_1c_4 - c_2c_3 \neq 0$, the following invariant function can be introduced

$$\frac{(z_4 - z_1)(z_2 - z_3)}{(z_2 - z_1)(z_4 - z_3)}.$$

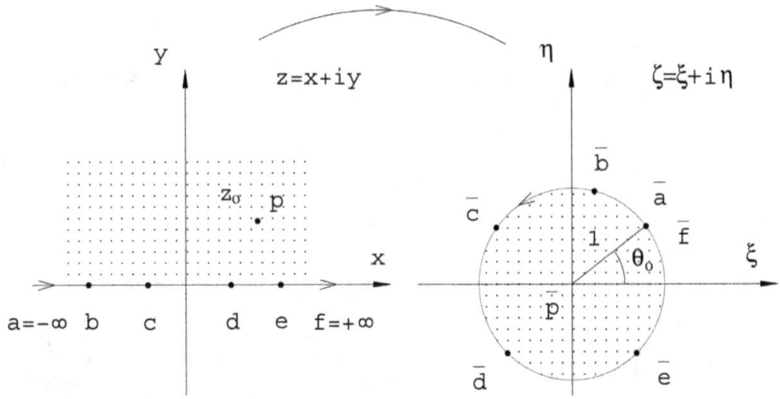

Fig. 1.9. The conformal mapping $\zeta = e^{i\theta_o}(z - z_o)/(z - \bar{z}_o)$.

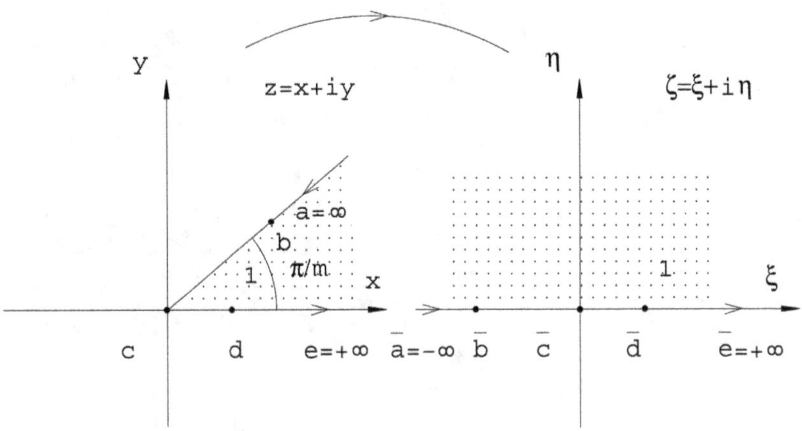

Fig. 1.10. The conformal mapping $\zeta = z^m$ with $m \geq 1/2$.

This bilinear transformation as depicted in Eq. (1.187) includes four basic operations, namely, scaling and rotation ($\zeta = cz$), translation ($\zeta = z + c$), and inversion ($\zeta = \frac{1}{z}$), and can map any circle or straight line (circle with an infinite radius) to a circle or a straight line.

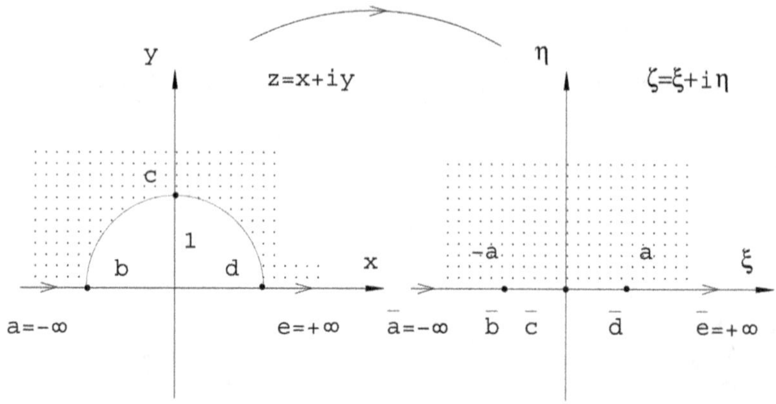

Fig. 1.11. The conformal mapping $\zeta = a/2(z+1/z)$.

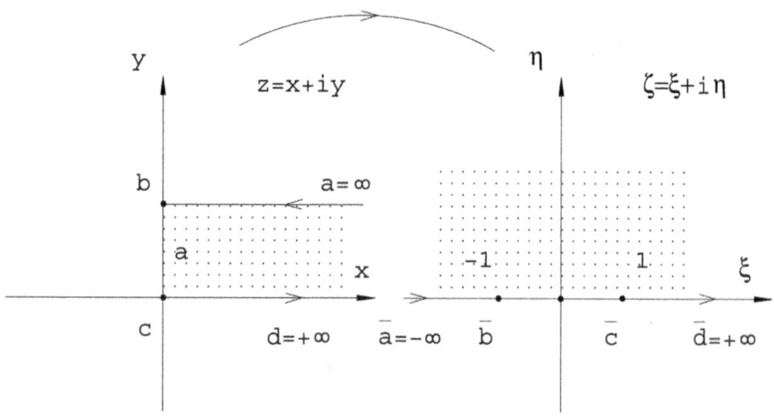

Fig. 1.12. The conformal mapping $\zeta = \cosh \pi z/a$.

In Fig. 1.10, an interior of a corner π/m is expanded to the entire upper plane with a power transformation. In Fig. 1.11, the exterior of a semi-circle is mapped to the entire upper plane with a so-called Joukowski transformation. In Fig. 1.12, a semi-infinite strip is transformed into the entire upper plane with a hyperbolic function.

More discussion on specific transformation is presented in Chapter 4. For a comprehensive study of this subject, the reader is referred to Hildebrand (1976). Finally, we present a useful theorem for conformal mapping named after Karl Herman Amandus Schwarz (1843–1921) and Elwin Bruno Christoffel (1829–1900).

Theorem 1.17 (Schwarz–Christoffel transformation). *The conformal mapping which maps the upper complex z plane to a polygon in the complex ζ plane is often expressed as $\zeta = c_1 \int \prod_{i=1}^{n} (z - x_i)^{\alpha_i/\pi - 1} dz + c_2$. The complex contants c_1 and c_2 determine the size, orientation, and position of the polygon, and α_i corresponds to the interior angles. Moreover, the points x_i along the real axis of the z plane are mapped to the vertices of the polygon in the ζ plane, and the order from $-\infty$ to $+\infty$ in the z plane corresponds to the counterclockwise order in the ζ plane. In addition, any three of the points x_i can be chosen at will, including the infinity.*

Chapter 2

Immersed Boundary Methods

Since its inception as the Ph.D. project of Professor Charlie Peskin in the 1970s, the immersed boundary (IB) methods have been introduced in numerous applications. In this chapter, we provide a review of various forms of the IB methods and their applications.

2.1 History and State-of-the-Art

Fluid–solid or fluid–structure interaction (FSI) plays a very important role in many engineering problems. In the past few decades, numerous research efforts have been directed to method developments for modeling of FSI systems (Belytschko, 1980; Liu *et al.*, 1986; Shyy *et al.*, 1996; Wang, 2008). Traditionally, staggered iterations are used to link available finite element codes for solids and structures with Computational Fluid Dynamics (CFD) codes (Bathe and Zhang, 2004; Bathe *et al.*, 1995b). Although this iterative procedure is convenient, complex dynamical system behaviors often get lost in the staggered iteration process between two different media (Zhang *et al.*, 2000; Tornberg and Shelley, 2004). In general, mesh updating and remeshing processes such as the arbitrary Lagrangian Eulerian (ALE) as described in Chapter 1 are computationally intensive and often topologically challenging (Tezduyar, 1991, 2001; Huerta and Liu, 1988; Liu and Ma, 1982; Zhang *et al.*, 2003).

Over the past six decades, marker and cell methods (1960s), finite difference methods for fluids (1970s), finite element methods for solids or structures (1970s), finite volume methods for fluids (1980s), finite element methods and spectral methods for fluids (1970s, 1980s), Lattice-gas methods (1990s), and boundary element methods or panel methods (1990s, 2000s) for hydrodynamics interacting with rigid or flexible objects have been developed and refined (Wei *et al.*, 2014; Triantafyllou *et al.*, 2000). Moreover, Stokesian Dynamics Brady and Bossis, (1980s) is found to be valid for Stokes flow with Reynold number below 1, especially useful for spherical particles, whereas the finite element method Joseph *et al.* (1990) is found to be effective for medium Reynolds numbers with high accuracy if used with mesh adaption. In addition, the fictitious Domain Method Glowinski *et al.* (1990) is effective for low to medium Reynolds numbers with many extra degrees of freedom. Moreover, front tracking, level set, and immersed interface methods Glim *et al.* (1990) have utilized specific treatment of interfaces. Procedures such as immersed interface methods (Li and LeVeque, 1994; LeVeque and Li, 1997), level set methods (Sethian, 1996; Osher and Fedkiw, 2002), front tracking methods (Glimm *et al.*, 2000; Tryggvason, 1988; Li *et al.*, 1996), dissipative particle dynamics (Pivkin *et al.*, 2011), and other finite element methods (FEM) and finite difference methods (FDM) techniques Shyy *et al.* (2000) have also been proposed to trace moving interfaces and to eliminate numerical problems associated with large motions of immersed objects (Farhat *et al.*, 2001; Shyy *et al.*, 1999; Kamakoti and Shyy, 2004; Lian and Shyy, 2005; Hughes *et al.*, 1981).

Similar to the fictitious domain method and the extended finite element method, immersed methods provide a direct coupling between immersed deformable solids and the surrounding fluid (Chessa and Belytschko, 2003; Wagner *et al.*, 2001; Glowinski *et al.*, 2001). In fact, the same principle can also be used to handle deformable solids immersed in another deformable solid medium (Wang, 2006). The immersed boundary (IB) method was first formulated by Professor Charles S. Peskin at the Courant Institute of Mathematical Science in New York University in the 1970s (Peskin, 1977). Back then he was a Ph.D. student at the Albert Einstein Medical School, following his undergraduate study at Harvard University. Since its inception, the IB method has been widely used in the modeling of biological systems (Peskin, 2002; McQueen and Peskin, 1985). For the past few decades, the application areas of the IB method include the heart and heart valves, swimming motions of aquatic organisms (Fauci and Peskin, 1988), wood pulp fiber dynamics (Stockie and Wetton, 1999), wave propagation in cochlea (Beyer, 1992), platelet aggregation (Fogelson and Peskin, 1988), biofilm processes (Dillon *et al.*, 1996), molecular interactions with fluids, and others. A very comprehensive overview of the theoretic background of the IB method and its applications for the past few decades has been presented in a review paper by Peskin (2002).

The main advantages of the IB method are its simplicity and geometric flexibility. It is a mixed Eulerian–Lagrangian scheme that combines the computational efficiency inherent in using a fixed Eulerian grid for the fluid motion with the ease of tracking the immersed boundary represented by a set of moving Lagrangian points. In essence, the influence of the immersed elastic boundary is relegated to an inhomogeneous forcing term that can be distributed to the fluid surrounding the FSI interface. If the immersed structure is narrow and thin, the displaced volume is negligible. The FSI interface is modeled as a set of material points linked with springs or fibers with merely axial tensions. However, a further development with the introduction of transverse forces, bending moments, and axial moment namely torsion will be presented in Chapter 5. This moving interface treatment facilitates the handling, without the use of ALE descriptions or adaptive meshing, of immersed boundaries or rather structures composed of nearly arbitrary shapes, sizes, and configurations. However, Beyer and LeVeque (1992) reported that Peskin's original method has been limited to first-order spatial accuracy by the Dirac delta function approximation. The high order adaptive mesh techniques have been presented by Griffith and his collaborators in Griffith and Peskin (2005) and Griffith and Luo (2017).

Stockie and Wetton (1999) analyzed the stiffness encountered in the IB method. It was found that very stiff "spring-like" links between boundary points require very restrictive time step selections. Thus, many recent efforts have been made to improve the accuracy and computational efficiency of the IB method. Noticeably, Peskin and Printz (1993) introduced a finite difference divergence operator to substitute the previous divergence derived from central difference schemes, which resulted in a dramatic improvement in overall volume conservation.

LeVeque and Li (1997) considered the jump in variables that are discontinuous across boundaries and presented a second-order accurate Cartesian grid method for solving elliptic equations whose solutions are not smooth across the interface due to discontinuous coefficients or singular source terms in the equation. The modified scheme incorporates the known jumps in the solution or its derivatives into finite difference schemes, and achieves second-order accuracy at all points on the uniform grid even for arbitrary interfaces.

McQueen and Peskin (1997) implemented a shared memory parallel vector computing scheme to compute the blood flow in the beating mammalian heart. Roma *et al.* (1999) presented an adaptive version of the immersed boundary method employing a hierarchical, nested adaptive mesh refinement for the improvement of resolution and accuracy. Accuracy enhancement is accomplished by using a sequence of nested, progressively finer rectangular grid patches which locally cover the immersed boundary vicinity and dynamically follow the immersed boundary motion. Implicit formulations of the immersed boundary method are also used to free the method from its time step restriction (Newren *et al.*, 2007).

Cortez and Minion (2000) replaced the Dirac delta function with a cut-off function, or blob, to regularize the force field exerted by the membranes on the fluid. Velocity field induced is directly computed on a regular Cartesian grid via a smoothed dipole potential. The results demonstrate better volume conservation and higher order convergence. In addition, Goldstein *et al.* (1993) used a feedback forcing method, which employed boundary body forces that allow the imposition of boundary conditions on a given surface not coinciding with the computational grid to compute the two-dimensional startup flow around a circular cylinder and three-dimensional plane.

Saiki and Biringen (1996) used the forcing method of Goldstein *et al.* (1993) to compute the flow around steady and rotating circular cylinders using fourth-order central difference approximations. Recently, Mohd-Yusof (1998) and Fadlun *et al.* (2000) derived an alternative formulation of forcing that does not affect the stability of the discrete-time equations. In Mohd-Yusof (1998), the new forcing was combined with B-splines to compute the laminar flow over a three-dimensional ribbed channel, showing substantial improvements to the previous formulations. Moreover, with respect to high Reynolds number flow, Briscolini and Santangelo (1989) used an immersed boundary approach referred to as the mask method to compute the unsteady two-dimensional flow around circular and square cylinders at Reynolds numbers up to 1000. Lai and Peskin (2000) presented a formally second-order accurate immersed boundary method with results showing less numerical viscosity and the potential to be used for high Reynolds number flows. Some approaches used in Briscoe and Panesar (1991); Saiki and Biringen (1996); Fadlun *et al.* (2000); Goldstein *et al.* (1993); Mohd-Yusof (1998), the so-called virtual boundary method, share the same spirit with the IB method but solve a rigid boundary problem rather than an elastic boundary problem.

Almost parallel to the development of the IB method, the computational mechanics in particular finite element methods (FEM) for solids and structures along with computational fluid dynamics (CFD) tools for fluid and solid mechanics also flourished. The researchers from Stanford, Berkeley, Northwestern University, University of Texas at Austin, and MIT have quickly implemented newly developed FEM and CFD techniques and applied to engineering applications. One of the emerging and very challenging systems is the FSI systems which involve complex fluid mechanics phenomena with moving boundaries, geometric and material nonlinearities of flexible structures or solids, as well as the intricate nonlinear dynamical coupling between of the two media.

FSI discipline originates from aero- and hydro-elasticity, which are often related to aeronautics and aerospace as well as nuclear industries. FSI systems, in particular those with highly deformable immersed structures or immersed solids, still pose unique challenges to applied mathematics and computational mechanics communities. The original work on FSI systems has been well documented by Bisplinghoff and Ashley (1962), Bisplinghoff *et al.* (1983), and Fung (1955). More recent accomplishments, following the tradition of analytical and experimental approaches, have been presented in Dowell *et al.* (1995), Dowell and Ilgamov (1988), Haszpra (1979), Blevins (1994), and Paidoussis (1998, 2004).

In practice, within the scope of nuclear, civil, aerospace, ocean, chemical, and mechanical engineering, there are many terminologies involved, i.e., flow-induced vibration, aeroelasticity, hydro-elasticity, fluid–structure interaction, and fluid–solid interaction. Typical problems include structure interaction with surface and sound waves, and vibrations and stabilities of cables, pipes, plates, and shells (Bathe *et al.*, 1995a). Each subject is by itself very extensive. For example, the aeroelasticity area also includes the sound vibration discussed in Junger and Feit (1972),

Cremer and Heckl (1973), and Howe (1998). Reviews on various finite element formulations for FSI systems are available in Morand and Ohayon (1995) and Bathe (1998).

Many techniques have been introduced to nonlinear computational fluid and solid mechanics. It was the advance of meshless or meshfree methods pioneered in 1990s by Liu and Belytschko at Northwestern University along with their teams that finally relieved the burden of remeshing and successive mesh generation and created the possibilities of having two distinct yet overlapping meshes for coupled fluid and solid domains, namely, Eulerian or arbitrary Lagrangian Eulerian description for fluids, and Lagrangian description for solids (Liu *et al.*, 1996a; 1996b; Li and Liu, 1996; Liu and Chen, 1995). In fact, this overlapping mesh idea is to some extent similar to the fictitious domain method by Glowinski and Pan Glowinski *et al.* (1999) and the overset method by Benek, Buning, and Steger in 1980s and 1990s (Benek *et al.*, 1983, 1985; Freitas and Runnels, 1999), yet in different contexts.

It was also identified that the kernel functions in meshless methods are directly related to the discretized delta function used in the IB method for the distribution of the solid forces and the interpolation of the surrounding fluid velocity. The hidden links between the IB method and the meshless finite element formulations kindled a recent surge of interests in the IB method from the computational mechanics community. In light of the recent shifts of research in both science and engineering fields directed toward the modeling of biological systems, a growing number of researchers are getting more and more interested in adopting ideas similar to the IB method for various multi-scale and multi-physics problems.

Because of continuing and growing interest in the IB method shown in both computational mechanics and applied mathematical sciences communities, a special mini-symposium has been organized by Fogelson and Wang in the Seventh World Congress (WCCM-VII) on Computational Mechanics held in Century Plaza Hotel in Los Angeles, California, USA, July 16–22, 2006 (Fogelson *et al.*, 2008). The success of the event becomes the impetus for researchers in both applied mathematics and engineering communities to explore new avenues to implement the key ideas in the IB methods for various fluid–structure interaction problems or in general problems with interfaces and multiple continua.

Recent advances and ideas include multi-scale and multi-physics modeling with IB methods, fundamental developments, mathematical theories and algorithms, high-performance computing, and various applications related to the IB method and its extensions.

An illustration of three different approaches in the representation of the same physical model involving fluid–solid or fluid–structure interactions (FSI) is depicted in Fig. 2.1. As shown in the left picture, the traditional approach to solving FSI problems is to represent the submerged solid with finite element methods (FEM) and the surrounding fluid with computational fluid dynamics (CFD). In this CFD/FEM approach, kinematic and dynamic matching along the FSI interface must be considered. In the immersed Boundary (IB) method (Peskin, 1977), the fluid mesh is independent of the motion of the immersed elastic boundary. It can be shown that the distribution of the internal forces and the interpolation of the motion of the immersed elastic boundary automatically represent the fluid–structure interaction (Peskin, 2002; Wang, 2003). Although fiber

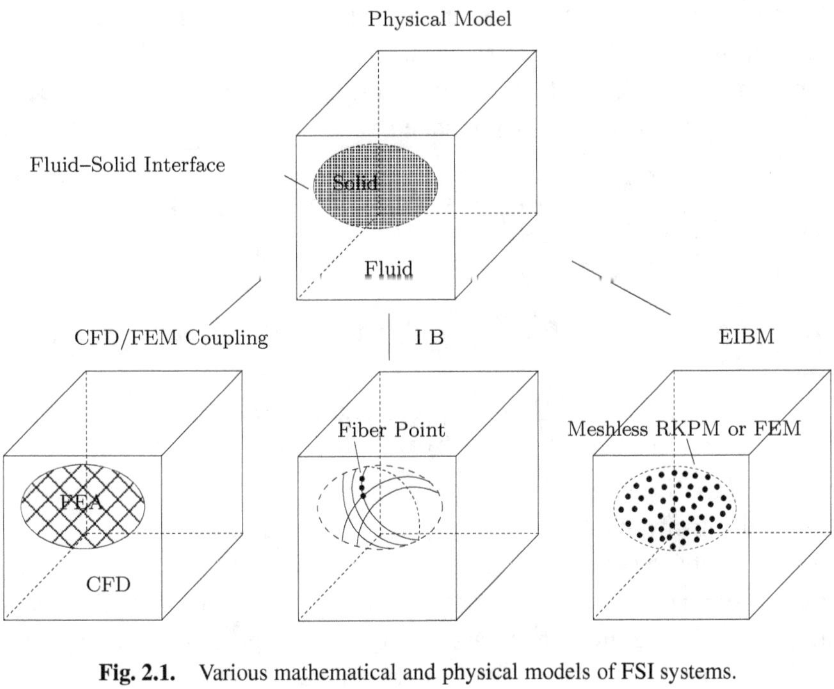

Fig. 2.1. Various mathematical and physical models of FSI systems.

network representations are consistent with some submerged structures, for example, the heart muscles, they in general do not yield as accurate stress and strain tensors as the continuum representations. The immersed continuum approaches will be addressed in Chapters 3 and 4.

2.2 Key IB Concepts

Consider a neutrally buoyant immersed flexible structure contained in a viscous incompressible fluid with periodic boundary conditions. Let $\mathbf{f}^s(\mathbf{r},t)$ stand for the elastic fiber point force, with the unit force per unit length as in the surface tension definition, at the Lagrangian parametric coordinate $\mathbf{r}(s) = (r_1(s), r_2(s), r_3(s))$ as a function of the arc length s,

$$\mathbf{f}^s(\mathbf{r},t) = \frac{\partial(T(\mathbf{r},t)\boldsymbol{\tau}(\mathbf{r},t))}{\partial s}, \tag{2.1}$$

where T represents the tension within the fiber and $\boldsymbol{\tau}$ is the unit tangent vector.

Denote $\mathbf{f}(\mathbf{x},t)$ as the corresponding effective body force in the fluid domain at the spatial position $\mathbf{x}$, and Ω as the domain of influence or support attached to every immersed elastic nodal point $\mathbf{X}(\mathbf{r},t)$, the governing equations of the FSI system involving a single neutrally buoyant smooth submerged elastic fiber Γ_s can be stated as

$$\rho\left(\frac{\partial \mathbf{v}}{\partial t} + \mathbf{v}\cdot\nabla\mathbf{v}\right) = -\nabla p + \mu\nabla^2\mathbf{v} + \mathbf{f}, \tag{2.2}$$

$$\nabla\cdot\mathbf{v} = 0, \tag{2.3}$$

$$\mathbf{f}(\mathbf{x},t) = \int_{\Gamma_s} \mathbf{f}^s(\mathbf{r},t)\delta(\mathbf{x}-\mathbf{X}(\mathbf{r},t))ds, \tag{2.4}$$

$$\frac{d\mathbf{X}(\mathbf{r},t)}{dt} = \int_{\Omega} \mathbf{v}(\mathbf{x},t)\delta(\mathbf{x}-\mathbf{X}(\mathbf{r},t))d\mathbf{x}, \tag{2.5}$$

where $\mathbf{f}^s(\mathbf{r},t)$ is given in Eq. (2.1); and $\mathbf{v}(\mathbf{x},t)$, $p(\mathbf{x},t)$, ρ, and μ represent the fluid velocity, pressure, density, and dynamic viscosity.

Note that at a typical fiber point k, the resultant elastic or internal force $\mathbf{f}_k^s$ is determined by the fiber configuration and material properties. Furthermore, if there are I arcs passing through the point k, assuming that each fiber has a constant tension T_i and an increment of the arc length Δs_i, the jth component of the nodal force $\mathbf{f}_k^s(\mathbf{x},t)$ is calculated as

$$f_{kj}^s(\mathbf{r},t) = \sum_{i=1}^{I} f_{kj}^{si}(\mathbf{r},t) = \sum_{i=1}^{I} \frac{\partial(T_i \tau_{kj}^i)}{\partial s} \Delta s_i, \tag{2.6}$$

where $\boldsymbol{\tau}_k^i$ represents the unit tangent vector of the ith arc at the point k.

It is clear that if a collection of immersed elastic fibers are used to represent various flexible structures, the line integral will be performed for all the fiber lines and the corresponding nodal forces will be distributed collectively as body forces in the fluid domain.

2.2.1 *Strong and weak formulations*

To better illustrate why the IB method works, we employ the familiar variational principles. Consider a fluid domain Ω enclosed with a sufficiently smooth boundary, $\partial\Omega = \Gamma_v \cup \Gamma_f$, as depicted in Fig. 2.2, where Γ_v stands for the Dirichlet boundary or essential boundary and Γ_f represents the Neumann boundary or natural boundary Γ_f. In addition, the entire background fluid domain Ω can be viewed as split into two parts, namely, the exterior portion Ω_{out} or Ω_e and the interior portion Ω_{in} or Ω_e. In addition, the immersed boundary Γ_s represents the entire interior portion boundary. Note that this enclosed immersed elastic boundary Γ_s (a line for two-dimensional cases and a surface for three-dimensional cases) as the generic representation of the immersed boundary, and the fluid domain Ω is subdivided into two regions: the interior region Ω_i and the exterior region Ω_e.

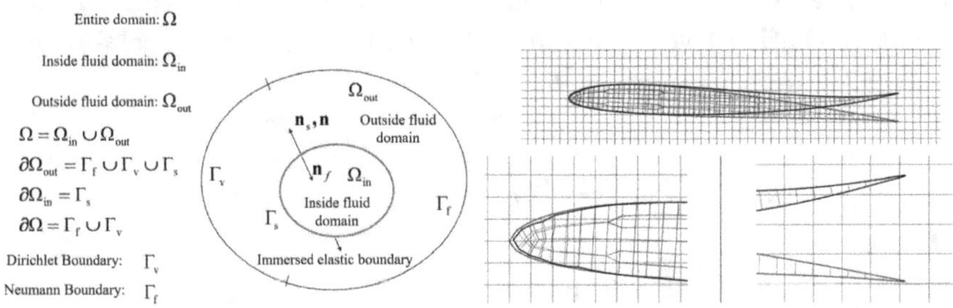

(a) IB Method (b) Two Sets of Meshes for Flexible Wings

Fig. 2.2. IB concept and a Lagrangian mesh over an Eulerian uniform mesh.

The boundaries of the interior and exterior regions can then be simply expressed as $\partial\Omega_i = \Gamma_s$ and $\partial\Omega_e = \Gamma_s \cup \Gamma_v \cup \Gamma_f$.

If we denote $\boldsymbol{\sigma}$ as the stress tensor, $\mathbf{v}$ as the velocity vector, and ρ as the density in the fluid domain, we establish the following governing set of equations or strong form:

$$\rho \dot{v}_i = \sigma_{ij,j} + f_i^e, \quad \text{in } \Omega_i (\text{or } \Omega \setminus \Omega_e), \tag{2.7}$$

$$\rho \dot{v}_i = \sigma_{ij,j} + f_i^e, \quad \text{in } \Omega_e, \tag{2.8}$$

$$[v_i] = 0, \quad \text{on } \Gamma_s, \quad \text{kinematic matching}, \tag{2.9}$$

$$[\sigma_{ij} n_j] = f_i^s + m\ddot{u}_i^s, \quad \text{on } \Gamma_s, \text{ dynamic matching}, \tag{2.10}$$

where the external body force $\mathbf{f}^e$ will be replaced by $\rho\mathbf{g}$, with $\mathbf{g}$ as the gravitational acceleration, $\mathbf{f}^s$ and m stand for the elastic force and the mass density of the immersed boundary Γ_s with the unit of force per unit length for two-dimensional cases and force per unit area for three-dimensional cases, $\mathbf{u}^s$ denotes the interface displacement, and the surface normal vector $\mathbf{n}$ is aligned with that of the interior fluid domain $\mathbf{n}^i$ and opposite to that of the exterior fluid domain $\mathbf{n}^e$.

At this point, we can derive a number of numerical approaches to solve Eqs. (2.7) to (2.10). A straightforward approach is to represent exterior and interior fluid domains with different meshes and to match them accordingly at interface Γ_s. This approach represents the traditional treatment of FSI problems, in which the solid mesh is coupled with the fluid mesh around the FSI interface (Bathe *et al.*, 1995b; Tezduyar, 2001).

Defining the Sobolev space $[\mathcal{H}^1_{0,\Gamma_v}(\Omega)]^d = \{\mathbf{w} | \mathbf{w} \in [\mathcal{H}^1(\Omega)]^d, \mathbf{w}|_{\Gamma_v} = \mathbf{0}\}$, named after Sergei Lvovich Sobolev (1908–1989), where d represents the spatial dimensions, we express Eqs. (2.7)–(2.10) in the variational form, or a so-called weak form: $\forall \mathbf{w} \in [\mathcal{H}^1_{0,\Gamma_v}(\Omega)]^d$

$$\int_{\Omega_e} w_i \left(\rho \frac{dv_i}{dt} - \rho g_i \right) d\Omega + \int_{\Omega_e} \frac{\partial w_i}{\partial x_j} \sigma_{ij} d\Omega$$

$$= \int_{\Gamma_f} w_i^f F_i^p d\Gamma + \int_{\Gamma_s} w_i^s \sigma_{ij} n_j^{out} d\Gamma = 0,$$

$$\int_{\Omega_i} w_i \left(\rho \frac{dv_i}{dt} - \rho g_i \right) d\Omega + \int_{\Omega_i} \frac{\partial w_i}{\partial x_j} \sigma_{ij} d\Omega = \int_{\Gamma_s} w_i^s \sigma_{ij} n_j^{in} d\Gamma = 0,$$

where the normal vector $\mathbf{n}^{\text{out}}$ for the exterior domain denoted as Ω_e or Ω_{out} must point to the outside of the domain, therefore align with the normal vector $\mathbf{n}_f$ as depicted in Fig. 2.2; whereas the normal vector $\mathbf{n}^{\text{in}}$ for the interior domain enclosed by the immersed boundary Γ_s and denoted as Ω_i or Ω_{in} must also point to the outside of the domain, therefore align with the normal vector $\mathbf{n}_s$ also depicted in Fig. 2.2.

Utilizing the kinematic boundary conditions for the immersed boundary Γ_s, we have $[v_i] = 0$, which essentially is equivalent to no-slip boundary and the velocity components at both sides of the immersed boundary are the same. Moreover, utilizing the dynamic boundary conditions for the immersed boundary Γ_s, we have

$$[\sigma_{ij}n_i] = f_i^s + m\frac{d^2u_i^s}{dt^2},$$

which essentially is equivalent to the force equilibrium in all directions and the force components exerted on the immersed boundary points from the surrounding must be balanced by the restoring force of the boundary along with its inertial force, or d'Alembert force, named after Jean-Baptiste le Rond d'Alembert (1717–1783).

Because, the immersed boundary is assumed to be volumeless, thus, by combining the exterior and interior domains together, employing the kinematic and dynamic boundary conditions, we have

$$\int_{\Omega_i} w_i[\rho(\dot{v}_i - g_i) - \sigma_{ij,j}]d\Omega + \int_{\Omega_e} w_i[\rho(\dot{v}_i - g_i) - \sigma_{ij,j}]d\Omega = 0, \quad (2.11)$$

which is essentially a direct expansion from the strong form depicted in Eqs. (2.7) (2.10).

In the variational form, $\mathbf{w} \in \left[\mathcal{H}^1_{0,\Gamma_v}(\Omega)\right]^d$ implies that the kinematic matching at the interface Γ_s written as Eq. (2.9) is satisfied for all $\mathbf{w}$. Furthermore, using integration by parts and the divergence theorem, introducing dynamic matching at the FSI interface Γ_s, and combining interior and exterior fluid domains with $\Omega_e \cup \Omega_i = \Omega$, Eq. (2.11) can be rewritten as: $\forall \mathbf{w} \in \left[\mathcal{H}^1_{0,\Gamma_v}(\Omega)\right]^d$

$$\int_{\Omega}[w_i\rho(\dot{v}_i - g_i) + w_{i,j}\sigma_{ij}]d\Omega + \int_{\Gamma_s} w_i^s(f_i^s + m\ddot{u}_i^s)d\Gamma - \int_{\Gamma_f} w_i f_i^{f,p} d\Gamma = 0. \quad (2.12)$$

In Eq. (2.12), the term involving the given surface traction $\mathbf{f}^{f,p}$ will remain the same as if the variational form is carried out for the entire fluid domain instead of the interior and exterior parts. Thus, the focus will be on the submerged interface Γ_s.

In Eq. (2.12), the external work comes from the external body force $\mathbf{f}^e$ or $\rho\mathbf{g}$, surface traction $\mathbf{F}^p$ at the Neumann boundary Γ_f, and elastic and inertial forces around the FSI interface Γ_s. Moreover, in Eq. (2.12), we do not stipulate the material derivative $\frac{d\mathbf{v}}{dt}$ and the stress $\boldsymbol{\sigma}$. Hence, the turbulent and the non-Newtonian fluid models can eventually be incorporated. Finally, the kinematic matching at the submerged interface Γ_s also implies that the submerged interface will move at the same velocity as that of the fluid–particles in the immediate vicinity.

In the IB method, we introduce the following two key equations:

$$f_i^{\text{fsi}} = -\int_{\Gamma_s} (f_i^s + m\ddot{u}_s)\delta(\mathbf{x} - \mathbf{x}^s)d\Gamma, \tag{2.13}$$

$$v_i^s = \int_{\Omega} v_i \delta(\mathbf{x} - \mathbf{x}^s)d\Omega, \tag{2.14}$$

where $\mathbf{f}^{\text{fsi}}$ is the so-called equivalent FSI body force.

In both Eqs. (2.13) and (2.14), the Dirac delta function is positioned at the current interface position $\mathbf{x}^s$. Before the discretization of the Dirac delta function, Eq. (2.14) can be simply interpreted as the evaluation of the fluid velocity at the submerged interface. In the discretized form, the Dirac delta function in Eq. (2.14) is equivalent to the shape function or kernel of the meshfree method.

Note that Γ_s represents the current configuration of the submerged interface, and nonlinear mechanics are employed to relate the elastic force $\mathbf{f}^s$ with the interfacial position $\mathbf{x}^s$ or the displacement $\mathbf{u}^s$. It is also clear that as long as we use the same delta function for both Eqs. (2.13) and (2.14), the virtual power input from the submerged elastic boundary (or the immersed boundary) to the fluid domain can be expressed as

$$\begin{aligned}\int_{\Omega} w_i f_i^{\text{fsi}} d\Omega &= -\int_{\Gamma_s}\int_{\Omega} w_i \delta(\mathbf{x} - \mathbf{x}^s)(f_i^s + m\ddot{u}_i^s)d\Omega d\Gamma \\ &= -\int_{\Gamma_s} w_i^s (f_i^s + m\ddot{u}_i^s)d\Gamma,\end{aligned}$$

which holds for all $\mathbf{w} \in \left[\mathcal{H}^1_{0,\Gamma_v}(\Omega)\right]^d$, and consequently the effect of the submerged elastic boundary can be simply replaced with the equivalent body force $\mathbf{f}^{\text{fsi}}$.

Finally, the governing equations (2.7)–(2.10) can be rewritten as

$$\rho \dot{v}_i = \sigma_{ij,j} + \rho g_i + f_i^{\text{fsi}}, \quad \text{in } \Omega; \tag{2.15}$$

and the variational equations (2.11) and (2.12) are modified as

$$\int_\Omega \left[w_i(\rho \dot{v}_i - \rho g_i - f_i^{\text{fsi}}) + w_{i,j}\sigma_{ij}\right] d\Omega - \int_{\Gamma_f} w_i f_i^{f,p} d\Gamma = 0. \tag{2.16}$$

Equation (2.16) provides us with the foundation of the key advantage of the IB method, namely, the independent Lagrangian solid mesh moves on top of a background Eulerian fluid mesh. We must also point out that the background fluid mesh could be a fixed Eulerian mesh or an ALE mesh with a prescribed mesh motion. In practice, such a mesh motion could follow moving structures or solids as well as conform to boundary deformation which will be elaborated in other chapters.

Again, an effective body force $\mathbf{f}^{\text{fsi}}$ is derived based on the equivalent effects of the immersed or submerged elastic boundary. Since we do not stipulate the stress and the material derivative of the entire fluid domain Ω, non-Newtonian and turbulent fluid models are not excluded from this formulation. In fact, the entire domain Ω could also be a continuum, fluid or solid as stipulated in Wang (2006) and Zhang (2017). However, when considering the bending moments and twisting moment or torque and shear forces in two transverse directions of the immersed boundary, the simple fiber or truss or cable model for the volumeless immersed boundary must be expanded to volumeless beams, codimension one structures or plates/shells, codimension two structures or beams/rods which will be addressed in Chapter 5.

2.2.2 *Mass and stiffness issues*

The key concepts of the immersed boundary (IB) method are illustrated in Fig. 2.3, namely, a Lagrangian mesh moving on top of a regular Eulerian mesh. As depicted in Fig. 2.3, an immersed particle or rather a point with mass has effect on the fluid flow so are immersed particles

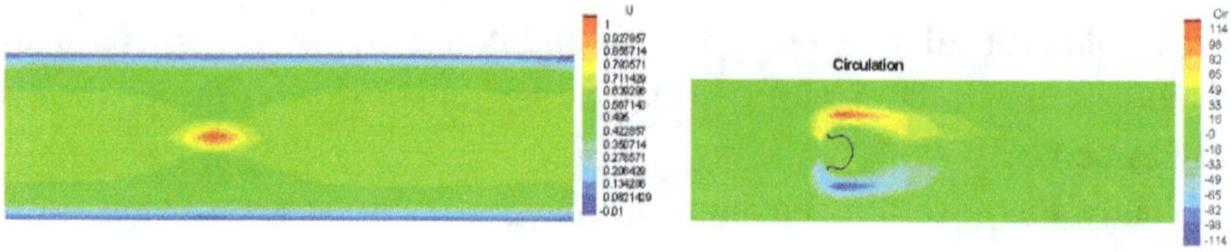

(a) (b)

Fig. 2.3. Illustrations of key concepts of the IB method. (a) Mass point and (b) points with stiffness.

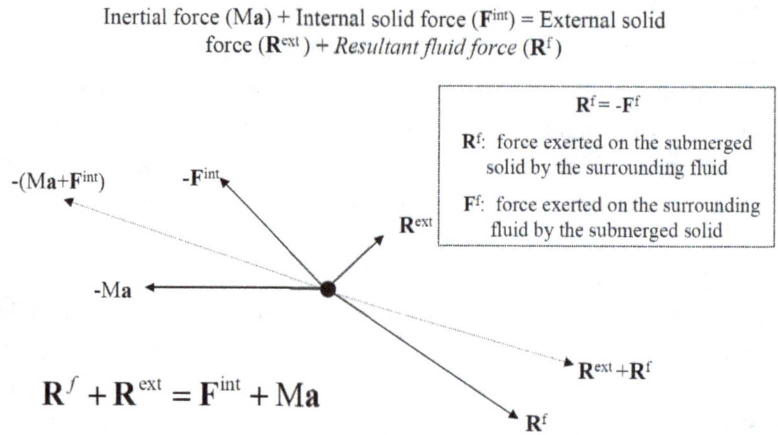

Fig. 2.4. Force balance at a typical submerged (immersed) point.

or points with or without finite mass connected with fibers (springs) and formed a submerged structure. Furthermore, also depicted in Fig. 2.3, the original IB method, the fluid channel is constructed with the immersed fiber tethered with a relatively stiffer spring. As a result, the force fields from these immersed springs will in effect form a physical wall. It is clear that if immersed massless particles are not connected with finite stiffness, the surrounding fluid will not experience the effect of their existence.

Let us consider a typical submerged or immersed point k as illustrated in Fig. 2.4, the nodal force at the point k, denoted as $\mathbf{f}_k$, is determined by tensions within the two adjacent fibers. If the immersed boundary has the same mass density as the surrounding fluid, i.e., is neutrally buoyant, at the submerged point k, there is no attached additional mass, and the resultant fluid force vector $\mathbf{R}_k^f$ and the external solid force vector $\mathbf{R}_k^{\text{ext}}$ must equal the resultant elastic or internal nodal force $\mathbf{f}_k$. Note that in Fig. 2.4, the Newton's third law is manifested in the negative sign of the force exerted on the surrounding fluid by the submerged solid $\mathbf{F}^f$.

However, if the mass density of the elastic structure ρ_s is different from the surrounding fluid density ρ_f, $\mathbf{R}_k^f$ and $\mathbf{R}_k^e$ must achieve a dynamic equilibrium with the resultant elastic nodal force and the inertial or d'Alembert's force. Denoting M and $\mathbf{a}_k$ as the elumped mass and the acceleration at the point k, we have

$$\mathbf{R}_k^f + \mathbf{R}_k^e = \begin{cases} \mathbf{f}_k, & \rho_s = \rho_f, \\ \mathbf{f}_k + \mathbf{M}\mathbf{a}_k, & \rho_s \neq \rho_f. \end{cases} \tag{2.17}$$

Again, this equilibrium depicted by Eq. (2.17) applies for both the boundary and the interior material points. Moreover, employing the discretized delta function, we can write the discretized form of Eqs. (2.4) and (2.5) as

$$\mathbf{f} = \sum_{k=1}^{K} \delta_h(\mathbf{x} - \mathbf{x}_k^s)(-\mathbf{R}_k^f), \tag{2.18}$$

$$\mathbf{v}_k^s = \sum_{\mathbf{x}_i \in \Omega_o^k} \delta_h(\mathbf{x}_i - \mathbf{x}_k^s)\mathbf{v}_i h^3, \tag{2.19}$$

where K is the total number of submerged points, Ω_o^k stands for the finite support domain of the discretized delta function surrounding and centered at the kth submerged point, and h is the grid size of the uniform background grid covering the entire fluid domain.

In the IB method, in order to satisfy energy conservation, namely, that the energy input to the fluid domain from the elastic boundary is the same as that from the equivalent body force, the same three-dimensional Dirac delta function $\delta(\mathbf{x} - \mathbf{X}(\mathbf{r}, t))$ must be used in both the distribution of the resultant nodal force and the interpolation of the velocity based on the surrounding fluid grid point velocities (Peskin, 2002), e.g., Eqs. (2.4) and (2.5).

The key attribute of the immersed methods presented in this section is to eliminate FSI interfaces and, consequently, fluid mesh distortion problems by distributing nodal structural forces, calculated with finite element

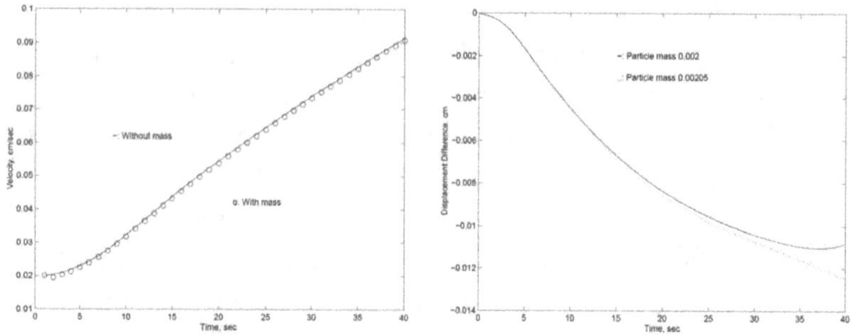

(a) stable (b) unstable

Fig. 2.5. Sensitivities of immersed particles with and without finite mass.

methods, through a delta function as equivalent body forces in the fluid domain. This approach enables the use of fast Fourier transform (FFT), parallel algorithm, and higher order schemes. In the original IB method, the time step selection is related to stiff spring supports from the boundary and the high elasticity moduli of the immersed solid.

In the original IB method, the immersed solids/structures are represented with a collection of nodes or material points linked through a stiffness matrix which do not occupy any volume and have no computational issue with the spectrum incompressible viscous fluid solver. Let M stand for the lumped mass, the inertial force $M\mathbf{a}$ and the internal solid force $\mathbf{F}^{\text{int}}$ have to be equal to the external force $\mathbf{R}^{\text{ext}}$ and the resultant force $\mathbf{R}^{f}$, which is exerted on the submerged solid point from the surrounding fluid (Kim *et al.*, 2003). As depicted in Fig. 2.5, with finite mass, the time step of the IB method must be judiciously chosen so that the numerical instability can be avoided (Kim and Peskin, 2007). This finite mass issue is essentially built in the spectrum solver for the incompressible viscous fluid which is inherent to the original IB method.

2.2.3 *Spectrum solver and RK2 scheme*

In this section, we present the original spectrum solver for viscous incompressible fluid implemented for the IB method which is initially based on uniform grids in space with grid size h. Later more sophisticated local adaptive meshing based on finite difference methods is utilized in Griffith *et al.* (2007) and Griffith and Patankar (2020).

Assume s is the axial arc length and the fibre can be spreading in the other two local coordinates q and r marked by the area S_o, thus, the positive of the fiber can be expressed as

$$\mathbf{x} = \mathbf{X}(q, r, s, t),$$

from which the unit tangent vector $\boldsymbol{\tau}$ can be expressed as

$$\boldsymbol{\tau}(q, r, s, t) = \frac{\partial \mathbf{X}/\partial s}{|\partial \mathbf{X}/\partial s|},$$

and the axial tensile force or tension $\mathbf{T}$ can be depicted as

$$T(q, r, s, t) = \sigma\left(\left|\frac{\partial \mathbf{X}}{\partial s}\right| - 1; q, r, s, t\right),$$

with the axial stiffness σ, or the product of the Young's modulus E and the cross-sectional area A of the immersed fiber.

Note that for a segment with $s \in [a, b]$, the net force experienced by the surrounding viscous fluid $\mathbf{f}^s$ can be expressed as

$$\mathbf{f}^s = \int_{S_o} T\boldsymbol{\tau} dqdr\bigg|_a^b = \int_a^b \int_{S_o} \frac{\partial}{\partial s}(T\boldsymbol{\tau}) dqdrds.$$

For the incompressible viscous fluid with period boundary conditions, it is very effective to utilize the spectrum solver based on FFT (Griffith and Luo, 2017). Again, with the immersed boundary Γ_s, the governing equations can be expressed as

$$\begin{aligned} \rho\left(\frac{\partial \mathbf{v}}{\partial t} + \mathbf{v}\cdot\nabla\mathbf{v}\right) &= -\nabla p + \mu\nabla^2\mathbf{v} + \mathbf{f}^{\mathrm{fsi}}, \\ \nabla\cdot\mathbf{v} &= 0. \end{aligned} \tag{2.20}$$

Moreover, the total force experienced by the surrounding viscous fluid can be distributed into the fluid domain with the same delta function or kernel $\delta(\mathbf{x} - \mathbf{X}(q, r, s, t))$ as an effective body force $\mathbf{f}^{\mathrm{fsi}}$, thus,

$$\mathbf{f}^{\mathrm{fsi}} = \int_{\Gamma_s}\int_{S_o} \frac{\partial}{\partial s}(T\boldsymbol{\tau})\delta(\mathbf{x} - \mathbf{X}(q, r, s, t)) dqdrds.$$

Furthermore, the immersed particle will move with the velocity interpolated through the surrounding fluid points with the same delta function or kernel $\delta(\mathbf{x}-\mathbf{X}(q,r,s,t))$ and yields

$$\frac{\partial \mathbf{X}}{\partial t}(q,r,s,t) = \mathbf{v}(\mathbf{X}(q,r,s,t),t) = \int_{\Gamma_s} \mathbf{v}(\mathbf{x},t)\delta(\mathbf{x}-\mathbf{X}(q,r,s,t))d\mathbf{x}.$$

Likewise, for the uniform spatial grids, since the same delta function or kernel $\delta_h^3(\mathbf{x}-\mathbf{X}(q,r,s,t))$ or for abbreviation δ_h^3 for three-dimensional (3D) cases is used for interpolation and application of velocity and force vectors, the rate of work done by fibers can be expressed as

$$\sum_{q,r,s} \mathbf{f}^s \cdot \frac{d\mathbf{X}}{dt}\Delta q\Delta r\Delta s = \sum_{q,r,s,\mathbf{x}} \mathbf{f}^s \cdot \mathbf{v}(\mathbf{x})\delta_h^3 h^3 \Delta q\Delta r\Delta s = \sum_{\mathbf{x}} \mathbf{f}^{\text{fsi}} \cdot \mathbf{v}(\mathbf{x})h^3.$$

In the explicit time integration scheme, the discretized governing equations are derived as follows:

$$\rho\left(\frac{\mathbf{v}^{n+1}-\mathbf{v}^n}{\Delta t} + \sum_{\alpha=1}^{3} v_\alpha^n D_\alpha^\pm \mathbf{v}^n\right) = -\mathbf{D}^o p^{n+1} + \mu \sum_{\alpha=1}^{3} D_\alpha^+ D_\alpha^- \mathbf{v}^{n+1} + \mathbf{f}^{n+1},$$
$$\mathbf{D}^o \cdot \mathbf{v}^{n+1} = 0,$$

with the central difference based gradient denoted as $\mathbf{D}^o = (D_1^o, D_2^o, D_3^o)^T$ and the central difference based Laplace operator denoted as $D_\alpha^+ D_\alpha^-$,

$$\mathbf{f}_i^{n+1} = -\sum_{k=1}^{I} (\mathbf{R}^f)_k^{n+1} \delta_h(\mathbf{x}_i - \mathbf{X}_k^n), \tag{2.21}$$

$$(\mathbf{R}^f)^{n+1} = \mathcal{N}(\mathbf{X}^n) + \int_{V_s} (\rho_s - \rho_f)\mathbf{N}^T\mathbf{N}\det(\mathcal{F})\ddot{\mathbf{U}}^n dV - (\mathbf{R}^{\text{ext}})^n, \tag{2.22}$$

$$\mathbf{X}_k^{n+1} = \mathbf{X}_k^n + \Delta t \sum_{\mathbf{x}_i \in \Omega} \mathbf{v}_i^{n+1} \delta_h(\mathbf{x}_i - \mathbf{X}_k^n)h^3, \tag{2.23}$$

where the operator $\mathcal{N}$ stands for the internal force calculation of Eq. (2.72), $\mathbf{D}^o$, $\mathbf{D}^+$ and $\mathbf{D}^-$ are center, forward, and backward difference operators, respectively; and the upwind difference operator D_α is written as

$$v_\alpha D_\alpha^\pm = \begin{cases} v_\alpha D_\alpha^+, & v_\alpha < 0, \\ v_\alpha D_\alpha^-, & v_\alpha > 0. \end{cases}$$

In fact, the convective term in Eqs. (2.20) can be rewrite with the sign function, namely, $\text{sgn}(v_\alpha) = 1$ with $v_\alpha > 0$, $\text{sgn}(v_\alpha) = -1$ with $v_\alpha < 0$, and

$\text{sgn}(v_\alpha) = 0$ with $v_\alpha = 0$:

$$v_\alpha D^{\pm}_{\alpha} = v_\alpha \left(\frac{1-\text{sgn}(v_\alpha)}{2} D^{+}_{\alpha} + \frac{1+\text{sgn}(v_\alpha)}{2} D^{-}_{\alpha} \right) = v_\alpha D^{o}_{\alpha} - \frac{|v_\alpha|}{2} h D^{+}_{\alpha} D^{-}_{\alpha},$$

using the identity

$$D^{+}_{\alpha} - D^{-}_{\alpha} = h D^{+}_{\alpha} D^{-}_{\alpha}.$$

In the first-order finite difference approach, the internal elastic force at time step $n+1$ can be simply expressed as

$$\frac{\mathbf{X}^{n+1}_k - \mathbf{X}^{n}_k}{\Delta t} = \sum_{\mathbf{x}} \mathbf{v}^n(\mathbf{x}) \delta_h (\mathbf{x} - \mathbf{X}^n_k) h^3,$$

and the axial tension

$$T^{n+1}_{k+1/2} = \sigma^{n+1}_{k+1/2} \left(\frac{|\mathbf{X}^{n+1}_{k+1} - \mathbf{X}^{n+1}_{k}|}{\Delta s} - 1 \right),$$

and the unit tangent vector

$$\boldsymbol{\tau}^{n+1}_{k+1/2} = \frac{\mathbf{X}^{n+1}_{k+1} - \mathbf{X}^{n+1}_{k}}{|\mathbf{X}^{n+1}_{k+1} - \mathbf{X}^{n+1}_{k}|},$$

along with the gradient of the axial tension

$$\mathcal{N}(\mathbf{X}^n) = -\frac{(T\boldsymbol{\tau})^{n+1}_{k+1/2} - (T\boldsymbol{\tau})^{n+1}_{k-1/2}}{\Delta s}.$$

To summarize, for this explicit version of the IB method, each time step proceeds as follows:

(1) At the beginning of time marching, the submerged solid configuration $\mathbf{X}^n$ and the fluid velocity $\mathbf{v}^n$ are given;
(2) Equation (2.22) is used to evaluate the resultant nodal forces $(\mathbf{R}^f)^{n+1}$, which includes the density difference and additional external forces;
(3) Equation (2.21) is used to transfer the nodal force $(\mathbf{R}^f)^{n+1}$ to the fluid lattice and the result of this computation is the Eulerian body force density $\mathbf{f}^{n+1}$;
(4) Equations (2.20) are solved simultaneously, often with the fast Fourier Transform, for the unknowns $(\mathbf{v}^{n+1}, p^{n+1})$;

(5) Once the new velocity field has been found, Eq. (2.23) is used to interpolate the velocity field at the material points, then move them accordingly;
(6) Thus, $\mathbf{X}^{n+1}$ and $\mathbf{v}^{n+1}$ are known, and this time step calculation is completed.

To have a better resolution in both space and time, a second-order Runge–Kutta (RK2) temporal scheme, named after Carl David Tolmé Runge (1856–1927) and Martin Wilhelm Kutta (1867–1944), is utilized. Consider the nature of the low spatial resolution in the original implementation of the IB method, a second-order Runge–Kutta method (RK2) is introduced as follows:

$$\frac{dy}{dt} = f(y),$$

with

$$\frac{y^{n+1/2} - y^n}{\Delta t/2} = f(y^n),$$

and

$$\frac{y^{n+1} - y^n}{\Delta t} = f(y^{n+1/2}).$$

A so-called skew symmetric difference scheme is introduced as

$$2\mathbf{v} \cdot \nabla\phi \simeq \mathbf{v} \cdot \mathbf{D}^o\phi + \mathbf{D}^o \cdot (\mathbf{v}\phi) = 2\mathbf{S}(\mathbf{v})\phi,$$

where the center difference operator is defined as $\mathbf{D}^o = \langle D_1^o, D_2^o, D_3^o \rangle$ along with the following identity

$$\begin{aligned} 2(\phi, \mathbf{S}(\mathbf{v})\phi) &= (\phi, \mathbf{v} \cdot \mathbf{D}^o\phi) + (\phi, \mathbf{D}^o \cdot (\mathbf{v}\phi)) \\ &= -(\mathbf{D}^o \cdot (\mathbf{v}\phi), \phi) + (\phi, \mathbf{D}^o \cdot (\mathbf{v}\phi)) = 0. \end{aligned}$$

Hence, the governing incompressible viscous fluid dynamics equations can be discretized as

$$\rho\left(\frac{\mathbf{v}^{n+1/2} - \mathbf{v}^n}{\Delta t/2} + S(\mathbf{v}^n)\mathbf{v}^n\right) + \mathbf{D}\bar{p}^{n+1/2} = \mu L(\mathbf{v}^{n+1/2}) + \mathbf{f}^{fsi,n+1/2},$$

$$\mathbf{D}\mathbf{v}^{n+1/2} = 0,$$

where the gradient operator and the Laplace operator are denoted as $\mathbf{D}$ and L.

As a consequence, the velocity of the kth immersed point can also be discretized as

$$\frac{\mathbf{X}_k^{n+1} - \mathbf{X}_k^n}{\Delta t} = \sum_{\mathbf{x}} \mathbf{v}^{n+1/2}(\mathbf{x})\delta_h(\mathbf{x} - \mathbf{X}_k^{n+1/2})h^3.$$

After the solution of $\bar{p}^{n+1/2}$ and $\mathbf{v}^{n+1/2}$ with the spectrum incompressible viscous fluid dynamics solver, we can complete the second portion with the solution of $p^{n+1/2}$ and $\mathbf{v}^{n+1}$ based on the second-order Runge–Kutta method (RK2) with the following discretization and the same spectrum solver

$$\rho\left(\frac{\mathbf{v}^{n+1} - \mathbf{v}^n}{\Delta t} + S(\mathbf{v}^{n+1/2})\mathbf{v}^{n+1/2}\right) + \mathbf{D}p^{n+1/2}$$
$$= \mu L\left(\frac{\mathbf{v}^{n+1} + \mathbf{v}^n}{2}\right) + \mathbf{f}^{fsi,n+1/2},$$
$$\mathbf{D}\mathbf{v}^{n+1} = 0.$$

For simplicity, assume the immersed boundary is neutrally buoyant with no additional external forces. It is important to note that the pressure unknowns $\bar{p}^{n+1/2}$ are intermediate and the pressure unknowns $p^{n+1/2}$ and the velocity unknowns $\mathbf{v}^{n+1}$ are staggered with a half of the time step Δt. Moreover, for the immersed solid or immersed structure point, we also have the following expression for the velocity of the immersed solid or immersed structure point

$$\frac{\mathbf{X}_k^{n+1} - \mathbf{X}_k^n}{\Delta t/2} = \sum_{\mathbf{x}} \mathbf{v}^n(\mathbf{x})\delta_h(\mathbf{x} - \mathbf{X}_k^n)h^3,$$

and the axial tension

$$T_{k+1/2}^{n+1/2} = \sigma_{k+1/2}^{n+1/2}\left(\frac{|\mathbf{X}_{k+1}^{n+1/2} - \mathbf{X}_k^{n+1/2}|}{\Delta s} - 1\right),$$

and the unit tangent vector

$$\boldsymbol{\tau}_{k+1/2}^{n+1/2} = \frac{\mathbf{X}_{k+1}^{n+1/2} - \mathbf{X}_k^{n+1/2}}{|\mathbf{X}_{k+1}^{n+1/2} - \mathbf{X}_k^{n+1/2}|},$$

along with the gradient of the axial tension

$$\mathbf{f}_k^{s,n+1/2} = \frac{(T\boldsymbol{\tau})_{k+1/2}^{n+1/2} - (T\boldsymbol{\tau})_{k-1/2}^{n+1/2}}{\Delta s},$$

and the corresponding equivalent FSI body force $\mathbf{f}^{\text{fsi}}$,

$$\mathbf{f}^{fsi,n+1/2}(\mathbf{x}) = \sum_k \mathbf{f}_k^{s,n+1/2} \delta_h\left(\mathbf{x} - \mathbf{X}_k^{n+1/2}\right)\Delta s.$$

The original IB method was implemented with a spectral solver for incompressible viscous fluid. As discussed in Section 1.3, spatial Fourier series are introduced in the spectral solver with a rectangular box domain with periodic boundary conditions. In this book, as presented in Eq. (3.172), we introduce the following notations with a discretized spatial function ϕ_j, the corresponding discrete Fourier transform (DFT) $\widehat{\phi_k}$ is denoted as

$$\widehat{\phi_m} = \frac{1}{N}\sum_{n=0}^{N-1} \phi_n e^{-i2\pi mn/N},$$

along with

$$\phi_m = \sum_{n=0}^{N-1} \widehat{\phi}_n e^{i2\pi mn/N}.$$

Consequently, the comparable difference operators are defined as

$$(D^+\phi)_j = \frac{\phi_{j+1} - \phi_j}{h} = \sum_{n=0}^{N-1} \widehat{\phi}_n e^{i2\pi jn/N} \frac{e^{i2\pi n/N} - 1}{h},$$

$$(D^-\phi)_j = \frac{\phi_j - \phi_{j-1}}{h} = \sum_{n=0}^{N-1} \widehat{\phi}_n e^{i2\pi jn/N} \frac{1 - e^{-i2\pi n/N}}{h},$$

$$(D^o\phi)_j = \frac{\phi_{j+1} - \phi_{j-1}}{2h} = \sum_{n=0}^{N-1} \widehat{\phi}_n e^{i2\pi jn/N} \frac{e^{i2\pi n/N} - e^{-i2\pi n/N}}{2h}.$$

For abbreviation, we have the DFT of all these operators,

$$\widehat{D}^+ = \frac{e^{i2\pi k/N} - 1}{h},$$

$$\widehat{D}^- = \frac{1 - e^{-i2\pi k/N}}{h},$$

$$\widehat{D}^o = \frac{e^{i2\pi k/N} - e^{-i2\pi k/N}}{2h}.$$

Similarly, with the Laplace operation $D^+D^- = D^+ - D^-$, we have

$$\widehat{D^+D^-} = \frac{e^{i2\pi k/N} + e^{-i2\pi k/N} - 2}{h^2} = -\frac{4\sin^2 \pi k/N}{h^2}.$$

Nodal force calculation is an important step in the IB method. The more accurate the nodal force distribution we calculate from the structure model, the better the simulation result we can expect. In previous structural models, elastic structures are normally discretized into certain patterns of networks that are composed of fibers. For example, in the simulation of hemodynamic interaction with heart muscles, the fiber network is generated in such a way that each fiber's direction is consistent with the muscle fiber direction. Nodal force at each fiber point is determined by tensions within fibers that are linked to this point. Therefore, in the IB method, nodal forces are calculated around a continuous arc and additional nodal points occupying the same location are introduced to represent the fiber crossing. Note that these fiber crossing points always move at at the same velocity and will never separate. In the three-dimensional simulation, a large number of redundant degrees of freedom could significantly increase the computation cost. In general, finite element methods for immersed continuum lead to better stress predictions.

If the fibers are assumed to be massless, inertial forces are neglected. Therefore, the resultant force $\mathbf{f}_k$ is entirely dependent upon the tension T_{k-1} and T_k and should be balanced by the force applied by the fluid. In other words, the force applied to the fluid is equal to the resultant force $\mathbf{f}_k$. This resultant force $\mathbf{f}_k$, or nodal force, is determined by the fiber configuration and material property. In the IB method, the program loops all fibers to evaluate the tension forces, then calculates the nodal force at every fiber point. For a typical connection at fiber point k, denoting two adjacent fibers as $k-1$ and k, we can either have constant tension or specify a linear elastic modulus K attribute to each fiber connection. For these two cases, the force calculation scheme is as follows:

(a) Constant tension T

$$\mathbf{f}_k = \frac{\partial(T\mathbf{t}_k)}{\partial s} = \frac{(T_{k+1/2}\mathbf{t}_{k+1/2} - T_{k-1/2}\mathbf{t}_{k-1/2})}{\Delta s} = \frac{T(\mathbf{t}_{k+1/2} - \mathbf{t}_{k-1/2})}{\Delta s}, \tag{2.24}$$

with

$$\mathbf{t}_{k-1/2} = \frac{\mathbf{x}_k^s - \mathbf{x}_{k-1}^s}{|\mathbf{x}_k^s - \mathbf{x}_{k-1}^s|} \quad \text{and} \quad \mathbf{t}_{k+1/2} = \frac{\mathbf{x}_{k+1}^s - \mathbf{x}_k^s}{|\mathbf{x}_{k+1}^s - \mathbf{x}_k^s|}, \tag{2.25}$$

where $\mathbf{t}_{k-1/2}$ and $\mathbf{t}_{k+1/2}$ represent the unit tangent vectors of fiber $k-1$ and k, respectively; likewise, the tension $T_{k-1/2}$ and $T_{k+1/2}$ within these fibers are constant and denoted as T.

(b) Linear elastic modulus K

$$\mathbf{f}_k = \frac{(K\frac{\partial \mathbf{x}}{\partial s})_{k+1/2} - (K\frac{\partial \mathbf{x}}{\partial s})_{k-1/2}}{\Delta s} = \frac{K(\mathbf{x}_{k+1}^s + \mathbf{x}_{k-1}^s - 2\mathbf{x}_k^s)}{(\Delta s)^2}. \tag{2.26}$$

Of course, as long as we represent the immersed objects with Lagrangian descriptions, the IB method will also be applicable for the modeling of one type of fluid immersed in another type of fluid. It also seems to be a natural extension to employ various finite element models for the nodal force calculation (Wang and Liu, 2004). In a test example, one viscous fluid modeled with a collection of fluid–particles is injected into another viscous fluid. Due to the surface tension effects, viscous fluid exiting the tube tends to form a droplet. Simulation results of one viscous fluid droplet squeezed into another viscous fluid are illustrated in Figs. 2.6 and 2.7.

In another simulation with a bubble floating to the water and air interface, the entire water with the free surface is modeled as the immersed particles as illustrated in Figs. 2.6 and 2.7. The three-dimensional fluid mesh size is $256 \times 4 \times 256$ along with $30,000$ immersed solid particles which will be used to construct the container and the bubble. Note that for the proper use of FFT, the mesh size must be a power of 2. Moreover, in the thickness direction, four layers are typically used for an effective two-dimensional physical and mathematical models. The computational time on SGI O2000 is around 8000 sec.

The bubble is modeled with and without additional mass and surface tension. Simulation results of one viscous fluid droplet squeezed into another viscous fluid are illustrated in Fig. 2.8. Similar surface tension effects have also been documented in Figs. 2.6–2.8. It is clear that the surface mass tends to drag the bubble down while the buoyancy tends to push the bubble up. Moreover, the hydrostatic pressure gradient along with

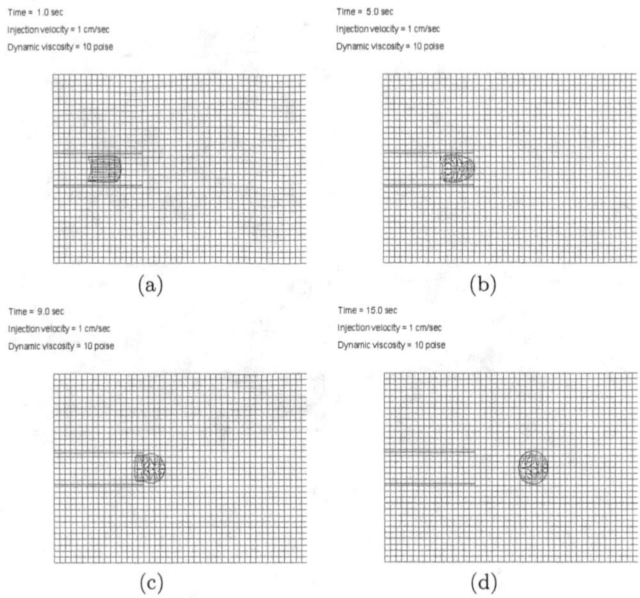

Fig. 2.6. One viscous fluid injected into another viscous fluid with surface tension.

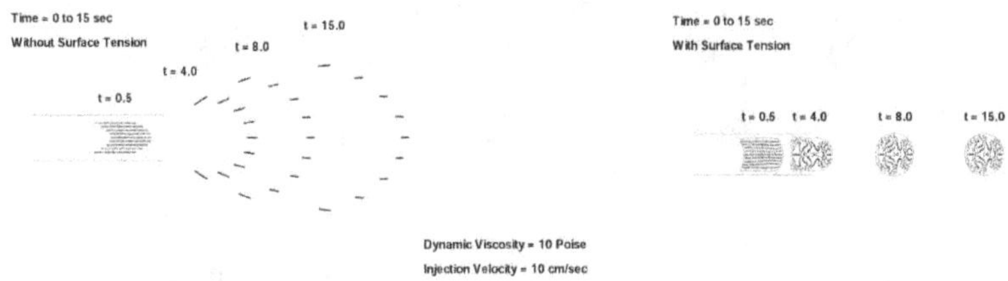

Fig. 2.7. Surface tension effects of one viscous fluid injected into another.

the dynamic pressure perturbations due to gravitational surface waves are properly obtained.

As shown in Figs. 2.2 and 2.9, various flexible structures as well as immersed points can be modeled by the IB method. In the example shown in Fig. 2.9, both the wall and valve fibers have a stiffness 0.2048 dyne/cm and the average steady flow velocity is 0.4 cm/sec. In this simulation, the three-dimensional fluid mesh size is $128 \times 128 \times 128$ along with $16,000$ immersed solid particles. Again for the proper use of FFT, the mesh size

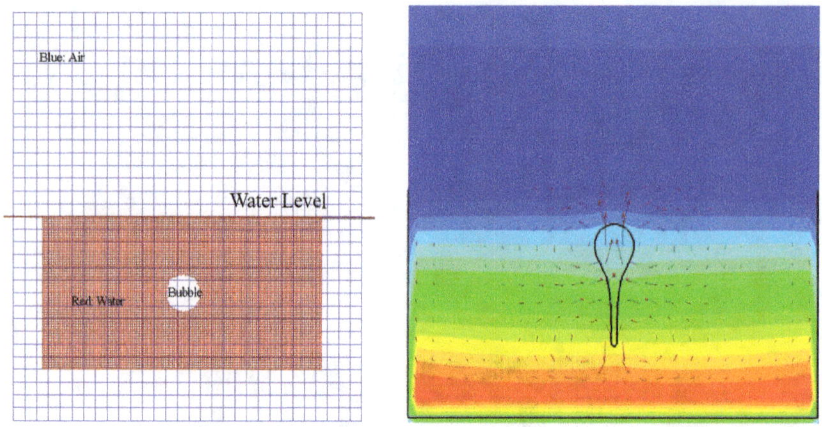

(a) (b)

Fig. 2.8. Buoyancy and bubble simulation with a free surface.

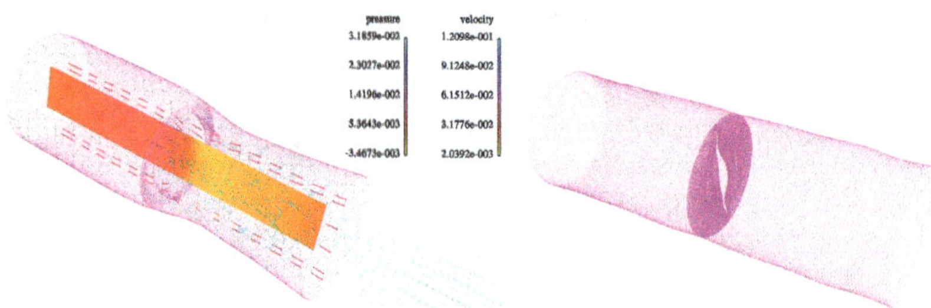

(a) (b)

Fig. 2.9. Simple immersed structures with immersed fibers. (a) Valve flow and (b) venous valve.

must be a power of 2. The computational time on SGI O2000 is around 7000 sec.

Oxygen-rich blood fills the left ventricle from the left atrium as shown in Fig 2.10. The heart muscle contracts and pumps the blood in the left ventricle to the systemic circulation. After collecting enough oxygen-poor blood from the right atrium, the right ventricle muscle contracts and pumps the blood to the pulmonary circulation. Four Chambers: right atrium (RA),

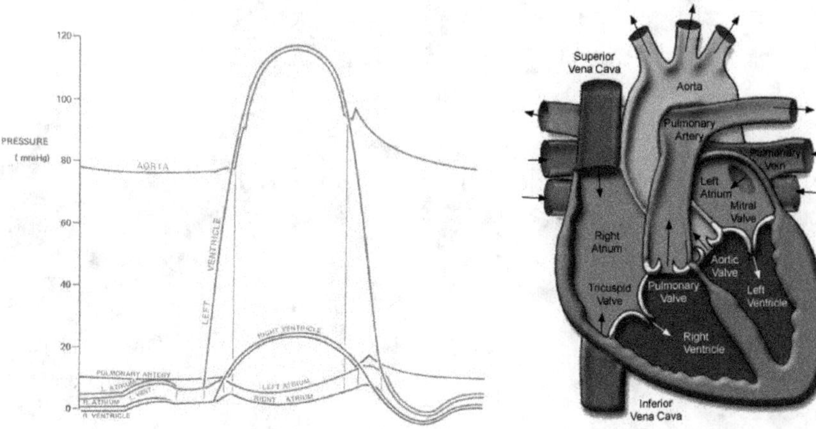

(a) Blood Pressures (b) 3D Heart Model

Fig. 2.10. Three-dimensional heart model (Peskin, 2012).

right ventricle (RV); left atrium (LA), left ventricle (LV). Four Valves: tricuspid valve (RA to RV), pulmonary valve (RV to PA); aortic valve (LV to aorta), mitral valve (LA to LV).

The ultimate goal of the IB method is to have a reasonable and realistic model of the three-dimensional human heart (McQueen and Peskin, 2001). As depicted in Fig. 2.10, also presented in Liu *et al.* (2006), human hearts have four chambers, namely right atrium (RA), right ventricle (RV); left atrium (LA), left ventricle (LV) along with four valves, namely, tricuspid valve (from RA to RV), pulmonary valve (from RV to PA); aortic valve (from LV to aorta), and mitral valve (LA to LV). In a typical cycle, oxygen-rich blood fills the left ventricle from the left atrium. The heart muscle contracts and pumps the blood in the left ventricle to the systemic circulation. After collecting enough oxygen-poor blood from the right atrium, the right ventricle muscle contracts and pumps the blood to the pulmonary circulation. Dynamics of these large and complex systems involving large deformation of heart muscles and valves. The IB method has been employed from the simple venous valve constructed with the immersed massless points as illustrated in Fig. 2.9 to the very complicated heart represented with immersed structures (McQueen and Peskin, 2001) as illustrated in Figs. 2.10 and 2.11.

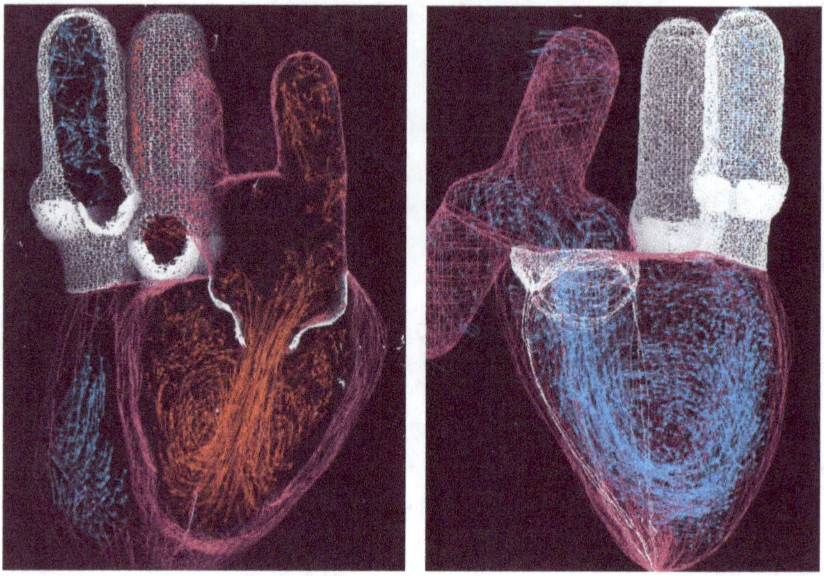

Fig. 2.11. Mitral valve and tricuspid valve with immersed fibers (Peskin, 2012).

2.2.4 *Delta functions*

In the IB method, all discretized delta functions are continuous and have finite support points. Furthermore, immersed boundary points should be two times as dense as the grid region near them where the equivalent forces are exerted.

Consider first a one-dimensional case, in accordance with the translation invariance (Peskin, 1977), for all r, where r is the parameter representing the position of the submerged boundary point and is scaled with respect to the grid size h. Thus, the one-dimensional discretized delta function or kernel can be expressed as $\delta_h(x) = \frac{1}{h}\phi(\frac{x}{h})$. In addition, to uniquely define the discretized delta function for all r, we also have $\sum_j (r-j)^m \phi(r-j) = 0$, where the selection of the mth moment depends on the number of support points. For instance, the discretized delta function with four support points is uniquely defined by the following:

(1) ϕ is a continuous function, with $\phi(r) = 0$ for $|r| \geq 2$;
(2) For all r, $\sum_{j \text{ even}} \phi(r-j) = \sum_{j \text{ odd}} \phi(r-j) = \frac{1}{2}$;

(3) For all r, $\sum_j (r-j)\phi(r-j) = 0$;
(4) For all r, $\sum_j \phi^2(r-j) = C$, where C is a numerical constant.

In the original IB method, this discretized delta function ϕ is uniquely defined and the constant C is determined to be $\frac{3}{8}$. Moreover, the discretized delta function employed in the IB method also has the following properties

(1) For all r_1 and r_2, $\sum_i \phi(r_1-i)\phi(r_2-i) \leq C$ with $C = \frac{3}{8}$;
(2) For all r, $\phi(-r) = \phi(r)$;
(3) For all r, $\frac{d\phi}{dr}$ is also continuous.

In general, for $0 < r < 1$, the discretized delta function $\phi(r-j)$ covers four non-zero support points. However, for the degenerate case of the 4-point discretized delta function centered at $r = 0$, we have five support points, namely, $r-j = -2,-1,0,1,2$. Moreover, the continuity issue of the derivative of the discretized delta function $\phi(r)$ with respect to r have been elaborated in Wang (2003). From the degenerate case, we can easily derive the constant C. Hence, we obtain the following four admissible branches of solutions for $0 < r < 1$,

$$\begin{aligned}
\phi(r) &= \frac{-(2r-3)+\sqrt{4r-4r^2+1}}{8}, \\
\phi(r-2) &= \frac{1}{2} - \phi(r), \\
\phi(r-1) &= -\frac{1}{4} + \frac{r}{2} + \phi(r), \\
\phi(r+1) &= \frac{3}{4} - \frac{r}{2} - \phi(r).
\end{aligned} \tag{2.27}$$

Therefore, in the three-dimensional case, one of the smoothed approximations to the delta function is given by

$$\delta_h(\mathbf{x}) = \frac{1}{h^3}\phi\left(\frac{x_1}{h}\right)\phi\left(\frac{x_2}{h}\right)\phi\left(\frac{x_3}{h}\right). \tag{2.28}$$

Note that for the delta function $\delta(x)$, with $\delta(x) = 0$, if $x \neq 0$, we have

$$\int_{-\infty}^{+\infty} \delta(x)dx = 1;$$

$$\int_{-\infty}^{+\infty} \delta(x)\phi(x)dx = \phi(0);$$

$$\int_{-\infty}^{+\infty} \delta(x)x^m dx = 0; \quad m = 1, 2, \ldots$$

$$\int_{-\infty}^{+\infty} \delta(x - x_1)\delta(x - x_2)dx = \delta(x_1 - x_2).$$

For the comparable discretized delta function $\delta_h(x)$ on the uniform grid with the grid size h, with $\delta_h(x) = 0$, if $|x| \geq 2h$, namely, with the support domain $[-2h, 2h]$, we have,

$$\sum_i \delta_h(x - ih)h = 1;$$
$$\sum_i \delta_h(x - ih)(x - ih)^m h = 0; \quad m = 1, 2, \ldots$$
$$\sum_i \delta_h(x_1 - ih)(x_2 - ih)^m h \neq \delta_h(x_1 - x_2).$$

Finally, using the discrete delta function $\delta_h(x)$, and its properties, the equivalent virtual power can be expressed as

$$\sum_{ijk} \mathbf{f}_{ijk} \cdot \mathbf{v}_{ijk} h^3 = \sum_{i,j,k;\alpha} \mathbf{F}_\alpha \cdot \mathbf{v}_{ijk} \delta_h(\mathbf{x}_{ijk} - \mathbf{X}_\alpha)\Delta s h^3 = \sum_\alpha \mathbf{F}_\alpha \cdot \mathbf{V}_\alpha \Delta s,$$

where the fluid velocity at grid point ijk is denoted as $\mathbf{v}_{ijk}$ and the corresponding equivalent immersed structure body force $\mathbf{f}_{ijk}$ is expressed as

$$\mathbf{f}_{ijk} = \sum_\alpha \mathbf{F}_\alpha \delta_h(\mathbf{x}_{ijk} - \mathbf{X}_\alpha)\Delta s;$$

in addition, the structure force at the structure node is denoted as α $\mathbf{F}_\alpha$ and the corresponding velocity $\mathbf{V}_\alpha$ is expressed as

$$\mathbf{V}_\alpha = \sum_{ijk} \mathbf{v}_{ijk} \delta_h(\mathbf{x}_{ijk} - \mathbf{X}_\alpha)h^3.$$

Of course, the second property with respect to the power will be proven rigorously in Theorem 2.1. It is interesting to note that the discretized delta function in Eq. (2.27) is very close to $(1 + \cos \pi r/2)/4$, with $r \in [-2, 2]$. Moreover, it is easy to confirm that the discretized delta function $\phi(r)$, with $r \in [-2, 2]$, defined in Eq. (2.27), has C^1 continuity (Wang, 2003). In this section, we discuss in detail the construction of the discretized delta function employed in the IB method and prove mathematically that such a function is C^1 and not C^2 continuous. A general discretized delta function with an odd number of support points is shown in Fig. 2.12. Assume ϕ_i,

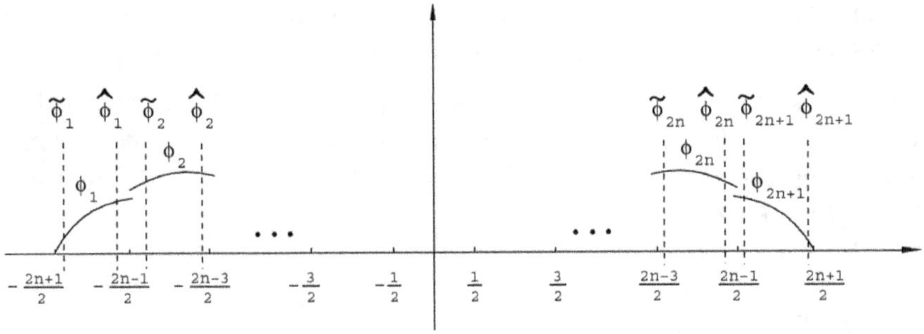

Fig. 2.12. A discretized delta function with an odd number of support points.

$1 \leq i \leq 2n+1$, are piecewise C^{∞}, and the global δ_h is written as $\phi(r)$ with $-(2n+1)/2 \leq r \leq (2n+1)/2$. For convenience, as illustrated in Fig. 2.12, assign $\hat{\phi}_k^m$ and $\tilde{\phi}_k^m$ as the limit values before the right station and after the left station of the kth branch of the discretized delta function, denoted as $\phi_k(r)$. For the global discretized delta function $\phi(r)$ to be C^m continuous, we must have

$$\hat{\phi}_k^{(m)} = \tilde{\phi}_{k+1}^{(m)}, \quad \text{for } 1 \leq k \leq 2n. \tag{2.29}$$

At this point, in order to preserve the properties of the delta function, we would like to enforce some essential properties for the corresponding discretized delta function.

Theorem 2.1. *For a non-negative integer m, the* mth *moment of the delta function is written as*

$$\int x^m \delta(x) dx = \begin{cases} 1, & m = 0, \\ 0, & m \geq 1. \end{cases} \tag{2.30}$$

With a uniform grid, the discretized mth *moment can be expressed either*

$$\sum_i (r-i)^m \phi(r-i) = 0, \tag{2.31}$$

or

$$\sum_i i^m \phi(r-i) = r^m. \tag{2.32}$$

Proof. Referring to Chapter 1, we can easily derive $\int \delta(x)dx = 1$ and $\int x^m \delta(x)dx = 0^m = 0$ for $m \geq 1$. Equation (2.31) is the direct application of Eq. (2.30) to a uniform grid. Suppose for any integer k, $0 \leq k \leq n-1$, Eq. (2.32) holds, i.e., $\sum_i i^k \phi(r-i) = r^k$. Employ the polynomial expansion $(r-i)^n = \sum_{j=0}^{n} r^j i^{n-j}(-1)^j C_n^j$, we can easily derive, from $\sum_i (r-i)^n \phi(r-i) = 0$,

$$\sum_i (r-i)^n \phi(r-i) = \sum_i i^n \phi(r-i) + \sum_i \sum_{j=1}^{n} r^j i^{n-j}(-1)^j C_n^j \phi(r-i) = 0. \tag{2.33}$$

Moreover, based on the premise, for $1 \leq j \leq n$, we also have $\sum_i i^{n-j}\phi(r-i) = r^{n-j}$. As a consequence, Eq. (2.33) is equivalent to

$$\sum_i i^n \phi(r-i) + \sum_{j=1}^{n} r^j r^{n-j}(-1)^j C_n^j = \sum_i i^n \phi(r-i) - r^n + (1-1)^n = 0, \tag{2.34}$$

from which we obtain $\sum_i i^n \phi(r-i) = r^n$, namely, Eq. (2.32) also holds for $k = n$.

Therefore, based on mathematical induction, Eq. (2.32) is valid for any non-negative integer m. Consider $m = 1$, we have $2n+2$ choices of implementing Eq. (2.32) which corresponds to $2n+2$ stations as shown in Fig. 2.12. Moreover, denote the station values as unknown vectors $\tilde{\mathbf{\Phi}} = \langle \tilde{\phi}_1, \tilde{\phi}_2, \ldots, \tilde{\phi}_{2n+1} \rangle$ and $\hat{\mathbf{\Phi}} = \langle \hat{\phi}_1, \hat{\phi}_2, \ldots, \hat{\phi}_{2n+1} \rangle$, we have

$$\tilde{\mathbf{A}}\tilde{\mathbf{\Phi}} = \mathbf{r} \quad \text{and} \quad \hat{\mathbf{A}}\hat{\mathbf{\Phi}} = \mathbf{r}, \tag{2.35}$$

with $\mathbf{r} = \langle -\frac{2n+1}{2}, -\frac{2n-1}{2}, \ldots, \frac{2n-1}{2}, \frac{2n+1}{2} \rangle$, and the $(2n+2) \times (2n+1)$ matrices $\tilde{\mathbf{A}}$ and $\hat{\mathbf{A}}$ are expressed as

$$\tilde{\mathbf{A}} = \begin{bmatrix} 0 & -1 & -2 & \cdots & -2n \\ 1 & 0 & -1 & \cdots & -2n+1 \\ 2 & 1 & 0 & \cdots & -2n+2 \\ \vdots & \vdots & \ddots & \vdots & \vdots \\ 2n+1 & 2n & 2n-1 & \cdots & 1 \end{bmatrix},$$

$$\hat{\mathbf{A}} = \begin{bmatrix} -1 & -2 & -3 & \cdots & -(2n+1) \\ 0 & -1 & -2 & \cdots & -2n \\ 1 & 0 & -1 & \cdots & -(2n-1) \\ \vdots & \vdots & \ddots & \vdots & \vdots \\ 2n & 2n-1 & 2n-2 & \cdots & 0 \end{bmatrix}.$$

It is obvious that both $\tilde{\mathbf{A}}$ and $\hat{\mathbf{A}}$ are rank-two matrices and the following conditions are dependent

$$\sum_{i=1}^{2n+1} \hat{\phi}_i = \sum_{i=1}^{2n+1} \tilde{\phi}_i = 1. \tag{2.36}$$

□

In fact, Eq. (2.36) can be easily derived by subtracting one row of the matrices $\tilde{\mathbf{A}}$ or $\hat{\mathbf{A}}$ from the other. In order to prove C^1 and not C^2 continuity of the discretized delta function, we must prove that Eq. (2.29) holds for $m = 1$ and not for $m = 2$. The condition (2.32), which is used for the derivation of the discretized delta function, suggests that properties of the Vandermonde matrix, named after Alexandre–Théophile Vandermonde (1735–1796), can be introduced. Similar to the derivation of the matrices $\tilde{\mathbf{A}}$ and $\hat{\mathbf{A}}$, we denote the station values as unknown vectors $\tilde{\mathbf{x}} =< \tilde{\phi}_2, \tilde{\phi}_3, \ldots, \tilde{\phi}_{2n+1} >$ and $\hat{\mathbf{x}} =< \hat{\phi}_1, \hat{\phi}_2, \ldots, \hat{\phi}_{2n} >$, and employing the condition (2.36), we derive

$$\mathbf{A}\tilde{\mathbf{x}} = \tilde{\mathbf{b}} \quad \text{and} \quad \mathbf{A}\hat{\mathbf{x}} = \hat{\mathbf{b}}, \tag{2.37}$$

with

$$\tilde{\mathbf{b}} = \mathbf{b}_o + \tilde{\mathbf{b}}_f \tilde{\phi}_1 \quad \text{and} \quad \hat{\mathbf{b}} = \mathbf{b}_o + \hat{\mathbf{b}}_f \hat{\phi}_{2n+1},$$

$$\mathbf{A} = \begin{bmatrix} 1 & 1 & 1 & \cdots & 1 \\ 0 & 1 & 2 & \cdots & 2n-1 \\ 0 & 1^2 & 2^2 & \cdots & (2n-1)^2 \\ \vdots & \vdots & \ddots & \vdots & \\ 0 & 1^{2n-1} & 2^{2n-1} & \cdots & (2n-1)^{2n-1} \end{bmatrix}, \tag{2.38}$$

and

$$\mathbf{b}_o = \begin{bmatrix} 1 \\ \frac{2n-1}{2} \\ \left(\frac{2n-1}{2}\right)^2 \\ \vdots \\ \left(\frac{2n-1}{2}\right)^{2n-1} \end{bmatrix}, \quad \tilde{\mathbf{b}}_f = \begin{bmatrix} -1 \\ 1 \\ -1 \\ \vdots \\ 1 \end{bmatrix}, \quad \text{and} \quad \hat{\mathbf{b}}_f = -\begin{bmatrix} 1 \\ 2n \\ (2n)^2 \\ \vdots \\ (2n)^{2n-1} \end{bmatrix}.$$

As a key observation, we identify that $\mathbf{b}_o$ can be considered as

$$\mathbf{b}_r = \langle 1, r, r^2, \ldots, r^{2n-1} \rangle,$$

which is evaluated at $r = (2n-1)/2$, whereas the following vectors $\mathbf{b}_1$ and $\mathbf{b}_2$ can be simply considered as the first and second derivatives of $\mathbf{b}_r$ evaluated at $r = (2n-1)/2$:

$$\mathbf{b}_1 = \left.\frac{\partial \mathbf{b}_r}{\partial r}\right|_{r=(2n-1)/2} = \left\langle 0, 1, 2\left(\frac{2n-1}{2}\right), \ldots, (2n-1)\left(\frac{2n-1}{2}\right)^{2n-2} \right\rangle,$$

$$\mathbf{b}_2 = \left.\frac{\partial^2 \mathbf{b}_r}{\partial r^2}\right|_{r=(2n-1)/2} = \left\langle 0, 0, 2, \ldots, (2n-1)(2n-2)\left(\frac{2n-1}{2}\right)^{2n-3} \right\rangle.$$

Therefore, the first and second derivatives of the vectors $\tilde{\mathbf{x}}$ and $\hat{\mathbf{x}}$ in Eqs. (2.37) can be evaluated from

$$\mathbf{A}\dot{\tilde{\mathbf{x}}} = \mathbf{b}_1 + \tilde{\mathbf{b}}_f \dot{\tilde{\phi}}_1 \quad \text{and} \quad \mathbf{A}\dot{\hat{\mathbf{x}}} = \mathbf{b}_1 + \hat{\mathbf{b}}_f \dot{\hat{\phi}}_{2n+1}, \tag{2.39}$$

$$\mathbf{A}\ddot{\tilde{\mathbf{x}}} = \mathbf{b}_2 + \tilde{\mathbf{b}}_f \ddot{\tilde{\phi}}_1 \quad \text{and} \quad \mathbf{A}\ddot{\hat{\mathbf{x}}} = \mathbf{b}_2 + \hat{\mathbf{b}}_f \ddot{\hat{\phi}}_{2n+1}. \tag{2.40}$$

Now, for the discretized delta function of the IB method, we also enforce the end support points, such that

$$\tilde{\phi}_1 = \hat{\phi}_{2n+1} = 0, \quad \dot{\tilde{\phi}}_1 = \dot{\hat{\phi}}_{2n+1} = 0, \quad \text{and} \quad \ddot{\tilde{\phi}}_1 = \ddot{\hat{\phi}}_{2n+1} = 0. \tag{2.41}$$

As a consequence, derived from Eq. (2.37), we obtain

$$\tilde{\mathbf{x}} = \hat{\mathbf{x}} = \mathbf{A}^{-1}\mathbf{b}_o. \tag{2.42}$$

It is clear that Eq. (2.42) is equivalent to Eq. (2.29) and the discretized delta function $\phi(r)$ is continuous. Note that in the IB method, for the derivation of discretized delta functions, an additional constraint is introduced:

$$\sum_i \delta_h^2(r-ih)=C, \tag{2.43}$$

where h stands for the grid size and C is a constant independent of the position r.

Equation (2.43) is provided based on the understanding of the translation invariance (Peskin, 1977). As a consequence, we also have the following additional constraint:

$$\sum_{i=1}^{2n+1}\tilde{\phi}_i^2=\sum_{i=1}^{2n+1}\hat{\phi}_i^2=C. \tag{2.44}$$

If the discretized delta function has C^2 continuity, by taking the first and second derivatives of Eq. (2.44), and employing the end conditions (2.41), we must have

$$\tilde{\mathbf{x}}^T\dot{\tilde{\mathbf{x}}}=\hat{\mathbf{x}}^T\dot{\hat{\mathbf{x}}}=0 \text{ and } \dot{\tilde{\mathbf{x}}}^T\dot{\tilde{\mathbf{x}}}+\tilde{\mathbf{x}}^T\ddot{\tilde{\mathbf{x}}}=\dot{\hat{\mathbf{x}}}^T\dot{\hat{\mathbf{x}}}+\hat{\mathbf{x}}^T\ddot{\hat{\mathbf{x}}}=0. \tag{2.45}$$

It is therefore clear, based on Eqs. (2.39) to (2.42), and (2.45), that the C^1 and not C^2 continuity condition is equivalent to

$$\mathbf{b}_o^T(\mathbf{A}^{-1})^T(\mathbf{A}^{-1})\mathbf{b}_1=0, \tag{2.46}$$

$$\mathbf{b}_o^T(\mathbf{A}^{-1})^T(\mathbf{A}^{-1})\mathbf{b}_2+\mathbf{b}_1^T(\mathbf{A}^{-1})^T(\mathbf{A}^{-1})\mathbf{b}_1\neq 0. \tag{2.47}$$

Let us first prove the equality (2.46). Considering a general Vandermonde matrix, we have

$$\det\begin{bmatrix} 1 & 1 & 1 & \cdots & 1\\ x_1 & x_2 & x_3 & \cdots & x_{2n}\\ x_1^2 & x_2^2 & x_3^2 & \cdots & x_{2n}^2\\ \vdots & \vdots & \vdots & \ddots & \vdots\\ x_1^{2n-1} & x_2^{2n-1} & x_3^{2n-1} & \cdots & x_{2n}^{2n-1}\end{bmatrix}=\prod_{1\le i<j}^{2n}(x_j-x_i). \tag{2.48}$$

Obviously, the matrix $\mathbf{A}$ is a Vandermonde matrix with $x_1=0$, $x_2=1,\ldots,x_j=j-1,\ldots,x_{2n}=2n-1$, and from Eq. (2.48), we have

$$\det\mathbf{A}=\prod_{0\le k<j}^{2n-1}(j-k).$$

Supposing we construct another Vandermonde matrix $\mathbf{A}_i$ by replacing the ith column of the matrix $\mathbf{A}$ with $\langle 1, r, r^2, \ldots, r^{2n-1} \rangle$, we obtain

$$\det \mathbf{A}_i = \frac{\prod_{0 \le k < j}^{2n-1} (j-k)}{\prod_{k \ne i-1}^{0,2n-1} (i-1-k)} \prod_{k \ne i-1}^{0,2n-1} (r-k).$$

Employing Cramer's rule, named after Gabriel Cramer (1704–1752), we then have

$$y_i = \frac{\det \mathbf{A}_i}{\det \mathbf{A}} = \frac{\prod_0^{2n-1} (r-k)}{\prod_{k \ne i-1}^{0,2n-1} (i-1-k)} \frac{1}{r-i+1}. \tag{2.49}$$

As a consequence, we also have

$$\frac{\partial y_i}{\partial r} = \sum_{j \ne i-1}^{0,2n-1} \frac{\prod_0^{2n-1} (r-k)}{\prod_{k \ne i-1}^{0,2n-1} (i-1-k)} \frac{1}{(r-i+1)(r-j)}, \tag{2.50}$$

$$\frac{\partial^2 y_i}{\partial r^2} = \sum_{j \ne i-1}^{0,2n-1} \sum_{l \ne i-1,j}^{0,2n-1} \frac{\prod_0^{2n-1} (r-k)}{\prod_{k \ne i-1}^{0,2n-1} (i-1-k)} \frac{1}{(r-i+1)(r-j)(r-l)}. \tag{2.51}$$

Therefore, employing Eqs. (2.49)–(2.51), we obtain

$$y_i \frac{\partial y_i}{\partial r} = \frac{\left[\prod_0^{2n-1} (r-k)\right]^2}{\left[\prod_{k \ne i-1}^{0,2n-1} (i-1-k)\right]^2} \frac{1}{(r-i+1)^2} \sum_{j \ne i-1}^{0,2n-1} \frac{1}{(r-j)},$$

$$y_i \frac{\partial^2 y_i}{\partial r^2} = \frac{\left[\prod_0^{2n-1} (r-k)\right]^2}{\left[\prod_{k \ne i-1}^{0,2n-1} (i-1-k)\right]^2} \frac{1}{(r-i+1)^2} \sum_{j \ne i-1}^{0,2n-1} \sum_{l \ne i-1,j}^{0,2n-1} \frac{1}{(r-j)(r-l)},$$

$$\frac{\partial y_i}{\partial r} \frac{\partial y_i}{\partial r} = \frac{\left[\prod_0^{2n-1} (r-k)\right]^2}{\left[\prod_{k \ne i-1}^{0,2n-1} (i-1-k)\right]^2} \frac{1}{(r-i+1)^2} \left[\sum_{j \ne i-1}^{0,2n-1} \frac{1}{(r-j)}\right]^2.$$

It is clear that the equality (2.46) is equivalent to

$$\sum_{i=1}^{2n} y_i \frac{\partial y_i}{\partial r} = 0. \tag{2.52}$$

Moreover, by replacing i with $2n-i+1$, we can easily confirm that

$$\prod_{k \ne i-1}^{0,2n-1} (i-1-k)^2 = \prod_{k \ne 2n-i}^{0,2n-1} (2n-i-k)^2. \tag{2.53}$$

Note that at $r = (2n-1)/2$, we have the following anti-symmetrical properties for i series, with $1 \le i \le 2n$ and j series, with $0 \le j < 2n-1$,

$$\begin{aligned} r-i+1 &= \frac{2n-2i+1}{2}, \\ r-(2n-i+1)+1 &= \frac{-2n+2i-1}{2}, \end{aligned} \tag{2.54}$$

and

$$\begin{aligned} r-j &= \frac{2n-2j-1}{2}, \\ r-(2n-j-1) &= \frac{-2n+2j+1}{2}. \end{aligned} \tag{2.55}$$

Hence, using the anti-symmetrical properties of (2.55), we can easily confirm that at $r = (2n-1)/2$, we have

$$\sum_{j=0}^{2n-1} \frac{1}{r-j} = 0, \tag{2.56}$$

$$\sum_{j\neq i-1}^{0,2n-1} \frac{1}{r-j} = -\frac{1}{r-i+1}. \tag{2.57}$$

Furthermore, by employing Eqs. (2.53) and (2.54), we derive

$$-\frac{1}{\prod_{k\neq i-1}^{0,2n-1}(i-1-k)^2}\frac{1}{(r-i+1)^3}-\frac{1}{\prod_{k\neq 2n-i}^{0,2n-1}(2n-i-k)^2}\frac{1}{(r-2n+i)^3}=0. \tag{2.58}$$

Based on Eq. (2.58), it is then straightforward to prove Eq. (2.52) by grouping terms i with $2n-i+1$, and consequently, Eq. (2.46). Employing similar procedures, we also derive

$$\sum_{j\neq i-1}^{0,2n-1}\sum_{l\neq i-1,j}^{0,2n-1}\frac{1}{(r-j)(r-l)} = \frac{2}{(r-i+1)^2} - \sum_{j=0}^{2n-1}\frac{1}{(r-j)^2}.$$

However, the anti-symmetrical properties will no longer hold for the cancelation of terms involving even powers of $(r-i+1)$ and $(r-j)$. Without much difficulty, we then also prove the inequality (2.47). Therefore, the discretized delta function in the IB method is C^1 and not C^2 continuous.

2.3 Extended IB Method

The idea of combining the IB method with the finite element methods stems from the proof that the nodal force calculation in the original IB method is equivalent to a traditional nonlinear finite element formulation for truss elements using the second Piola–Kirchhoff stress and the Green–Lagrangian strain (Kim *et al.*, 2003), named after Augustin–Louis Cauchy (1789–1857), George Green (1793–1841), Giulio Fagnano de Santi–Piola (1765–1840), and Gustav Robert Kirchhoff (1824–1887),

Instead of representing the immersed structures or immersed solids with fibers as illustrated in Fig. 2.9, the goal of the extended IB method is to represent the immersed incompressible solid with nonlinear finite element formulations. In addition, these incompressible solids have finite mass and volume. The initial attempt to connect a traditional linear elasticity model with the IB method can be traced back to Sulsky and Brackbill (1991), in which a stress function is transferred to the background fluid grid. In recent finite element extensions of the IB method, a more direct connection between fluid and solid domains is accomplished by employing the internal nodal forces calculated in the context of finite element methods. In these new attempts, submerged solids can experience large displacements and deformations. We must point out that, contrary to the traditional method of tracking the FSI interface, both the interior and the FSI interface of the submerged solid are modeled as submerged material points in contact with the background fluid, namely, the background fluid is everywhere.

Of course, the physical interpretation of finite element methods can be simplified as replacing a continuous medium with a collection of nodes or material points linked through a stiffness matrix. If the structure is best approximated with a fiber network, the stiffness matrix can be derived from truss elements or a finite difference scheme, as used in the original IB method. Otherwise, a traditional finite element formulation must be used to represent two- or three-dimensional elastic structures interacting with the surrounding fluids. However, the compatibility and compressibility issues of finite volume, finite mass, and finite elasticity of immersed solids/structures will be addressed in Chapters 3–5.

In the extended immersed boundary method (EIBM) as proposed in Wang and Liu (2004), instead of the volumeless immersed elastic boundary, we consider the immersed elastic solid which occupies a finite volume within the fluid domain. In particular, we replace the kinematic and dynamic matching of the fluid–solid interface and the effect of the submerged solid with a judiciously chosen collection of equivalent nodal forces calculated in the context of meshless finite element formulations. The employment of the finite element representation enables a realistic stress analysis for the submerged solid subject to large deformations. Finally, in EIBM, we also replace the discretized delta function commonly used in the IB method with the kernel functions in meshless reproducing kernel particle method (RKPM). With this improvement, using the same finite support domain, we could improve the discretized delta function from C^1 to C^n continuity, where n is chosen according to the required smoothness and resolution of such functions, and more importantly, enables the use of non-uniform meshing in the fluid domain. These extensions may eventually open doors to the modelling of complex biological fluid involving molecular, cellular, and flexible vessel-flow interactions. Note that in EIBM, traditional incompressible viscous spectrum fluid dynamics solver is still used. Further developments using the finite element based fluid solvers that have more direct coupling with finite element methods for solids and structures will be employed in the immersed finite element method (IFEM) which will be elaborated in Chapter 3.

2.3.1 *Meshless kernels*

In immersed methods, the nonlinear mapping from the fluid mesh to the solid mesh is essential. At a typical solid node j, with a finite support domain Ω_j, the discretized form of the constraint of the velocities of the immersed solid and the corresponding fluid occupying the same solid domain can be expressed as

$$\mathbf{v}_j^s = \sum_i \mathbf{v}_i \phi_i(\mathbf{x}_i - \mathbf{x}_j^s), \quad \forall \mathbf{x}_i \in \Omega_j, \tag{2.59}$$

where $\phi_i(\mathbf{x}_i - \mathbf{x}_j^s)$ is the kernel function centered at the solid node j, represented with $\mathbf{x}_j^s$.

It is very important to realize that the material points of the submerged solid domain will move in the entire domain, therefore, even if we do not adjust the size of the support domain attached to these material points, Eq. (2.59) represents a nonlinear mapping simply denoted as

$\mathcal{M}$. Note that for the second view of immersed incompressible solids, namely dry, impermeable, and completely separated from the surrounding fluid, within the solid domain, we have the stress difference $\sigma^s_{ij} - \sigma^f_{ij}$ or $-(p^s - p^f)\delta_{ij} + (\tau^s_{ij} - \tau^f_{ij})$. In order to use the definition of σ^f_{ij}, we must also map the unknown pressure from the fluid mesh denoted with node i to the solid mesh denoted with node j. In this case, it is beneficial to use the continuous pressure mixed finite element formulation for both fluid and solid domains (Wang and Bathe, 1997a; Bao *et al.*, 2001). Consequently, like Eq. (2.59), we have

$$p^f_j = \sum_i p_i \phi_i(\mathbf{x}_i - \mathbf{x}^s_j), \quad \forall \mathbf{x}_i \in \Omega_j. \tag{2.60}$$

It turns out that such discretized mapping using various kernel functions has been studied recently in the meshless finite element methods. For example, reproducing kernel particle method (RKPM) was proposed as an alternative or enhancement to various numerical procedures including finite element methods (Liu *et al.*, 1995, 1996a). Unlike the discretized delta function originally proposed in the IB method (Peskin, 2002), the kernel functions in the meshless methods can handle non-uniform meshing, marking an important improvement for the increase of local resolutions near the interfaces. In the IB method, all discretized delta functions are continuous and have finite support points within a uniform background mesh. For non-uniform meshing in the fluid domain, different kernel functions must be introduced. As illustrated in Liu *et al.* (1996b), both wavelet and smooth particle hydrodynamics (SPH) methods belong to a class of reproducing kernel methods where the reproduced function $u^R(x)$ is derived as

$$u^R(x) = \int_{-\infty}^{+\infty} u(y)\phi(x-y)dy, \tag{2.61}$$

with a projection operator or a window function $\phi(x)$.

The reproducing condition requires that up to nth order polynomial be reproduced, i.e.,

$$x^n = \int_{-\infty}^{+\infty} y^n \phi(x-y)dy. \tag{2.62}$$

From the convolution theorem in the Fourier transform domain, Eq. (2.61) can be expressed as

$$\hat{u}^R(\omega) = \hat{u}(\omega)\hat{\phi}(\omega),$$

where ω is the natural frequency or wave number of the functions $u(x)$ and $\phi(x)$.

Supposing $\hat{\phi}(\omega)$ is a perfect rectangular window function, a so-called ideal low-pass filter function, which corresponds to the sinc function in the functional domain, spatial or temporal, $\hat{u}(\omega)$ contains signals within $|\omega| \leq \omega_c$, where ω_c is the cut-off or band-limited frequency or wave number defined by the rectangular window function. Nevertheless, it has been recognized that the ideal low-pass filter does not have a compact support in the functional domain and is therefore not useful as an interpolation function in computational mechanics. Therefore, as presented in Liu *et al.* (1996b), a correction function $C(x;x-y)$ is introduced in the finite computation domain Ω, i.e., the domain of influence or support:

$$C(x;x-y) = \sum_{k=0}^{n} \beta_k(x)(x-y)^k,$$

and the consequent modified window function ϕ is constructed as

$$\bar{\phi}(x;x-y) = \sum_{k=0}^{n} \beta_k(x)(x-y)^k \phi(x-y). \tag{2.63}$$

Define $m_k(x)$ as the kth moment of the window function $\phi(x)$, with $i = \sqrt{-1}$, we have

$$m_k(x) = \int_{\Omega} (x-y)^k \phi(x-y)dy = i^k \hat{\phi}^{(k)}(0), \quad k = 0,1,2,\ldots,n. \tag{2.64}$$

Similarly, define $\bar{m}_k(x)$ as the kth moment of $\bar{\phi}(x)$, and we have

$$\bar{m}_k(x) = \int_{\Omega} (x-y)^k \bar{\phi}(x-y)dy = i^k \hat{\bar{\phi}}^{(k)}(0), \quad k = 0,1,2,\ldots,n. \tag{2.65}$$

In general, with a proper construction of the correction function with respect to the selected window functions, such as the scaling function, wavelet, or spline family, we will be able to enforce

$$\hat{\bar{\phi}}(0) = 1, \quad \hat{\bar{\phi}}^{(1)}(0) = 0, \ldots, \quad \hat{\bar{\phi}}^{(n)}(0) = 0. \tag{2.66}$$

In essence, the window function is required to be flatter at $\omega = 0$ in the Fourier domain as the order n of reproducing gets higher in Eq. (2.62). This implies that as $n \to \infty$, $\hat{\bar{\phi}}(\omega)$ becomes flatter at $\omega = 0$ and approaches an ideal filter. Therefore, based on Eqs. (2.64)–(2.66), we can set up the moment equations to solve for β_k,

$$\begin{bmatrix} m_0(x) & m_1(x) & \cdots & m_n(x) \\ m_1(x) & m_2(x) & \cdots & m_{n+1}(x) \\ \vdots & \vdots & \ddots & \vdots \\ m_n(x) & m_{n+1}(x) & \cdots & m_{2n}(x) \end{bmatrix} \begin{bmatrix} \beta_o(x) \\ \beta_1(x) \\ \vdots \\ \beta_n(x) \end{bmatrix} = \begin{bmatrix} 1 \\ 0 \\ \vdots \\ 0 \end{bmatrix}. \tag{2.67}$$

Furthermore, we can also define a dilation parameter a and introduce

$$\bar{\phi}_a(x-y) = \frac{1}{a}\bar{\phi}\left(\frac{x-y}{a}\right).$$

Thus, by changing the value of the dilation parameter a, usually by a factor of 2, a sequence of low-pass filters is defined. Note that the dilation parameter a is directly linked to mesh density. The difference between two such low-pass filters defines a high-pass filter. Therefore, we can perform a multi-resolution analysis with the projection operator P_a expressed as

$$P_a u(x) = \int_\Omega u(y)\bar{\phi}_a(x-y)dy.$$

Based on the resolution of the projection operator $P_{a/2^n}$, a hierarchical representation of a function $u(x)$ is defined as

$$u(x) = \lim_{n\to\infty} P_{a/2^n}u(x), \ldots, P_a u(x), \ldots, \lim_{n\to\infty} P_{2^n a}u(x) = \oslash.$$

Furthermore, we also define a complementary projection operator Q_{2a}, such that the higher scale projected solution $P_a u$ can be represented by the following sum of low and high scale projections:

$$P_a u(x) = P_{2a}u(x) + Q_{2a}u(x),$$

where the Q_{2a} projection, which can be viewed as a peeled off scale or the rate of variation of $P_a u(x)$, is simply given as a wavelet projection:

$$Q_{2a}u(x) = \int_\Omega u(y)\bar{\psi}_{2a}(x-y)dy,$$

with $\bar{\psi}_{2a}(x-y) = \bar{\phi}_a(x-y) - \bar{\phi}_{2a}(x-y)$.

In accordance with the reproducing property, we have

$$\int_\Omega y^k \bar{\psi}_{2a}(x-y)dy = 0, \quad \text{with } k = 0, 1, \ldots, n,$$

where n can be considered as the order of the polynomial based wavelet.

The discretized reconstruction of the delta function for non-uniform spacing can be written as

$$\bar{\phi}_{aJ}(x) = C(x;x - x_J)a^{-1}\phi\left(\frac{x - x_J}{a}\right)\Delta x_J.$$

For uniform spacing, if the basis functions are chosen to be $1, x, x^2$, and the cubic spline as the window function, the modified window function satisfying the moment reproducing conditions is given as

$$\delta_h(r) = \bar{\phi}_h(x) = \frac{1}{h}\bar{\phi}(r) = \frac{1}{h}$$

$$\times\left(\frac{27}{17} - \frac{30}{17}r^2\right)\begin{cases} \frac{1}{6}(r+2)^3, & -2 \le r < -1, \\ \frac{2}{3} - r^2\left(1 + \frac{r}{2}\right), & -1 \le r < 0, \\ \frac{2}{3} - r^2\left(1 - \frac{r}{2}\right), & 0 \le r < 1, \\ -\frac{1}{6}(r-2)^3, & 1 \le r < 2, \\ 0, & \text{otherwise}. \end{cases} \tag{2.68}$$

Likewise, to reproduce $1, x, x^2, x^3, x^4$ terms, the following correction function can be introduced:

$$\frac{170010}{80347} - \frac{429450}{80347}r^2 + \frac{178290}{80347}r^4.$$

In Fig. 2.13, we provide a comparison of the function and Fourier domains of the modified window functions with the original discretized delta function. It is clear, as we increase the reproducing order n, with the same finite support region, that the kernel functions approach more the ideal low pass filter. In the extended immersed continuum method (EIBM) or the extended IB method, we employ the RKPM kernel functions. Instead of using the discretized delta function, it is possible to interpolate the solid point velocity within a single finite element Boffi *et al.* (2007). The mathematical implication, with respect to the convergence behaviors and accuracy, is still to be discovered. In addition, as shown in Fig. 2.14 the translation invariance is no longer satisfied for RKPM kernel functions. Numerical evidence also suggests that a restriction based on the translation invariance could be relaxed.

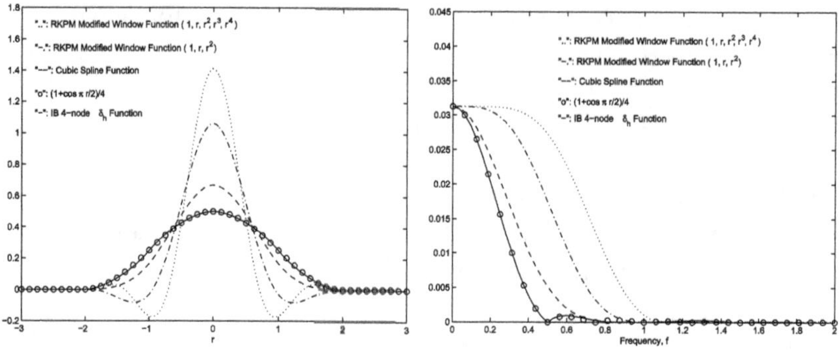

(a) Spatial Function (b) Spectrum

Fig. 2.13. Various discretized delta functions and their respective spectra.

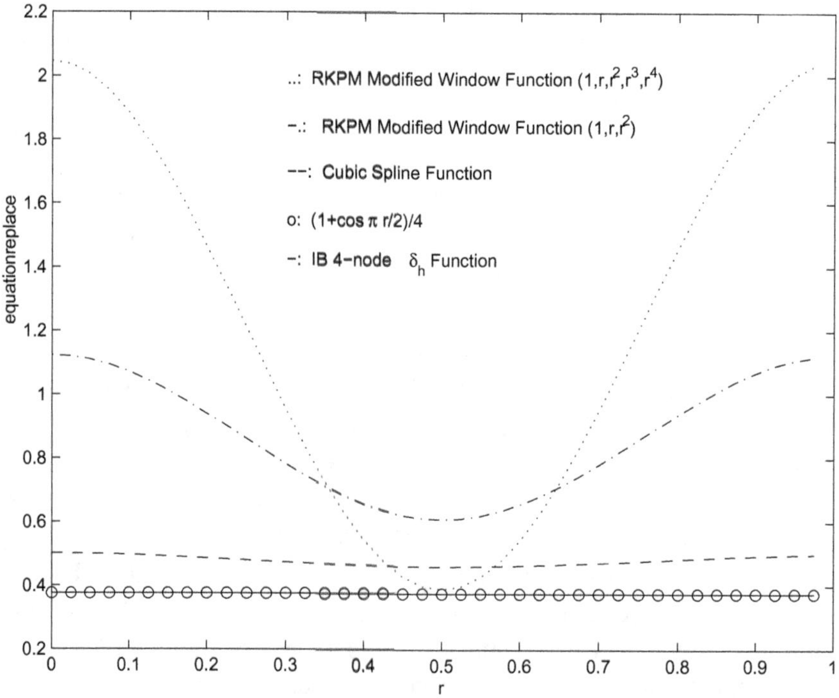

Fig. 2.14. Various discretized delta functions with respect to translation invariance.

2.3.2 *Key improvements*

The goal of the extended IB method is to expand the immersed boundary formulation to the submerged solid with finite volume. As illustrated in Fig. 2.2, the solid is represented with Lagrangian nodes and the background fluid is depicted with Eulerian nodes. In essence, the influence of the immersed elastic boundary or the submerged elastic body is relegated to inhomogeneous forcing terms which can be distributed onto the surrounding fluid domain. Therefore, just as in the IB method, in EIBM, fluid–structure interaction problems are solved by using independent grids for the submerged solid and the surrounding fluid. This moving interface treatment facilitates the handling, without the use of arbitrary Lagrangian-Eulerian (ALE) or adaptive meshing (Belytschko *et al.*, 2000; Huerta and Liu, 1988; Liu, 1981; Liu *et al.*, 1988; Tezduyar, 2001), of submerged structures composed of nearly arbitrary shapes, sizes, and configurations. Furthermore, the benefit in improving the computation efficiency emerges if Eqs. (2.2) and (2.3) are discretized on a regular Eulerian lattice and the discretized fast Fourier Transform is employed. It is clear that Eq. (2.5) represents the equation of motion of the immersed elastic boundary or the submerged elastic body. The fact that the immersed material points move at the local fluid velocity implies the equivalence to the no-slip condition along the fluid–structure interface (Peskin, 2002).

The governing equations for the extended IB method with a submerged structure are the combination of the viscous incompressible Navier–Stokes equation, named after Claude–Louis Navier (1785–1836), written as,

$$\rho\left(\frac{\partial \mathbf{v}}{\partial t}+\mathbf{v}\cdot\boldsymbol{\nabla}\mathbf{v}\right)=-\boldsymbol{\nabla}p+\mu\nabla^2\mathbf{v}+\mathbf{f}, \tag{2.69}$$

$$\boldsymbol{\nabla}\cdot\mathbf{v}=0, \tag{2.70}$$

along with the equivalent force exerted on the surrounding fluid from the immersed solids or immersed structures which consists of the buoyancy, additional elasticity, and additional external force expressed as

$$\mathbf{f}=-\int_{V_s}\mathbf{R}^f\delta(\mathbf{x}-\mathbf{X})\det(\mathcal{F})dV, \tag{2.71}$$

$$\mathbf{R}^f = \int_{V_s} \mathbf{S}\frac{\partial \boldsymbol{\varepsilon}}{\partial \mathbf{X}} dV + \int_{V_s} (\rho_s - \rho_f)\mathbf{N}^T\mathbf{N}\det(\mathcal{F})\ddot{\mathbf{U}}dV - \mathbf{R}^{\text{ext}}, \quad (2.72)$$

$$\frac{d\mathbf{X}}{dt} = \int_{V_f} \mathbf{v}\delta(\mathbf{x}-\mathbf{X})dV, \quad (2.73)$$

where V_s represents the original solid configuration employed in the total Lagrangian formulation.

Equations (2.69) and (2.70) are the governing equations for the incompressible viscous fluid, Eq. (2.72) has to be calculated for each material point, and Eqs. (2.71) and (2.73) distribute the solid force vector to the fluid and the fluid velocity on the solid material using the delta function. Note that Eq. (2.71) and the inertia force term are carried out in the current configuration of the submerged solid and hence a change of coordinate based on the total Lagrangian formulation is needed.

Furthermore, the internal nodal force vector ${}^t\mathbf{F}$ at the current time t with reference to the initial time o between node 1 and 2 can be represented as

$$ {}^t\mathbf{F} = {}^0A\,{}^0L\frac{{}^tP\,{}^0L}{{}^0A\,{}^tL}\begin{bmatrix} \dfrac{1}{{}^0x_1^1 - {}^0x_1^2}\left(1 + \dfrac{{}^tu_1^1 - {}^tu_1^2}{{}^0x_1^1 - {}^0x_1^2}\right) \\ \dfrac{1}{{}^0x_1^1 - {}^0x_1^2}\dfrac{{}^tu_1^1 - {}^tu_1^2}{{}^0x_1^1 - {}^0x_1^2} \\ \dfrac{-1}{{}^0x_1^1 - {}^0x_1^2}\left(1 + \dfrac{{}^tu_1^1 - {}^tu_1^2}{{}^0x_1^1 - {}^0x_1^2}\right) \\ \dfrac{-1}{{}^0x_1^1 - {}^0x_1^2}\dfrac{{}^tu_1^1 - {}^tu_1^2}{{}^0x_1^1 - {}^0x_1^2} \end{bmatrix}. $$

In total Lagrangian representation of truss elements with large displacements, as described in (Bathe, 1996), ${}^tP$ is the axial force within the truss element at time t; ${}^0A$ and ${}^0L$ stand for the initial cross-sectional area and the initial length, respectively; ${}^0\mathbf{x}$ and ${}^t\mathbf{x}$ represent the initial and the current positions of node k, respectively; thus ${}^t\mathbf{u}$ is the displacement for node k, denoted as ${}^t\mathbf{u} = {}^t\mathbf{x} - {}^0\mathbf{x}$. Therefore, for nonlinear truss element mesh or fiber network, as depicted in Fig. 2.15, we have, in isoparametric finite

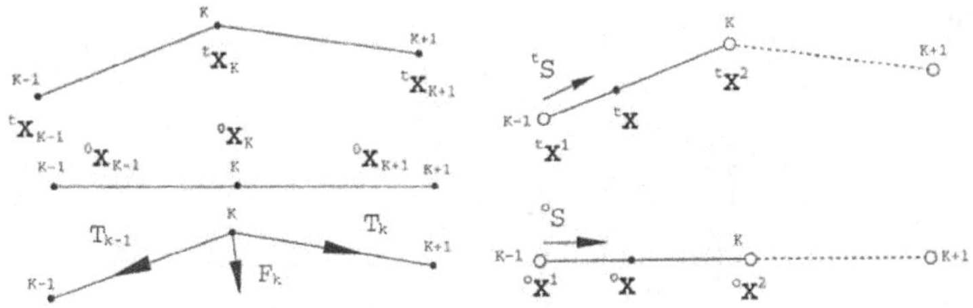

Fig. 2.15. Total Lagrangian representation of truss elements with large displacements.

element formulations, the initial ($t = 0$) and the current ($t = t$) coordinates expressed as

$$
{}^{0}\mathbf{x} = \frac{1-r}{2}{}^{0}x^{1} + \frac{1+r}{2}{}^{0}x^{2},
$$
$$
{}^{t}\mathbf{x} = \frac{1-r}{2}{}^{t}x^{1} + \frac{1+r}{2}{}^{t}x^{2}.
$$

Note that in the total Lagrangian kinematic descriptions, the original position ${}^{0}\mathbf{x}$ is often denoted as $\mathbf{X}$. Thus, the increment of the displacement at time t can be expressed as $\mathbf{u}$, with ${}^{t}\mathbf{u} = {}^{t}\mathbf{x} - {}^{0}\mathbf{x}$. For constant tension within the immersed boundary, we have ${}^{t}P = T$, whereas for constant elastic modulus, we have ${}^{t}P = K|\frac{\partial \mathbf{x}}{\partial s}|$ with the arc length s. Furthermore, the Cauchy stress tensor at time t can also be denoted as

$$
{}^{t}\boldsymbol{\tau} = \begin{bmatrix} {}^{t}\tilde{\tau} & 0 \\ 0 & 0 \end{bmatrix},
$$

where the Cauchy normal stress component within the axial direction of the local coordinate system is denoted as ${}^{t}\tilde{\tau}$, which can also be expressed as the function of the second Piola–Kirchhoff stress normal component in the same direction ${}^{t}_{0}S_{11}$ as follows:

$$
{}^{t}\tilde{\tau} = \frac{{}^{t}\rho}{{}^{0}\rho}\left(\frac{{}^{0}L + \Delta L}{{}^{0}L}\right)^{2} {}^{t}_{0}S_{11} = \frac{{}^{t}P}{{}^{t}A}.
$$

Based on mass conservation, we have

$$
{}^{0}\rho\,{}^{0}L\,{}^{0}A = {}^{t}\rho\,{}^{t}L\,{}^{t}A.
$$

Moreover, if the density of the immersed solids/structures does not change, namely, ${}^{0}\rho = {}^{t}\rho$, we have

$$
{}^{0}L\,{}^{0}A = {}^{t}L\,{}^{t}A,
$$

as a consequence, we then have

$$ {}^{t}_{0}S_{11} = \frac{{}^{t}P}{{}^{0}A} \frac{{}^{0}L}{{}^{0}L + \Delta L}. $$

Again, in the total Lagrangian description, the increment of the Green–Lagrangian strain which includes the geometric nonlinearity, can be expressed as

$$ {}_{0}e_{11} = \frac{\partial u_1}{\partial {}^{0}x_1} + \frac{\partial {}^{t}u_1}{\partial {}^{0}x_1} \frac{\partial u_1}{\partial {}^{0}x_1} + \frac{\partial {}^{t}u_2}{\partial {}^{0}x_1} \frac{\partial u_2}{\partial {}^{0}x_1}, $$

with

$$ \frac{\partial {}^{t}u_1}{\partial {}^{0}x_1} = \frac{{}^{t}u_1^1 - {}^{t}u_1^2}{{}^{0}x_1^1 - {}^{0}x_1^2} \text{ and } \frac{\partial {}^{t}u_2}{\partial {}^{0}x_1} = \frac{{}^{t}u_2^1 - {}^{t}u_2^2}{{}^{0}x_1^1 - {}^{0}x_1^2}. $$

Thus the interpolation matrix which relates the linear strain components with the displacements, as defined in Bathe (1996), can be expressed as

$$ {}^{t}_{0}\mathbf{B}_L^T = \begin{bmatrix} \dfrac{1}{{}^{0}x_1^1 - {}^{0}x_1^2} \left(1 + \dfrac{{}^{t}u_1^1 - {}^{t}u_1^2}{{}^{0}x_1^1 - {}^{0}x_1^2} \right) \\ \dfrac{1}{{}^{0}x_1^1 - {}^{0}x_1^2} \dfrac{{}^{t}u_1^1 - {}^{t}u_1^2}{{}^{0}x_1^1 - {}^{0}x_1^2} \\ \dfrac{-1}{{}^{0}x_1^1 - {}^{0}x_1^2} \left(1 + \dfrac{{}^{t}u_1^1 - {}^{t}u_1^2}{{}^{0}x_1^1 - {}^{0}x_1^2} \right) \\ \dfrac{-1}{{}^{0}x_1^1 - {}^{0}x_1^2} \dfrac{{}^{t}u_1^1 - {}^{t}u_1^2}{{}^{0}x_1^1 - {}^{0}x_1^2} \end{bmatrix}^T. $$

Finally, in the extended IB method, using the finite element representations,

$$ \tau_{ij} = \frac{1}{\det {}^{t}_{0}\mathbf{X}} {}^{t}_{0}x_{i,m} {}^{t}_{0}S_{mn} {}^{t}_{0}x_{j,n}, $$

where the deformation gradient ${}^{t}_{0}\mathbf{X}$ is defined as

$$ {}^{t}_{0}X_{ij} = \frac{\partial {}^{t}x_i(r,t)}{\partial {}^{0}x_j(r,t)}, $$

with the parameter r in isoparametric finite element formulations.

In finite element notations, the force equilibrium is expressed as the balance between the internal force $\mathbf{F}$ and the external force $\mathbf{R}$. In the virtual work, we have

$$\delta_0 e_{11} = {}_0^t\mathbf{B}_L^T \delta\mathbf{u},$$

thus

$${}^t\mathbf{F} = \int_{{}^0V} {}_0^t\mathbf{B}_L^T {}_0^t\mathbf{S} d^0V.$$

Without loss of generality, consider the initial position as horizontal, namely, ${}^0x_2^1 = {}^0x_2^2 = 0$, we have

$${}^0x_1^1 - {}^0x_1^2 = -{}^0L, \quad {}^tx_1^1 - {}^tx_1^2 = -{}^tL\cos\theta, \quad \text{and} \quad {}^tx_2^1 - {}^tx_2^2 = -{}^tL\sin\theta,$$

where the polar angle is denoted as θ and moreover the following expression matching with the physics

$${}^t\mathbf{F} = {}^tP \begin{bmatrix} -\cos\theta \\ -\sin\theta \\ \cos\theta \\ \sin\theta \end{bmatrix}.$$

Similarly, in the explicit time integration scheme, the discretized governing equations are derived as follows:

$$\rho\left(\frac{\mathbf{v}^{n+1} - \mathbf{v}^n}{\Delta t} + \sum_{\alpha=1}^{3} v_\alpha^n D_\alpha^{\pm}\mathbf{v}^n\right) = -\mathbf{D}^o p^{n+1} + \mu \sum_{\alpha=1}^{3} D_\alpha^+ D_\alpha^- \mathbf{v}^{n+1} + \mathbf{f}^{n+1},$$
$$\mathbf{D}^o \cdot \mathbf{v}^{n+1} = 0,$$

with

$$\begin{aligned}
\mathbf{f}_i^{n+1} &= -\sum_{k=1}^{I} (\mathbf{R}^f)_k^{n+1} \delta_h(\mathbf{x}_i - \mathbf{X}_k^n), \\
(\mathbf{R}^f)^{n+1} &= \mathcal{N}(\mathbf{X}^n) + \int_{V_s} (\rho_s - \rho_f)\mathbf{N}^T\mathbf{N}\det(\mathcal{F})\ddot{\mathbf{U}}^n dV - (\mathbf{R}^{\text{ext}})^n, \\
\mathbf{X}_k^{n+1} &= \mathbf{X}_k^n + \Delta t \sum_{\mathbf{x}_i \in \Omega} \mathbf{v}_i^{n+1} \delta_h(\mathbf{x}_i - \mathbf{X}_k^n) h^3.
\end{aligned}$$

For this explicit version of the extended IB method, each time step proceeds as follows:

(1) At the beginning of time marching, the submerged solid configuration $\mathbf{X}^n$ and the fluid velocity $\mathbf{v}^n$ are given;
(2) The resultant nodal forces $(\mathbf{R}^f)^{n+1}$ are evaluated with the finite element formulations;
(3) The nodal force $(\mathbf{R}^f)^{n+1}$ is transferred to the fluid lattice and the result of this computation is the Eulerian body force density $\mathbf{f}^{n+1}$;
(4) Again, the governing equations for incompressible viscous fluid are solved simultaneously, often with the fast Fourier Transform, for the unknowns $(\mathbf{v}^{n+1}, p^{n+1})$;
(5) Once the new velocity field has been found, Eq. (2.23) is used to interpolate the velocity field at the material points, then move them accordingly; and
(6) Thus, $\mathbf{X}^{n+1}$ and $\mathbf{v}^{n+1}$ are known, and this time step calculation is completed.

In EIBM, the same spectrum solver as presented in Section 2.2.3 for incompressible viscous fluid is employed. Traditionally, the immersed boundary formulation is attached to an explicit time marching scheme. Therefore, if the boundary is too stiff, an excessively small time step is often required. Likewise, in the extended IB method, if the submerged structure has a relatively large elastic modulus, it is also required to have a sufficiently small time step. Mayo and Peskin (1993) explored a partially implicit scheme, the derivation of which is based on the Jacobi iterative procedure for linear systems of equations. Further work in employing the implicit scheme to the extended IB method is needed. Moreover, in EIBM, a more direct connection between the fluid and solid domains is accomplished by employing the internal nodal forces calculated in the context of finite element methods and hence submerged solids can experience large deformations. Equation (2.6) will be automatically included in the finite element assemblage process. In particular, both the interior and the fluid–structure interface of the submerged structure are modelled as submerged material points, where the nodal forces are calculated using the finite element methods.

The physical interpretation of finite element methods can be simplified as replacing a continuous medium with a collection of nodes or material points linked through a stiffness matrix. Therefore, if the structure is approximated with an equivalent fiber network, the stiffness matrix can be derived from truss elements or a finite difference scheme as used in the IB method. In fact, we have demonstrated in Section 2.3.2 that the nodal force calculation in the immersed boundary is equivalent to a traditional nonlinear finite element formulation for truss elements using the second Piola–Kirchhoff stress and the Green–Lagrangian strain (Wang, 2003). In the extended IB method, we employ finite element formulations that have been proven to be versatile in handling deformable objects with both material and geometrical nonlinearities.

In EIBM, as in the IB method, an equivalent force density term is added to the fluid equations to account for the presence of the solid. The force density term is calculated from the constitutive equation of the solid material. The background fluid grid covers the entire fluid domain which includes the domain occupied by the submerged solid. Consequently, both the viscous and elastic stress terms are calculated in the submerged solid domain within which there should have been, in accordance with the physical and mathematical models, only the elastic stress term. Nevertheless, in general, the elastic stress term is much larger than the viscous term. A detailed discussion on this subject can be found in Patankar *et al.* (2003), Griffith and Patankar (2020), and Griffith *et al.* (2007).

The new concept employed in EIBM, seemingly a logical improvement on the immersed boundary method, constitutes a brand new approach to dealing with fluid–structure interaction problems as depicted in details in Chapters 3 and 4. In EIBM, the submerged structure can include structures that have finite volumes and also fibers that occupy no volume, just as in the IB method. In fact, it can also be shown explicitly that the judiciously chosen force density term provides an equivalent effect to the kinematic and dynamic matching at the fluid–structure interface and consequently to the fluid–structure interactions.

Note that although the entire submerged solid domain is decomposed into a collection of submerged material points, due to the distributed nodal forces, with a sufficiently dense solid mesh, the surrounding fluid will not penetrate the structure interior and the fluid–structure interface will be automatically defined by the submerged material points enclosing the solid domain. The physical interpretation of finite element methods can be simplified as replacing a continuous medium with a collection of nodes or material points linked through a stiffness matrix. Therefore, if the structure is approximated with an equivalent fiber network, the stiffness matrix can be derived from truss elements or a finite difference scheme as used in the immersed boundary method. In fact, it has been shown that the nodal force calculation in the immersed boundary is equivalent to a traditional nonlinear finite element formulation for truss elements using the second Piola–Kirchhoff stress and the Green–Lagrangian strain (Wang, 2003).

Furthermore, if the surrounding fluid is viscous and incompressible, the immersed solid must be incompressible in immersed methods. In addition, there are two views of immersed solids. The first view, matching the original understanding of the immersed boundary method, is that the immersed solid is wet and permeated with the same fluid as the surrounding. Therefore, the elasticity forces will be the additional force due to the solid portion of the immersed solid. This understanding is realistic for some biological system modeling, since tissues are mostly FSI systems and elastic parts are contributed by elastomer, collagen, or other solid constituents. The second view of the immersed solid is more in tune with traditional FSI systems in which the immersed solid is dry, impermeable, and completely separated from the surrounding fluid. In this context, if we were to use the immersed methods, the immersed solid must be incompressible to match with the surrounding incompressible fluid. Hence, similarities of immersed methods to the fictitious domain as presented in Glowinski *et al.* (2001) and Zhou and Wei (2006) will be elaborated in Chapter 4.

In the initial versions of the finite element formulations of immersed methods (Boffi *et al.*, 2007; Boffi and Gastaldi, 2003; Wang and Liu, 2004; Zhang *et al.*, 2004), the immersed solids are assumed to be incompressible. Furthermore, if the immersed solids are impermeable, the additional elastic forces in comparison with the viscous counterparts calculated with the material properties of the surrounding fluid are very large, hence both views for immersed solids yield similar results. In the extended immersed boundary method, the fluid solver is still the same spectrum solver with

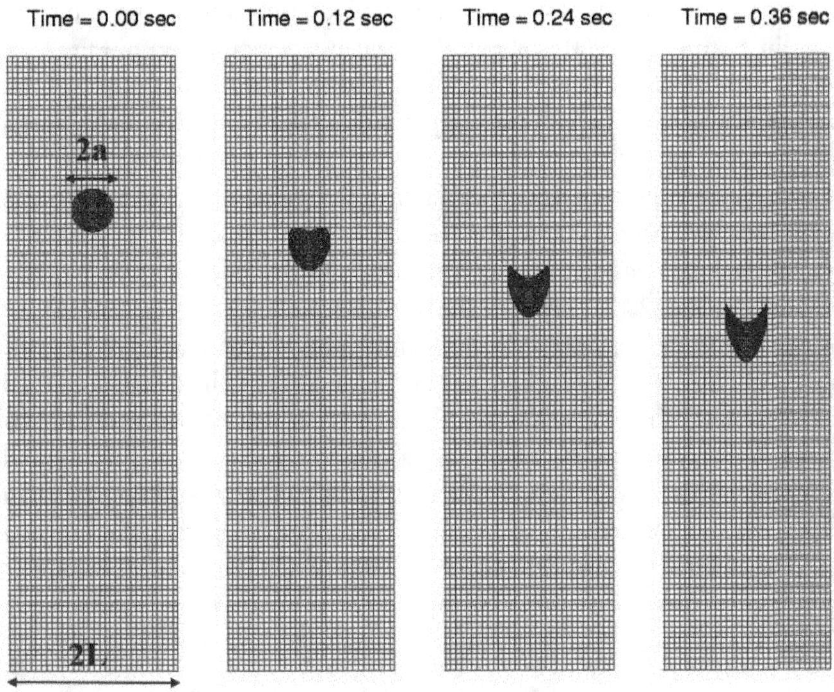

Fig. 2.16. Independent immersed disk mesh over background uniform fluid mesh.

periodic boundary conditions based on Chorin's algorithm, named after Alexandre Joel Chorin (1938-), and FFT solver (Peskin, 1977; Chorin and Marsden, 1998; Chorin, 1967). Since the background fluid mesh is uniform, the same discretized delta function as derived in the immersed boundary method is employed. In the immersed finite element method, the fluid solver is replaced with the stabilized finite element formulation. In this case, the background fluid mesh is unstructured. Therefore, the discretized delta function is replaced with the kernel functions for meshless methods (Li and Liu, 1999; Liu and Chen, 1995).

Finally, in the simulation of a deformable disk falling in a viscous channel, with a background fluid domain mesh 256×4 and a solid mesh with 1602 nodes, as shown in Figs. 2.16 and 2.18, the terminal velocity matches fairly well with the theory, in particular those for hard disks. With FFT based spectrum solver for the incompressible viscous fluid, the computation time on SGI O2000 is about 2000 sec. As demonstrated in Fig. 2.16, the immersed disk has a typical finite element mesh independent of the background uniform fluid mesh. In Figs. 2.18, comparable phenomena similar to those of the terminal velocity for a falling rigid disk in viscous fluid have been predicted with different solid stiffness, mesh density, and delta function. Moreover, with EIBM, it is now convenient to use traditional finite element methods to derive the shear stresses and the pressure as shown in Fig. 2.17. However, the immersed solid is strictly incompressible to satisfy the compatibility requirements of both fluid and solid media. This crucial issue will be addressed in the immersed finite element method and its many variations in Chapter 3. In the same spirit of the extended IB method, the newly proposed immersed finite element method Wang and Liu (2004); Zhang *et al.* (2004), as elaborated in Chapter 3, has marked a clear step forward to combine the much-developed computational fluid and solid mechanics with the concept of the IB method. The implementation of

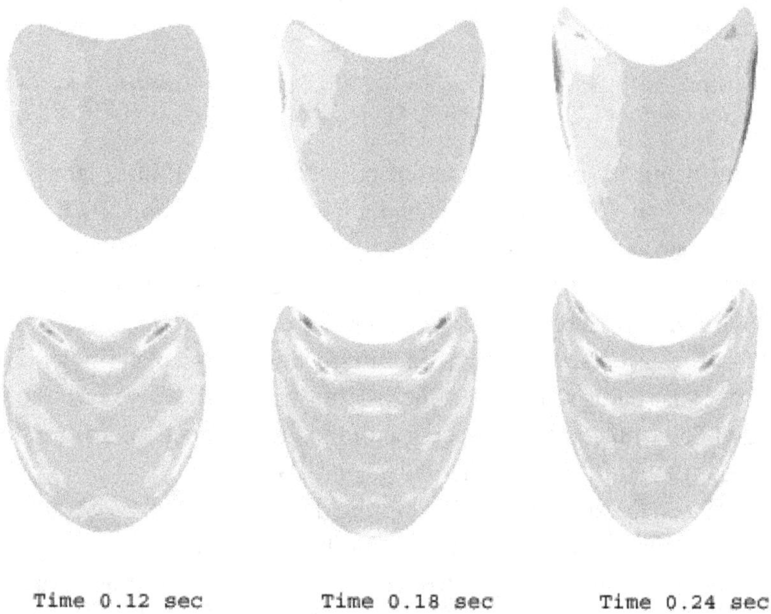

Fig. 2.17. Shear stress and pressure wave predicted with the extended IB method.

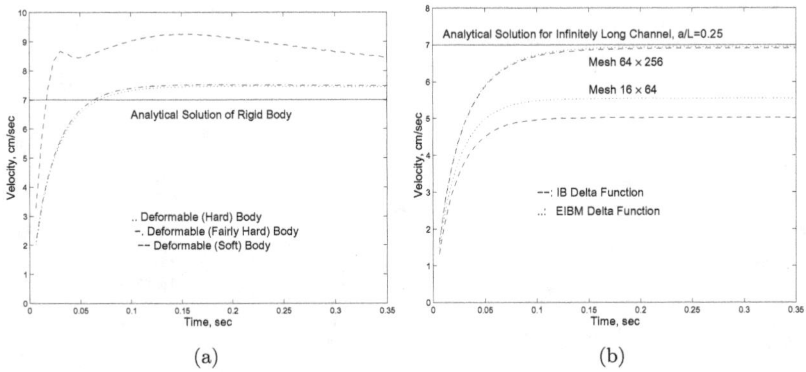

Fig. 2.18. Almost rigid (hard) disk terminal velocity test with EIBM. (a) Different Moduli and (b) different meshes and kernels.

unstructured meshes for background fluids and the adoption of the kernels for meshless finite element methods also open doors for the modeling of many complex FSI systems.

Chapter 3

Immersed Finite Element Methods

3.1 Inception and Developments

For the past few decades, tremendous research efforts have been directed to the development of modeling and simulation approaches for fluid–structure interaction problems. Methods (Tezduyar, 1991, 2001) developed by Tezduyar and his coworkers are widely used in the simulations of fluid–particle (rigid) and fluid–structure interactions. To accommodate the complicated motions of fluid–structure boundaries, we often use adaptive meshing or the arbitrary Lagrangian Eulerian (ALE) techniques (Huerta and Liu, 1988; Hughes *et al.*, 1981; Liu, 1981; Liu and Gvildys, 1986; Liu and Ma, 1982). Recently, such approaches have also been adopted by Hu *et al.* in the modeling of fluid–particle (rigid) systems (Hu *et al.*, 2001). Nevertheless, mesh update or remeshing algorithms can be time consuming and expensive and a detailed discussion on this issue is presented in the work of Johnson and Tezduyar (1999). In 1999, Glowinski *et al.* developed the distributed Lagrange multiplier (DLM) method for the modeling of viscous flows interacting with rigid particles (Glowinski *et al.*, 2001). Their finite element formulations enable the use of a fixed structured fluid grid along with fast and efficient solvers. However, none of the aforementioned methods deals with the interaction between fluid and *flexible* solid particles with large deformations.

In the 1970s, Peskin developed the Immersed Boundary (IB) method (Peskin, 1977) to study flow patterns around heart valves. The mathematical formulation of the IB method employs a mixture of Eulerian and Lagrangian descriptions for fluid and solid domains. In particular, the entire fluid domain is represented by a uniform background grid, which can be solved by finite difference methods with periodic boundary conditions; whereas the submerged structure is represented by a fiber network. The interaction between fluid and structure is accomplished by distributing the nodal forces and interpolating the velocities between Eulerian and Lagrangian domains through a smoothed approximation of the Dirac delta function. The advantage of the IB method is that the fluid–structure interface is tracked automatically, which removes the costly computations due to various mesh update algorithms.

Nevertheless, one major disadvantage of the IB method is the assumption of the fiber-like one-dimensional immersed structure, which may carry mass, but occupies no volume in the fluid domain. This assumption also limits accurate representations for immersed flexible solids which may occupy finite volumes within the fluid domain. Furthermore, uniform fluid grids also set limitations in resolving fluid domains with complex shapes and boundary conditions.

In the proposed immersed finite element method (IFEM), we will eliminate the aforementioned drawbacks of the IB method and adopt parts of the work on the extended immersed boundary method (EIBM) as elaborated in Section 2.3 (Wang and Liu, 2004). With finite element formulations for both fluid and solid domains, the submerged structure is solved more realistically and accurately in comparison with the corresponding fiber network representation. The fluid solver is based on a stabilized equal-order finite element formulation applicable to problems involving moving boundaries (Tezduyar, 1991, 2001; Hughes *et al.*, 1986). This stabilized formulation prevents numerical oscillations without introducing excessive numerical dissipations. Moreover, in the IFEM, the background fluid mesh does not have to follow the motion of the flexible fluid-structure interfaces and thus it is possible to assign sufficiently refined fluid mesh within the region around which the immersed deformable structures in general occupy and move.

Unlike the Dirac delta functions in the IB method which yield C^1 continuity (Peskin, 2002; Wang, 2003), the discretized delta function in IFEM is the C^n shape function often employed in the meshfree reproducing kernel particle method (RKPM). Due to the higher order smoothness in the RKPM delta function, the accuracy is increased in the coupling procedures between fluid and solid domains (Wang and Liu, 2004). Furthermore, the RKPM shape function is also capable of handling nonuniform fluid grids.

The immersed finite element method (IFEM) is a volume-based finite element fluid–structure interaction numerical method. Finite elements are not only used as a way of discretization, but the entire IFEM formulation is derived from the weak form, or principle of virtual work, where the total work in the entire fluid–structure coupled system is maintained.

Even though the concept of "immersed" is similar to that of the immersed boundary (IB) method (Peskin, 2002) where the solid is immersed in the fluid and the two domains are constructed independently, the major distinction of the two is that the solid is not just an interface as it is in the IB method, it actually occupies a volume in the computational domain. As such, the formulation for the IFEM must be volume-based rather than point-based in the IB method. Similarly, the coupling scheme is derived through the balance of the total work in the entire computational space rather than a direct elastic force evaluation on each discrete point.

In IFEM, the fluid and solid domains are defined on two sets of meshes independently and governed by their respective governing equations. The solid domain Ω^s is completely immersed in its surrounding fluid Ω^f, depicted in Fig. 3.1. Since the solid domain is immersed in the fluid domain, the solid boundary Γ^s is also the common interface intersecting with the fluid domain, called the fluid–structure interface $\Gamma^{fsi} \equiv \Gamma^s$.

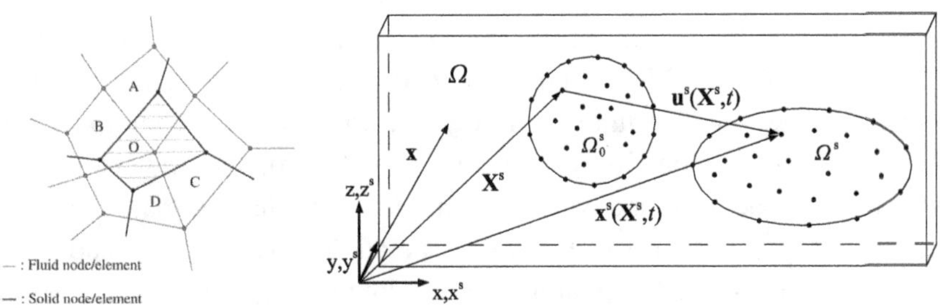

(a) (b)

Fig. 3.1. Communication between Lagrangian solid and Eulerian fluid nodes. (a) Fluid and solid points. (b) Coordinate systems.

The fluid domain and the solid domain together form the entire computational domain, $\Omega = \Omega^f \cup \Omega^s$. All variables associated with the solid are denoted with either superscript or subscript s, while all of those associated with the fluid are denoted with either superscript or subscript f. The key assumption of the IFEM algorithm is that the fluid exists everywhere in the entire computation domain, similar to the IB method. However, unlike the connected discrete points in representing the solid boundary in IB, the solid in IFEM occupies volume in the computational space. The part of the fluid that overlaps with the solid domain is called the artificial fluid, $\bar{\Omega}$, since it does not physically exist. Note that by definition and implementation, the *artificial* fluid domain $\bar{\Omega}$ and the solid domain Ω^s do occupy the same space. The solid domain gets updated at every time step from its initial Ω_0^s. As a result, the artificial domain also moves in the entire computational space. In fact, it is very important to understand that incorporating the dynamics of the solid into the coupled system, the inertial force from the acceleration must be included. In the original IB method, the dynamics of the boundary is not considered but rather is carried by the dynamics of the fluid. The solid boundary has simply been solved as a quasi-static deformation.

The mathematical studies of the immersed finite element method (IFEM) have been presented in the works of Kim and Liu (2006) and Liu *et al.* (2007a). The original IFEM has been widely used to investigate various biomedical problems (Zhang *et al.*, 2004; Zhang and Gay, 2007; Wang and Zhang, 2010). However, it is not the method of choice here due to its limitation in handling fluid and solid having large density differences and large Reynolds number flows as it is the case in the laryngeal flow simulation (Liu *et al.*, 2004a, 2006, Liu and Liu, 2006a; Gay *et al.*, 2006; Zhang, 2014). In the modified IFEM (mIFEM), as elaborated in Section 6.2, the algorithm is entirely re-formulated to accommodate for large Reynolds number flows or where the solid inertia is dominant in fluid-structure interactions, and when the solid and fluid have disparate density difference such as that between soft tissue and air. The mIFEM addresses these problems by re-formulating the coupling scheme and solving for the solid dynamics rather than imposing it. A more detailed derivation of the algorithm can be found in the work of Wang and Bathe (1997b). It has been rigorously derived and verified with 2D and 3D studies (Wang and Bathe, 1997a, 1997b; Wang and Zhang, 2015).

3.2 Kinematics and Formulations

3.2.1 *Coordinate systems*

Let us consider an incompressible three-dimensional deformable structure in Ω^s completely immersed in an incompressible fluid domain Ω^f. Together, the fluid and the solid occupy a domain Ω, but they do not intersect:

$$\Omega^f \cup \Omega^s = \Omega \quad \text{and} \quad \Omega^f \cap \Omega^s = \oslash.$$

In contrast to the IB formulation, the solid domain can occupy a finite volume in the fluid domain. Since we assume both fluid and solid to be incompressible, the union of two domains can be treated as one incompressible continuum with a continuous velocity field. In the computation, the fluid spans over the entire domain Ω, thus an Eulerian fluid mesh is adopted; whereas a Lagrangian solid mesh is constructed on top of the Eulerian fluid mesh. The coexistence of fluid and solid in Ω^s requires some considerations when developing the momentum and continuity equations.

As shown in Fig. 3.1, in the computational fluid domain Ω, the fluid grid is represented by the time-invariant position vector $\mathbf{x}$, while the material points of the structure in the initial solid domain Ω_0^s and the current solid domain Ω^s are represented by $\mathbf{X}^s$ and $\mathbf{x}^s(\mathbf{X}^s,t)$, respectively. The superscript or subscript s is used in the solid variables to distinguish the fluid and solid domains.

In the fluid calculations, the velocity $\mathbf{v}$ and the pressure p are the unknown fluid field variables; whereas the solid domain involves the calculation of the nodal displacement $\mathbf{u}^s$, which is defined as the difference of the current and the initial coordinates: $\mathbf{u}^s = \mathbf{x}^s - \mathbf{X}^s$. The velocity $\mathbf{v}^s$ is the material derivative of the displacement, i.e., $\mathrm{d}\mathbf{u}^s/\mathrm{d}t$. The Eulerian coordinates in the computational fluid domain Ω are described with the time invariant position vector $\mathbf{x}$; the solid positions in the initial configuration Ω_0^s and the current configuration Ω^s are represented by $\mathbf{X}^s$ and $\mathbf{x}^s(\mathbf{X}^s,t)$, respectively. The displacement of the solid is defined as $\mathbf{u}^s = \mathbf{x}^s - \mathbf{X}^s$. The definitions of the spatial coordinate, displacement, velocity, and acceleration in both fluid and solid domains are summarized in Table 3.1. Since displacement and acceleration are not explicitly used in the fluid calculations, they are not defined under the column of the fluid domain.

Table 3.1. IFEM nomenclature.

	Fluid Domain (Ω)	**Solid Domain (Ω^s)**
Spatial coordinate	$\mathbf{x}$	$\mathbf{x}^s$
Displacement	–	$\mathbf{u}^s = \mathbf{x}^s - \mathbf{X}^s$
Velocity	$\mathbf{v}$	$\mathbf{v}^s = \mathrm{d}\mathbf{u}^s/\mathrm{d}t = \dot{\mathbf{u}}^s$
Acceleration	–	$\mathbf{a}^s = \mathrm{d}^2\mathbf{u}^s/\mathrm{d}t^2 = \ddot{\mathbf{u}}^s$

3.2.2 *Equations of motions*

In this section, we start by introducing the equations of motion for a continuum that contains both fluid and solid. The governing equations of both domains are then extracted from the equations of motion along with a delta function for the coupling between fluid and solid domains. The detailed mathematical derivation and the proof of equivalency between strong and weak forms have been illustrated in the work of Liu *et al.* (2007a).

In a continuum, the inertial force of a particle is balanced with the derivative of the Cauchy stress σ and the external force $\mathbf{f}^{\text{ext}}$ exerted on the continuum:

$$\rho\frac{\mathrm{d}v_i}{\mathrm{d}t} = \sigma_{ij,j} + f_i^{\text{ext}}. \tag{3.1}$$

If the solid density ρ^s is different from the fluid density ρ^f, i.e., $\rho = \rho^s$ in Ω^s and $\rho = \rho^f$ in Ω^f, we can divide the inertial force into two components within the solid domain Ω^s:

$$\rho\frac{\mathrm{d}v_i}{\mathrm{d}t} = \begin{cases} \rho^f\frac{\mathrm{d}v_i}{\mathrm{d}t}, & \mathbf{x}\in\Omega/\Omega^s, \\ \rho^f\frac{\mathrm{d}v_i}{\mathrm{d}t} + (\rho^s - \rho^f)\frac{\mathrm{d}v_i}{\mathrm{d}t}, & \mathbf{x}\in\Omega^s. \end{cases} \tag{3.2}$$

Note that to illustrate the point of addition and subtraction of these two sub-domains, we represent the fluid domain Ω^f as the entire domain Ω minus the solid domain Ω^s, namely, Ω/Ω^s. Moreover, in this section, the external force f_i^{ext} is illustrated as the gravitational force and likewise, we obtain

$$f_i^{\text{ext}} = \begin{cases} \rho^{\text{f}} g_i, & \mathbf{x}\in\Omega/\Omega^s, \\ \rho^{\text{f}} g_i + (\rho^{\text{s}} - \rho^{\text{f}})g_i, & \mathbf{x}\in\Omega^s. \end{cases} \tag{3.3}$$

Since the computational fluid domain is the entire domain Ω, we ignore the hydrostatic pressure and rewrite Eq. (3.3) as

$$f_i^{\text{ext}} = \begin{cases} 0, & \mathbf{x} \in \Omega/\Omega^s, \\ (\rho^{\text{s}} - \rho^{\text{f}})g_i, & \mathbf{x} \in \Omega^s. \end{cases} \tag{3.4}$$

Furthermore, the derivative of the Cauchy stress in Eq. (3.1) can also be decomposed as

$$\sigma_{ij,j} = \begin{cases} \sigma_{ij,j}^f, & \mathbf{x} \in \Omega/\Omega^s, \\ \sigma_{ij,j}^f + \sigma_{ij,j}^s - \sigma_{ij,j}^f, & \mathbf{x} \in \Omega^s. \end{cases} \tag{3.5}$$

Note that in Eq. (3.5) the fluid stress in the solid domain in general is much smaller than the corresponding solid stress (Patankar *et al.*, 2003). Let us now define the fluid–structure interaction (FSI) force within the domain Ω^s as $f_i^{fsi,s}$:

$$f_i^s = -(\rho^s - \rho^f)\frac{\mathrm{d}v_i}{\mathrm{d}t} + \sigma_{ij,j}^s - \sigma_{ij,j}^f + (\rho^{\text{s}} - \rho^{\text{f}})g_i, \quad \mathbf{x} \in \Omega^s. \tag{3.6}$$

Naturally, the interaction force f_i^s in Eq. (3.6) is calculated with the Lagrangian description. Moreover, a Dirac delta function δ is used to distribute the interaction force from the solid domain onto the computational fluid domain:

$$f_i^{\text{fsi}}(\mathbf{x},t) = \int_{\Omega^s} f_i^s(\mathbf{X}^s,t)\,\delta(\mathbf{x} - \mathbf{x}^s(\mathbf{X}^s,t))\,d\Omega. \tag{3.7}$$

Hence, the governing equation for the fluid can be derived by combining the fluid terms and the interaction force as

$$\rho^f \frac{\mathrm{d}v_i}{\mathrm{d}t} = \sigma_{ij,j}^f + f_i^{\text{fsi}}, \quad \mathbf{x} \in \Omega. \tag{3.8}$$

Since we consider the entire domain Ω to be incompressible, we only need to apply the incompressibility constraint once in the entire domain Ω:

$$v_{i,i} = 0. \tag{3.9}$$

To delineate the Lagrangian description for the solid and the Eulerian description for the fluid, we introduce different velocity field variables v_i^s and v_i to represent the motions of the solid in the domain Ω^s and the fluid within the entire domain Ω. The coupling of both velocity fields is accomplished with the Dirac delta function:

$$v_i^s(\mathbf{X}^s,t) = \int_{\Omega} v_i(\mathbf{x},t)\,\delta(\mathbf{x}-\mathbf{x}^s(\mathbf{X}^s,t))\,d\Omega. \tag{3.10}$$

Finally, with the variables in the solid domain Ω^s defined with the Lagrangian description and the fluid domain Ω with the Eulerian description, we can summarize the derivations of the governing equations as well as the interpolation and distribution processes as follows:

$$\rho^f \frac{\mathrm{d}v_i}{\mathrm{d}t} = \sigma^f_{ij,j} + f_i^{\mathrm{fsi}}, \quad \text{in } \Omega, \tag{3.11a}$$

$$v_{i,i} = 0, \quad \text{in } \Omega, \tag{3.11b}$$

$$f_i^{\mathrm{fsi}}(\mathbf{x},t) = \int_{\Omega^s} f_i^s(\mathbf{X}^s,t)\,\delta(\mathbf{x}-\mathbf{x}^s(\mathbf{X}^s,t))\,d\Omega, \tag{3.11c}$$

$$f_i^s = -(\rho^s-\rho^f)\frac{\mathrm{d}v_i^s}{\mathrm{d}t} + \sigma^s_{ij,j} - \sigma^f_{ij,j} + (\rho^s-\rho^f)g_i, \quad \text{in } \Omega^s, \tag{3.11d}$$

$$v_i^s(\mathbf{X}^s,t) = \int_{\Omega} v_i(\mathbf{x},t)\,\delta(\mathbf{x}-\mathbf{x}^s(\mathbf{X}^s,t))\,d\Omega, \tag{3.11e}$$

where the velocity in the solid domain is simply denoted as $\mathbf{v}^s$ instead of $\mathbf{v}$.

3.3 Governing Equations and Discretizations

In this section, we discuss in detail the strong and weak forms of governing equations for both fluid and solid domains. In particular, using the kinematic and dynamic matching along the fluid–structure interface, for brevity, we exclude the fluid–structure interface boundary integral terms in the corresponding weak forms for both fluid and solid domains.

3.3.1 *Strong form*

This section explains in detail the three-dimensional (3D) finite element fluid solver used in the IFEM formulation. The derivations include the strong form, namely, the Navier–Stokes equations, the Petrov–Galerkin weak form, and the corresponding discretization.

For the Eulerian description of the fluid, the material time derivative $\mathrm{d}v_i/\mathrm{d}t$ is expressed as the time derivative and the convective terms:

$$\frac{\mathrm{d}v_i}{\mathrm{d}t} = v_{i,t} + v_j v_{i,j}. \tag{3.12}$$

By substituting Eq. (3.12) into Eq. (3.11a), the governing equations for the fluid domain Ω are written as

$$\rho^f(v_{i,t} + v_j v_{i,j}) = \sigma^f_{ij,j} + f_i^{\text{fsi}}, \tag{3.13a}$$

$$v_{i,i} = 0. \tag{3.13b}$$

Moreover, the Cauchy stress tensor σ^f_{ij} is also defined as

$$\sigma^f_{ij} = -p\delta_{ij} + 2\mu\tau_{ij},$$

$$\tau_{ij} = \frac{1}{2}(v_{i,j} + v_{j,i}),$$

where μ is the dynamic viscosity and p is the pressure.

In addition to these equations of motion, we stipulate the following boundary conditions:

$$v_i = g_i, \quad \mathbf{x} \in \Gamma_{g_i},$$

$$\sigma^f_{ij} n_j = h_i, \quad \mathbf{x} \in \Gamma_{h_i},$$

where g_i and h_i are given functions.

For each spatial degree of freedom i, we assume that the fluid boundary Γ can be partitioned as $\Gamma = \Gamma_{g_i} \bigcup \Gamma_{h_i}$, with $\Gamma_{g_i} \bigcap \Gamma_{h_i} = \oslash$, which implies that the Dirichlet and Neumann boundaries span the entire fluid surface with no intersection. Similarly, the solid domain Ω^s is solved using the Lagrangian description. The material time derivative $\mathrm{d}v_i^s/\mathrm{d}t$ can therefore be directly introduced as $\ddot{u}_i^s$ and Eq. (3.6) yields

$$(\rho^s - \rho^f)\ddot{u}_i^s = \sigma^s_{ij,j} - \sigma^f_{ij,j} + (\rho^s - \rho^f)g_i - f_i^s. \tag{3.14}$$

The force balance equation in Eq. (3.14) is the linear momentum equation in the updated Lagrangian description. The interaction force $\mathbf{f}^s$ can be treated as an additional body force in the solid continuum which is also reiterated and elaborated in Chapter 4. Obviously, if the densities are the same for both fluid and solid, and if the solid behaves as the fluid, the entire computational domain would behave as the fluid with $f_i^s = 0$.

3.3.2 *Petrov–Galerkin weak form*

For the fluid domain, transforming the strong form of the conservation equations into the weak form requires two test functions: the velocity test function δv_i and the pressure test function δp for the momentum equation and the continuity equation, respectively. To reduce or eliminate the numerical oscillations in the velocity and pressure solutions from the standard Galerkin method for incompressible materials, we employ the test functions $\delta \tilde{v}_i$ and $\delta \tilde{p}$ along with the stabilization parameters τ^m and τ^c. The algorithm follows the stabilized equal-order finite element formulation developed in the works of Tezduyar (2001) and Hughes *et al.* (1986):

$$\begin{aligned}\delta \tilde{v}_i &= \delta v_i + \tau^m v_k \delta v_{i,k} + \tau^c \delta p_{,i},\\ \delta \tilde{p} &= \delta p + \tau^c \delta v_{i,i},\end{aligned}$$

where τ^m and τ^c are the scalar stabilization parameters dependent of the computational grid, time step size, and flow variables.

The weak form of the continuity equation (3.13b) is obtained by multiplying the pressure test function $\delta \tilde{p}$ and integrating over Ω:

$$\int_\Omega (\delta p + \tau^c \delta v_{i,i}) v_{j,j}\, d\Omega = 0. \tag{3.15}$$

Similarly, the weak form of the momentum equation (3.13a) is achieved by multiplying the velocity test function $\delta \tilde{v}_i$:

$$\begin{aligned}&\int_\Omega (\delta v_i + \tau^m v_k \delta v_{i,k} + \tau^c \delta p_{,i}) \left[\rho^f (v_{i,t} + v_j v_{i,j}) - f_i^{\mathrm{fsi}}\right] d\Omega - \int_\Omega \delta v_i \sigma_{ij,j}^f\, d\Omega\\ &\quad - \sum_e \int_{\Omega_e} (\tau^m v_k \delta v_{i,k} + \tau^c \delta p_{,i}) \sigma_{ij,j}^f\, d\Omega = 0.\end{aligned} \tag{3.16}$$

Applying integration by parts and the divergence theorem, we can rewrite Eq. (3.16) as

$$\begin{aligned}&\int_\Omega (\delta v_i + \tau^m v_k \delta v_{i,k} + \tau^c \delta p_{,i}) \left[\rho^f (v_{i,t} + v_j v_{i,j}) - f_i^{\mathrm{fsi}}\right] d\Omega - \int_{\Gamma_{h_i}} \delta v_i h_i \mathrm{d}\Gamma\\ &\quad + \int_\Omega \delta v_{i,j} \sigma_{ij}^f d\Omega - \sum_e \int_{\Omega_e} (\tau^m v_k \delta v_{i,k} + \tau^c \delta p_{,i}) \sigma_{ij,j}^f d\Omega = 0,\end{aligned} \tag{3.17}$$

where individual finite elements are denoted with the subscript e and the stabilization supplementary terms are evaluated at the element interiors.

The detailed illustrations of the stabilized finite element formulations are available in the works of Tezduyar (1991) and Yang *et al.* (2017). Let us assume that there is no traction applied on the fluid boundary, i.e., $\int_{\Gamma_{h_i}} \delta v_i h_i d\Gamma = 0$, following Zhang *et al.* (2002), the final weak form yields

$$\int_{\Omega} (\delta v_i + \tau^m v_k \delta v_{i,k} + \tau^c \delta p_{,i}) \left[\rho^f (v_{i,t} + v_j v_{i,j}) - f_i^{\text{fsi}}\right] d\Omega + \int_{\Omega} \delta v_{i,j} \sigma_{ij}^f d\Omega$$
$$- \sum_e \int_{\Omega_e} (\tau^m v_k \delta v_{i,k} + \tau^c \delta p_{,i}) \sigma_{ij,j}^f d\Omega + \int_{\Omega} (\delta p + \tau^c \delta v_{i,i}) v_{j,j} d\Omega = 0, \tag{3.18}$$

where τ^m and τ^c are scalar stabilization parameters that are in general functions of computational grid, the time step size, and the flow variables.

The nonlinear systems are solved with the Newton–Raphson method. Moreover, to improve the computation efficiency, we also employ the GMRES iterative algorithm and compute the residuals based on matrix-free techniques (Saad and Schultz, 1986; Zhang *et al.*, 2002).

To develop the weak form, we multiply the momentum equation, Eq. (3.14), by a test function δu_i and integrate over the current configuration Ω^s:

$$\int_{\Omega^s} \delta u_i \left[(\rho^s - \rho^f) \ddot{u}_i^s - \frac{\partial \sigma_{ij}^s}{\partial x_j} - (\rho^s - \rho^f) g_i + f_i^s \right] d\Omega^s = 0. \tag{3.19}$$

Note that for brevity we ignore the fluid stress within the solid domain. The transformation of the weak form from the updated Lagrangian to the total Lagrangian description is to change the integration domain from Ω^s to Ω_0^s. Since we consider incompressible fluid and solid, and the Jacobian determinant is 1 in the solid domain, the transformation of the weak form to total Lagrangian description yields

$$\int_{\Omega_0^s} \delta u_i \left[(\rho^s - \rho^f) \ddot{u}_i^s - \frac{\partial P_{ij}}{\partial X_j} - (\rho^s - \rho^f) g_i + f_i^s \right] d\Omega_0^s = 0, \tag{3.20}$$

where the first Piola–Kirchhoff stress P_{ij} is defined as $P_{ij} = J F_{ik}^{-1} \sigma_{kj}^s$ and the deformation gradient F_{ij} as $F_{ij} = \partial x_i / \partial X_j$.

Using integration by parts and the divergence theorem, we can rewrite Eq. (3.20) as

$$\int_{\Omega_0^s} \delta u_i(\rho^s - \rho^f)(\ddot{u}_i^s - g_i)\, d\Omega_0^s + \int_{\Omega_0^s} \delta u_{i,j} P_{ij}\, d\Omega_0^s + \int_{\Omega_0^s} \delta u_i f_i^s\, d\Omega_0^s = 0. \tag{3.21}$$

Note again that the boundary integral terms on the fluid-structure interface for both fluid and solid domains will cancel each other and for brevity are not included in the corresponding weak forms.

3.3.3 *Finite element discretizations*

The discretization process using the finite elements follows the procedures in the work of Belytschko *et al.* (2000). The velocity and pressure functions, v_i and p, along with the test functions δv_i and δp, are interpolated as

$$v_i^h = \sum_I N_I d_{iI}, \quad \delta v_i^h = \sum_I N_I c_{iI}, \quad p^h = \sum_I N_I p_I, \quad \text{and} \quad \delta p^h = \sum_I N_I q_I,$$

where N_I is the nodal shape function at node I.

Substituting the interpolations into Eq. (3.18), we derive

$$\begin{aligned}
&\sum_I \int_\Omega (c_{iI} N_I + \tau^m v_k^h c_{iI} N_{I,k} + \tau^c q_I N_{I,i}) \left[\rho^f (v_{i,t}^h + v_j^h v_{i,j}^h) - f_i^{\text{fsi},h}\right] d\Omega \\
&\quad + \sum_I \int_\Omega c_{iI} N_{I,j} \sigma_{ij}^{\text{f},h} d\Omega - \sum_I \sum_e \int_{\Omega_e} (\tau^m v_k^h c_{iI} N_{I,k} + \tau^c q_I N_{I,i}) \sigma_{ij,j}^{\text{f},h} d\Omega \\
&\quad + \sum_I \int_\Omega (q_I N_I + \tau^c c_{iI} N_{I,i}) v_{j,j}^h d\Omega = 0.
\end{aligned}$$

Combining the test function vectors $\mathbf{c}_i$ and $\mathbf{q}$, we have four set of equations with respect to four unknowns at each node I:

$$\mathbf{c}_i^T \mathbf{r}_i^w = 0 \quad \text{and} \quad \mathbf{q}^T \mathbf{r}^q = 0,$$

where the residual vectors r_{iI}^{w} and r_I^q are given as follows:

$$
\begin{aligned}
r_{iI}^{w} &= \int_{\Omega}(N_I+\tau^m v_k^h N_{I,k})\left[\rho^f(v_{i,t}^h+v_j^h v_{i,j}^h)-f_i^{\text{fsi},h}\right]d\Omega \\
&\quad +\int_{\Omega} N_{I,j}\sigma_{ij}^{\text{f},h}d\Omega-\sum_e\int_{\Omega_e}\tau^m v_k^h N_{I,k}\sigma_{ij,j}^{\text{f},h}d\Omega+\int_{\Omega}\tau^c N_{I,i}v_{j,j}^h d\Omega, \\
r_I^q &= \int_{\Omega}\tau^c N_{I,i}\left[\rho^f(v_{i,t}^h+v_j^h v_{i,j}^h)-f_i^{\text{fsi},h}\right]d\Omega \\
&\quad -\sum_e\int_{\Omega_e}\tau^c N_{I,i}\sigma_{ij,j}^{\text{f},h}d\Omega+\int_{\Omega}N_I v_{j,j}^h d\Omega,
\end{aligned}
$$

from which the pressure and velocity unknowns are solved iteratively in each time step.

With respect to immersed solids or structures, the discretization of the equilibrium equation (3.21) is done by introducing the finite element interpolations of the test and trial functions:

$$
\delta u_i^h=\sum_I N_I\delta u_{iI} \quad \text{and} \quad u_i^h=\sum_J N_J u_{iJ}. \tag{3.22}
$$

By substituting the interpolation functions into Eq. (3.21), we can obtain the discretized form:

$$
\begin{aligned}
&\sum_I\sum_J\int_{\Omega_0^s}N_I\delta u_{iI}(\rho^s-\rho^f)N_J\ddot{u}_{iJ}^s\,d\Omega_0^s+\sum_I\int_{\Omega_0^s}N_{I,j}\delta u_{iI}P_{ij}\,d\Omega_0^s \\
&\quad -\sum_I\int_{\Omega_0^s}N_I\delta u_{iI}(\rho^s-\rho^f)g_i\,d\Omega_0^s+\sum_I\int_{\Omega_0^s}N_I\delta u_{iI}f_i^s\,d\Omega_0^s=0.
\end{aligned}
$$

Using the arbitrariness of δu_{iI}, we derive

$$
\begin{aligned}
&\sum_J\int_{\Omega_0^s}N_I(\rho^s-\rho^f)N_J\ddot{u}_{iJ}^s\,d\Omega_0^s+\int_{\Omega_0^s}N_{I,j}P_{ij}\,d\Omega_0^s \\
&\quad -\int_{\Omega_0^s}N_I(\rho^s-\rho^f)g_i\,d\Omega_0^s+\int_{\Omega_0^s}N_I f_i^s\,d\Omega_0^s=0.
\end{aligned} \tag{3.23}
$$

Therefore, at a discrete submerged point I represented with the position vector $\mathbf{x}_I^s(\mathbf{X}_I^s,t)$, we have the following force equilibrium:

$$
f_{iI}^{\text{inert}}+f_{iI}^{\text{int}}-f_{iI}^{\text{ext}}+f_{iI}^{s}=0, \tag{3.24}
$$

where f_{iI}^{inert}, f_{iI}^{int}, f_{iI}^{ext}, and f_{iI}^{s} stand for the inertial, internal, external, and fluid-structure interaction forces at node I, respectively.

In the original IFEM implementation, we consider a three-dimensional almost incompressible hyperelastic material model with the Mooney–Rivlin material description (Wang and Liu, 2004), in which the elastic energy potential W is given as

$$W = C_1(J_1 - 3) + C_2(J_2 - 3) + \frac{\kappa}{2}(J_3 - 1)^2, \tag{3.25}$$

where C_1, C_2, and κ are the material constants and J_1, J_2, and J_3 are functions of the invariants of the Cauchy–Green deformation tensor $\mathbf{C}$ defined as $C_{ij} = F_{im}F_{jm}$.

For structures with large displacements and deformations, the second Piola–Kirchhoff stress S_{ij} and the Green–Lagrangian strain E_{ij} are used in the total Lagrangian formulation:

$$S_{ij} = \frac{\partial W}{\partial E_{ij}} \quad \text{and} \quad E_{ij} = \frac{1}{2}(C_{ij} - \delta_{ij}).$$

The first Piola–Kirchhoff stress P_{ij} can then be obtained from the second Piola–Kirchhoff stress S_{ij}:

$$P_{ij} = S_{ik}F_{jk},$$

where the definition of P_{ij} is employed to derive the internal force f_{iI}^{int}.

Note that if the solid material model is purely incompressible as constrained in Eqs. (3.10) and (3.9), we should have the invariant $I_3 = 1$ and its function $J_3 = 1$ based on the definition of $\det(\mathbf{C}) = 1$ for incompressible materials. Therefore, the third term in the definition of W should be 0. However, the divergence free property is not strictly satisfied in the process of interpolating the velocities, which induces accumulated numerical errors in the fluid and solid calculations and creates possible artificial stress due to the volume change of the solid. Therefore, in the implementation, we assign the bulk modulus κ to be sufficiently small so that the last term in Eq. (3.25) has nearly no contribution in the energy potential. A more elaborate study on this subject has been discussed in the work of Liu *et al.* (2007a).

3.4 Algorithms and Validations

3.4.1 *Explicit schemes*

The governing equations of IFEM in discretized form (except the Navier–Stokes equations, for convenience) are summarized as follows:

$$f_{iI}^{s} = -f_{iI}^{\text{inert}} - f_{iI}^{\text{int}} + f_{iI}^{\text{ext}}, \quad \text{in } \Omega^{s}, \tag{3.26a}$$

$$f_{iJ}^{\text{fsi}} = \sum_{I} f_{iI}^{s}(\mathbf{X}^{s},t)\phi_{I}(\mathbf{x}_{J}-\mathbf{x}_{I}^{s}), \quad \mathbf{x}_{I}^{s} \in \Omega_{\phi J}, \tag{3.26b}$$

$$\rho^{f}(v_{i,t}+v_{j}v_{i,j}) = \sigma_{ij,j}+\rho g_{i}+f_{i}^{\text{fsi}}, \quad \text{in } \Omega, \tag{3.26c}$$

$$v_{j,j} = 0, \quad \text{in } \Omega, \tag{3.26d}$$

$$v_{iI}^{s} = \sum_{J} v_{iJ}(t)\phi_{J}(\mathbf{x}_{J}-\mathbf{x}_{I}^{s}), \quad \mathbf{x}_{J} \in \Omega_{\phi I}. \tag{3.26e}$$

An outline of the IFEM algorithm with a semi-explicit time integration is illustrated as follows:

Given the structure configuration $\mathbf{x}^{s,n}$ and the fluid velocity $\mathbf{v}^{n}$ at time step n,

(1) evaluate the nodal interaction forces $\mathbf{f}^{s,n}$ for solid material points, using Eq. (3.26a),
(2) distribute the material nodal force onto the fluid lattice, from $\mathbf{f}^{s,n}$ to $\mathbf{f}^{\text{fsi},n}$, using the delta function as in Eq. (3.26b),
(3) solve for the fluid velocities $\mathbf{v}^{n+1}$ and the pressure p^{n+1} implicitly using Eqs. (3.26c) and (3.26d),
(4) interpolate the velocities in the fluid domain onto the material points, i.e., from $\mathbf{v}^{n+1}$ to $\mathbf{v}^{s,n+1}$ as in Eq. (3.26e), and
(5) update the positions of the structure using $\mathbf{u}^{s,n+1} = \mathbf{v}^{s,n+1}\Delta t$ and go back to Step (1).

Note that even though the fluid is solved fully implicitly, the coupling between fluid and solid is explicit. If we rewrite the fluid momentum equation (for clarity only discretized in time), we have

$$\rho^{f}\left(\frac{v_{i}^{m+1}-v_{i}^{m}}{\Delta t}+v_{j}^{m+1}v_{i,j}^{m+1}\right) = \sigma_{ij,j}^{\text{f},m+1}+f_{i}^{fsi,m}.$$

It is clear that the interaction force is not updated during the iteration, i.e., the solid equations are calculated with values from the previous time step. For a fully implicit coupling, this force must be a function of the current fluid velocity and the term $f_i^{\text{fsi},m+1}$ should be included into the linearization of the fluid equations.

3.4.2 *Delta functions*

The coupling between the fluid and the structure consists of two parts: the interpolation of the fluid grid velocity $\mathbf{v}(\mathbf{x},t)$ onto the structural nodal velocity $\mathbf{v}^s(\mathbf{X}^s,t)$ and the distribution of the solid domain force $\mathbf{f}^s(\mathbf{X}^s,t)$ onto the equivalent fluid domain force $\mathbf{f}^{\text{fsi}}(\mathbf{x},t)$. Note again that the fluid variables are depicted in the Eulerian coordinate $\mathbf{x}$; whereas the solid variables are represented in the Lagrangian coordinate $\mathbf{X}^s$.

In this section, we describe the construction of the function δ that is used in the interactions. Both Dirac delta functions in the IB method (Peskin, 2002) and EIBM (Wang and Liu, 2004) are implemented for uniform fluid meshes. However, the reproducing kernel particle method (RKPM) delta function employed in IFEM can handle both uniform and nonuniform grids (Liu *et al.*, 1995). With this improvement, we are able to deal with fluid domains with arbitrary geometries as discussed in Chapter 2.

To satisfy the consistency condition, a correction function $C(x;x-y)$ is introduced in the finite domain of influence or support so that the window function yields

$$\int_\Omega C(x;x-y)\frac{1}{r}\phi\left(\frac{x-y}{r}\right)d\Omega = 1,$$

where r is the dilation parameter or refinement of the window function.

The additional constant, r^{-1}, scales the window function so that the integral over the domain of support equals one. The discretized reconstruction of the delta function for non-uniform spacing can be written as

$$\phi_I(x) = C(x;x-x_I)\frac{1}{r}\phi\left(\frac{x-x_I}{r}\right)\Delta x_I. \tag{3.27}$$

The window function is used as the delta function for the interpolation of the nodal velocities and the distribution of the nodal forces between the fluid and solid domains. To ensure the conservation of energy, the delta functions in both distribution and interpolation procedures must be the same at each time step.

In particular, for each solid point I at its current position $\mathbf{x}_I^s(t)$, we construct an influence domain $\Omega_{\phi I}$ that covers a subset of the fluid grid nodes J positioned at $\mathbf{x}_J$. To define the influence domain for each solid point I, a search algorithm is needed at each time step during the calculation of the RKPM shape functions. When non-uniform fluid mesh is used, the size of the influence domain for each solid point is dependent of the definition of the RKPM shape function in Eq. (3.27) and fluid mesh density.

Therefore, in the interpolation process from the fluid onto the solid grid, the discretized form of Eq. (3.10) can be written as

$$v_{iI}^s = \sum_J v_{iJ}(t)\phi_J(\mathbf{x}_J - \mathbf{x}_I^s), \quad \mathbf{x}_J \in \Omega_{\phi I}. \tag{3.28}$$

Here, the solid velocity $\mathbf{v}_I^s$ at node I can be calculated by gathering the velocities at fluid nodes within the influence domain $\Omega_{\phi I}$. A dual procedure takes place in the distribution process from the solid onto the fluid grid. The discretized form of Eq. (3.7) is expressed as

$$f_{iJ}^{\text{fsi}} = \sum_I f_{iI}^s(\mathbf{X}^s, t)\phi_I(\mathbf{x}_J - \mathbf{x}_I^s), \quad \mathbf{x}_I^s \in \Omega_{\phi J}. \tag{3.29}$$

By interpolating the fluid velocities onto the solid particles in Eq. (3.28), the fluid within the solid domain is bounded to solid material points. This ensures not only the no-slip boundary condition on the surface of the solid, but also stops automatically the fluid from penetrating the solid, provided the solid mesh is at least two times denser than the surrounding fluid mesh. This heuristic criterion is based on the numerical evidence and needs further investigation.

3.4.3 *Validations*

In this section, we also present the simulations on particulate flows with both rigid and deformable spheres using the IFEM method. The algorithm requires no mesh update or remeshing and the structures are not restricted to only rigid cases. The demonstrated problems are as follows: a rigid sphere and a soft sphere dropping between two planes and multiple soft spheres dropping in a square channel. The sphere flexibility is controlled by adjusting the material properties. Naturally, as the modulus increases, the sphere behaves more like a rigid body.

For a test example with a rigid sphere dropping in a viscous channel, the material moduli are set to be high. The channel has a size of $2 \times 2 \times 8$ cm.

Table 3.2. (a) Properties used for one rigid sphere dropping in a channel and (b) properties used for one soft sphere dropping in a channel.

(a) One rigid sphere dropping in a channel			
Fluid	4987 nodes 26030 elements	$\rho^{\mathrm{f}} = 1$ g/cm^3 $\mu = 0.01$ dyne/cm$^2 \cdot$ s	
Solid	997 nodes 864 elements	$\rho^{\mathrm{s}} = 2.02$ g/cm^3 $D = 0.4$ cm $g = 980$ cm/s^2	$C_1 = 2.93 \times 10^5$ dyne/cm^2 $C_2 = 1.77 \times 10^5$ dyne/cm^2
(b) One soft sphere dropping in a channel			
Fluid	4861 nodes 22557 elements	$\rho^{\mathrm{f}} = 1$ g/cm^3 $\mu = 10$ dyne/cm$^2 \cdot$ s	
Solid	997 nodes 864 elements	$\rho^{\mathrm{s}} = 2.02$ g/cm^3 $D = 0.5$ cm $g = 980$ cm/s^2	$C_1 = 2.93 \times 10^2$ dyne/cm^2 $C_2 = 1.77 \times 10^2$ dyne/cm^2

The diameter of the sphere is 0.4 cm. The properties for both fluid and solid as well as the discretization for each mesh are summarized in Table 3.2(a).

As illustrated in Fig. 3.2, the use of the non-uniform radial-based gradient fluid mesh from the submerged sphere provides the unique ability and the advantage of IFEM over the IB method as well as EIBM. The interaction force $\mathbf{f}^{\mathrm{fsi}}$ is calculated using Eq. (3.26a). As shown in Figs. 3.3–3.4, the pressure and velocities in the fluid domain are influenced by the movement of the sphere and the vortices and recirculation are clearly captured near the sphere.

Since its inception, IFEM has been employed in the solutions of many FSI systems. To verify the accuracy of the IFEM algorithm, we compared the solution of terminal velocity with the experimental data. It is explained in the work of Clift *et al.* (1978) that for a particle moving with steady terminal velocity U_T in a gravitational field, the drag force F_D balances the difference between the weight and buoyancy:

$$F_D = g(\rho^s - \rho^f)(\pi/6)D^3, \tag{3.30}$$

and as a consequence, the drag coefficient C_D is calculated as

$$C_D = 4(\rho^s - \rho^f)gD/3\rho^f U_T^2 = 4\rho^f(\rho^s - \rho^f)gD^3/3\mu^2 \,\mathrm{Re}_T^2, \tag{3.31}$$

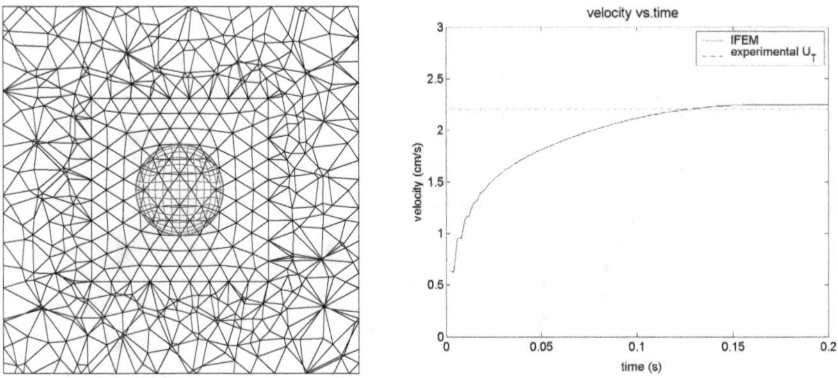

(a) (b)

Fig. 3.2. A gradient fluid mesh and terminal velocity validation with experiments. (a) Fluid and solid meshes. (b) Terminal velocity.

where Re_T is the Reynolds number at the terminal velocity.

According to the work of Clift *et al.* (1978), since both the drag coefficient and Reynolds number are dependent of the terminal velocity U_T, an iterative procedure is needed. It is more convenient to express Re as a function of the so-called *best number* N_D, which is defined as

$$N_D = C_D\,\mathrm{Re}_T^2 = 4\rho^f(\rho^s - \rho^f)gD^3/3\mu^2. \tag{3.32}$$

Finally, the Reynolds number can be expressed as a function of N_D. In this case, for $580 \le N_D \le 1.55 \times 10^7$, the correlation for Re as a function of N_D is expressed as

$$\log_{10}\mathrm{Re} = -1.81391 + 1.34671W - 0.12427W^2 + 0.006344W^3, \tag{3.33}$$

with $W = \log_{10} N_D$.

Now reset the material moduli for the soft deformable solid as $C_1 = 2.93 \times 10^2$ dyne/cm^2 and $C_2 = 1.77 \times 10^2$ dyne/cm^2. As expected from the simulation, the sphere experiences a higher pressure at the bottom and a lower pressure on the top as it tries to immerse itself deeper into the fluid. To demonstrate the advantage of immersed methods, we model a case with 20 soft spheres dropping in a channel. These same-sized spheres are placed randomly near the top of the channel. The spheres experience the gravitational and internal forces as well as the interaction forces with the surrounding fluid.

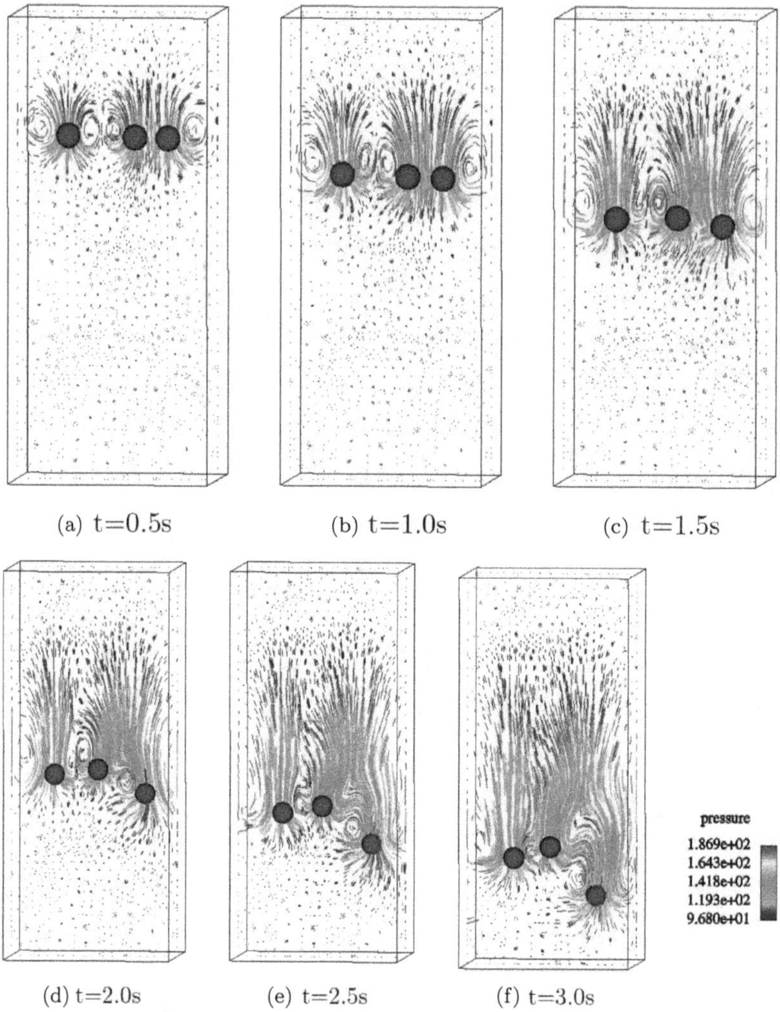

Fig. 3.3. Theree spheres dropping in a channel. Fluid velocity and pressure distributions at (a) t = 0.5 s. (b) t = 1.0 s. (c) t = 1.5 s. (d) t = 2.0 s. (e) t = 2.5 s. (f) t = 3.0 s.

Using the parameters shown in the above example, the experimental terminal velocity is found to be 2.203 cm/s with the terminal Reynolds number 88.11. The calculated terminal velocity from the IFEM solution, as shown in Fig. 3.2, is found to be 2.24412 cm/s, which deviates only 1.87%

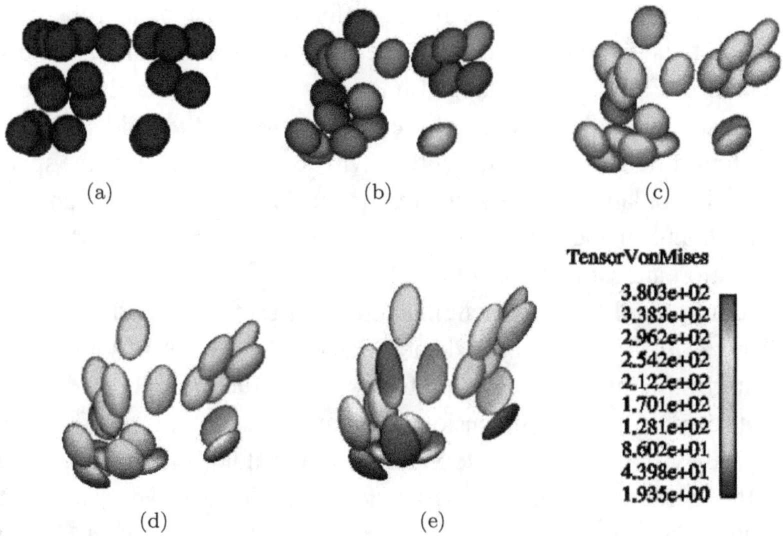

Fig. 3.4. Stress distribution and deformation of 20 spheres dropping in a channel. (a) $t = 0.0$ s. (b) $t = 1.0$ s. (c) $t = 2.0$ s. (d) $t = 3.0$ s. (e) $t = 4.0$ s.

Table 3.3. Properties for 20 soft spheres dropping in a channel.

Fluid	4861 nodes 22557 elements	$\rho^f = 1$ g/cm^3 $\mu = 10$ dyne/cm$^2 \cdot$ s	
Solid	20×997 nodes 20×864 elements	$\rho^s = 2.02$ g/cm^3 $D = 0.5$ cm $g = 980$ cm/s^2	$C_1 = 2.93 \times 10^1$ dyne/cm^2 $C_2 = 1.77 \times 10^1$ dyne/cm^2

from the experimental value. This example shows a soft sphere dropping in a viscous fluid channel. The sphere flexibility is controlled by the parameters C_1 and C_2 as defined in Section 3.2.2 and listed in Table 3.2(b). For the case of 20 soft deformable spheres dropping in a viscous channel example solves 20 soft spheres dropping in a channel. These same-sized spheres are placed randomly near the top of the channel. The spheres experience the gravitational and internal forces as well as the interaction forces with the surrounding fluid. The properties used in this problem can be found in Table 3.3.

The movements of the spheres at different time steps and the deformations of the spheres along with the pressure distributions are clearly presented in Fig. 3.4, the enlarged images of the spheres and at different time steps. Due to the absence of the deformation of the rigid sphere, the interaction force $\mathbf{f}^s$ is calculated with gravity and acceleration of the sphere. In the fluid domain, however, the pressure and velocities are influenced by the movement of the sphere, as shown in Fig. 3.3, and the vortices and recirculation are captured near the sphere.

The immersed finite element method (IFEM), as proposed in the work of Zhang *et al.* (2004) has been demonstrated to be capable of solving complex fluid–structure interaction problems. The structural models in IFEM are not restricted to one-dimensional volumeless structures such as fibers; instead, they may occupy finite volumes in the fluid. Finite elements are used to discretize the submerged solid and as a consequence, the calculated stress are more realistic and accurate. In addition, in IFEM, different material laws can be assigned to the submerged solid.

The higher order RKPM delta function can handle nonuniform grids and is particularly useful for fluid domains with arbitrary geometries and boundaries. Finally, unlike the periodic boundary conditions implemented in most IB applications, essential and natural boundary conditions in the context of finite element methods are relatively easy to implement in IFEM.

In the simple setup of a pulsatile flow in a channel with a square cross section, a shell-like flexible structure similar to the heart valve is installed. The shell-like structure is made of incompressible rubber. As shown in Fig. 3.5, the shell-like structure deflects in a pulsatile flow through a square cross-sectional channel. Moreover, as depicted in Fig. 3.5, the simulation results are very much the same as the experimental observation at individual time steps. In the experiment conducted by ABIOMED,

Fig. 3.5. Structure deflecting in a pulsatile flow channel.

an innovative medical technology company based in Danvers, MA, water was pulsed through a column with a square cross section (5 × 5 cm) at a frequency of 1 Hz. In the simulation, both velocity vectors of the fluid flow and stress tensors of the flexible structure are captured for the entire fluid and solid domains, whereas the preliminary experimental observation can only provide us with the dynamic behavior of the structure and overall flow patterns. This brings up an important point: For the actual valve design, simulation tools are very much required since many areas of the valve are difficult to visualize and only limited amounts of data can be experimentally collected.

In finite element formulations for immersed methods (Gay *et al.*, 2006; Liu *et al.*, 2006; Wang, 2006, 2007), various meshless kernels or radial-based interpolation functions are employed for unstructured background meshes (Liu *et al.*, 1995). This mesh-free delta function provides not only a higher order smoothness but also the ability to handle non-uniform fluid grids. The advantage of the immersed finite element methods is to allow independent Lagrangian mesh of the immersed continuum to move on top of a fixed background mesh which does not have to be uniform as in the spectral solver used in the original immersed boundary method. In our further demonstrations, a more complicated flexible wing can also be modeled with IFEM, as shown in Fig. 3.6, with a comparable simulation with traditional ALE formulations, as shown in Fig. 3.7. As elaborated in the works of Wang *et al.* (2009a) and Zhang *et al.* (2004), with an attach angle of 14°,

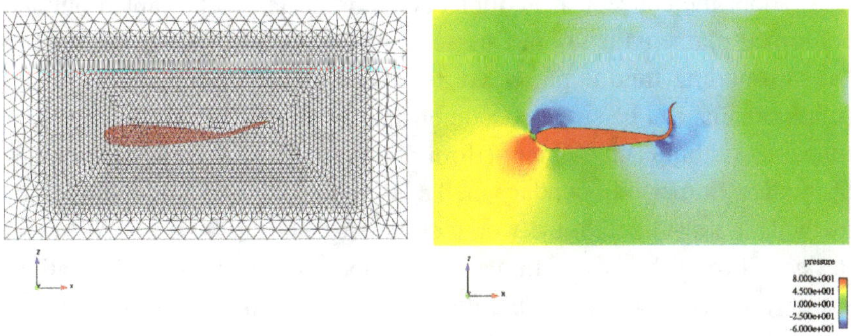

Fig. 3.6. Flexible airfoil IFEM simulation.

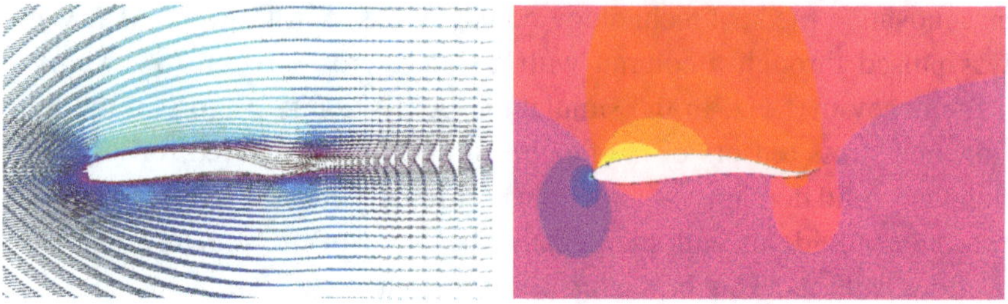

(a) (b)

Fig. 3.7. Flexible airfoil ADINA ALE simulation. (a) Velocity. (b) Pressure.

within a Reynolds number between 200 and 400, an airfoil with a length 4 cm, with an incoming flow velocity 10.3 cm/s, with a density of 1 g/cm^3, dynamics viscosity 0.2 to 0.1 dyne/cm^2 s or g/cm s, along with 17181 fluid elements and 5860 solid nodes. Note the advantage of the immersed finite element method with an unstructured dense fluid grid around the flexible airfoil.

Again, it is necessary to combine both immersed solid domain Ω_s and exterior fluid domain Ω_f into a complete computation domain Ω. Since this complete domain stays the same regardless of the position of the immersed continuum, a fixed background mesh becomes possible. Moreover, the same delta function must be used in both the distribution of the resultant nodal force and the interpolation of the solid velocity based on the surrounding fluid velocities. Hence, to match with the original topology, we must adopt the concept of fictitious domains. It has been pointed out and confirmed mathematically in Chapter 2 that as long as the power input to the surrounding fluid is preserved, the communication between fluid and solid domains can be simply accomplished by finite element interpolation functions within each element (Boffi and Gastaldi, 2003). Recent studies of implicit immersed boundary methods in the context of finite difference methods are available in the works of Newren *et al.* (2007) and Mori and Peskin (2008). The key of immersed methods is the energy conservation, namely, the energy input to the fluid domain from the immersed solid or immersed structure is the same as that from the equivalent body force.

In fact, this treatment in the immersed boundary or immersed continuum methods can be viewed as the synchronization of the fluid motion with the solid motion within the immersed solid domain Ω_s, namely, $\mathbf{v}^s = \mathbf{v}^f$. A wonderful comparison of both implicit and explicit approaches for material point methods has also been presented in the work of Sulsky and Kaul (2004). Similar formulations have already been extended to electrokinetics and mass and heat transfers (Fogelson *et al.*, 2008; Liu *et al.*, 2007a, 2007b).

Chapter 4

Coupled Continua

In this section, we explore the situation of a compressible solid with finite mass and elasticity immersed in a compressible fluid. Although in this situation, both bulk moduli can be high, the formulation must be geometrically consistent as well as matching with the physics. In fact, the interaction could be one continuum inside another continuum and does not have to be limited to FSI systems. Mixed finite element formulations are used along with mixed elements satisfying the Inf-Sup conditions. The key for the successful modeling is the matrix-free Newton–Krylov iteration procedure based on the Newton–Raphson iteration, named after Isaac Newton PRS (1642–1726) and Joseph Raphson (1668–1715), and the construction of the so-called orthogonal Krylov subspaces, named after Aleksey Nikolaevich Krylov (1863–1945).

Because of the indirect and implicit formulation of the equivalent fluid-structure or fluid–solid interaction body force $\mathbf{f}^{fsi}$, the exact Jacobian matrix, named after Carl Gustav Jacob Jacobi (1804–1851), formulated in the Newton–Raphson iterative procedures for nonlinear equations, cannot be established analytically or explicitly. It is essential to have the simultaneous convergence of key variables governing both continua as illustrated in Fig. 4.1. Note from a topological point of view, the immersed continuum if impenetrable, does produce a hole in the surrounding, not background, continuum as depicted in Fig. 4.1. An addition and subtraction of the equivalent fluid domain occupying the topological hole left by the immersed continuum must be considered in order to have the same effects as the original IB method with immersed fibers that have no volumes. Furthermore, in order to adopt the two separate kinematic descriptions, namely, the Lagrangian description for the immersed continua, named after Joseph–Louis Lagrange (1736–1813), and the Euerlian description for the background continua, named after Leonhard Euler (1707–1783), fictitious immersed continua must be employed to topologically compensate the holes within the background continua occupied by immersed continua as illustrated in Figs. 4.1 and 4.2 which depict the core concept and immersed continua approaches. Background fluids with Eulerian or ALE description are in communication with immersed solids through meshless kernel and unstructured mesh. With ALE descriptions for background fluid, it is possible to add another layer of complexity with background fluid coupled with flexible structures on the boundary, similar to that of a collapsible tube with internal flexible blood cells which will be explored in Chapter 6.

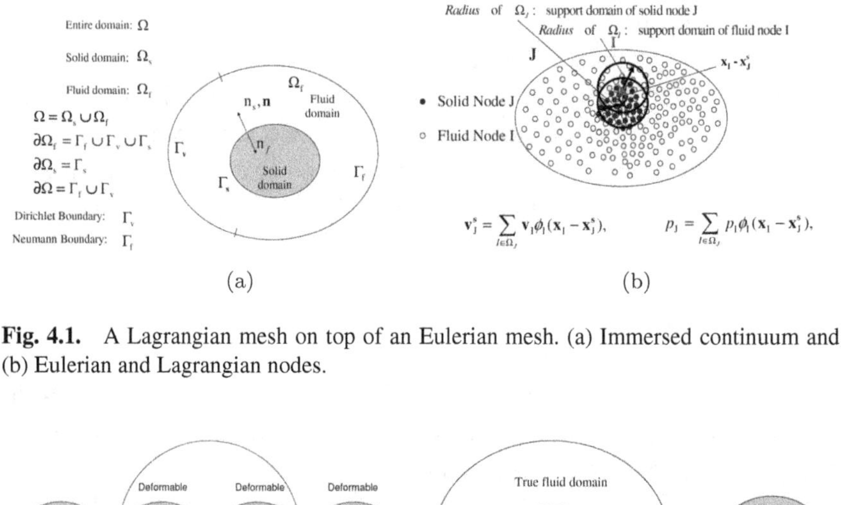

Fig. 4.1. A Lagrangian mesh on top of an Eulerian mesh. (a) Immersed continuum and (b) Eulerian and Lagrangian nodes.

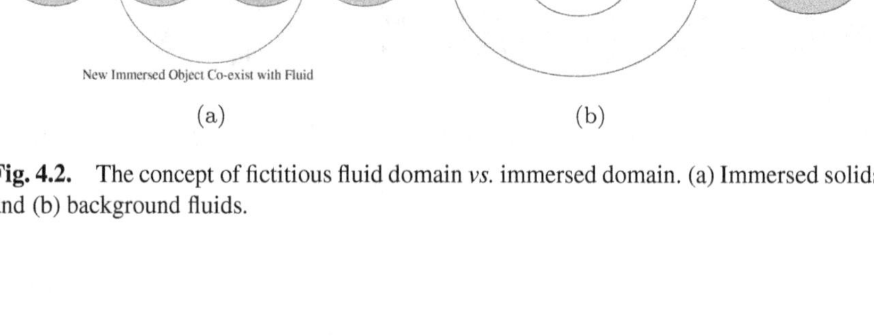

Fig. 4.2. The concept of fictitious fluid domain *vs.* immersed domain. (a) Immersed solids and (b) background fluids.

4.1 Immersed Continua

The initial attempt to connect a traditional elasticity model with the immersed boundary method can be traced back to Sulsky and Brackbill (1991), in which a stress function is transferred to the fluid grid. As illustrated in the converged solution, the pressure unknowns for the immersed continua and the background continua must converge simultaneously with the displacement or velocity unknowns for both continua. In fact, the pressure unknowns are directly related to the volumetric changes and are related to the geometrical compatibilities whereas the displacement or velocity unknowns for the immersed continua and the background continua must satisfy the kinematic and dynamic boundary conditions along the interfaces in addition to the geometric compatibility conditions. Furthermore, also illustrated in the converged solutions with mixed finite element formulations for both fluids and solids, as shown in Figs. 4.3–4.5, the rate of convergence of the pressure unknowns are one order of magnitude lower than the displacement or velocity unknowns. This issue is due to the so-called Inf-Sup conditions for mixed displacement-pressure elements that have been elaborated in

```
 time step
   1   0.1564344650402309E+00   0.1564344650402309E+00
uf us pf ps    1   0.22690E-04   0.41492E-04   0.18809E-01   0.25455E+00
uf us pf ps    2   0.38812E-06   0.29509E-06   0.44533E-03   0.54820E-02
uf us pf ps    3   0.84238E-10   0.35385E-10   0.15255E-07   0.21101E-05
iteration dis. increment  pre.increment   lambda increment

 time step
   2   0.3090169943749474E+00   0.3090169943749474E+00
uf us pf ps    1   0.30205E-04   0.58749E-04   0.18102E-01   0.26171E+00
uf us pf ps    2   0.24444E-04   0.54535E-04   0.24321E-02   0.49779E-01
uf us pf ps    3   0.51401E-07   0.22541E-07   0.22905E-04   0.22451E-03
uf us pf ps    4   0.77258E-12   0.14687E-11   0.23592E-09   0.10960E-06
iteration dis. increment  pre.increment   lambda increment

 time step
   3   0.4539904997395467E+00   0.4539904997395467E+00
uf us pf ps    1   0.69094E-04   0.16581E-03   0.14654E-01   0.27672E+00
uf us pf ps    2   0.55662E-06   0.13497E-05   0.20055E-03   0.22952E-02
uf us pf ps    3   0.46562E-10   0.25301E-10   0.18101E-07   0.58024E-06
iteration dis. increment  pre.increment   lambda increment

 time step
   4   0.5877852522924731E+00   0.5877852522924731E+00
uf us pf ps    1   0.73341E-04   0.19456E-03   0.13530E-01   0.29402E+00
uf us pf ps    2   0.52377E-05   0.14885E-04   0.55545E-03   0.11425E-01
uf us pf ps    3   0.18568E-08   0.16315E-08   0.10133E-05   0.51892E-05
iteration dis. increment  pre.increment   lambda increment
```

Fig. 4.3. Monolithic Newton–Krylov matrix-free implicit IB method.

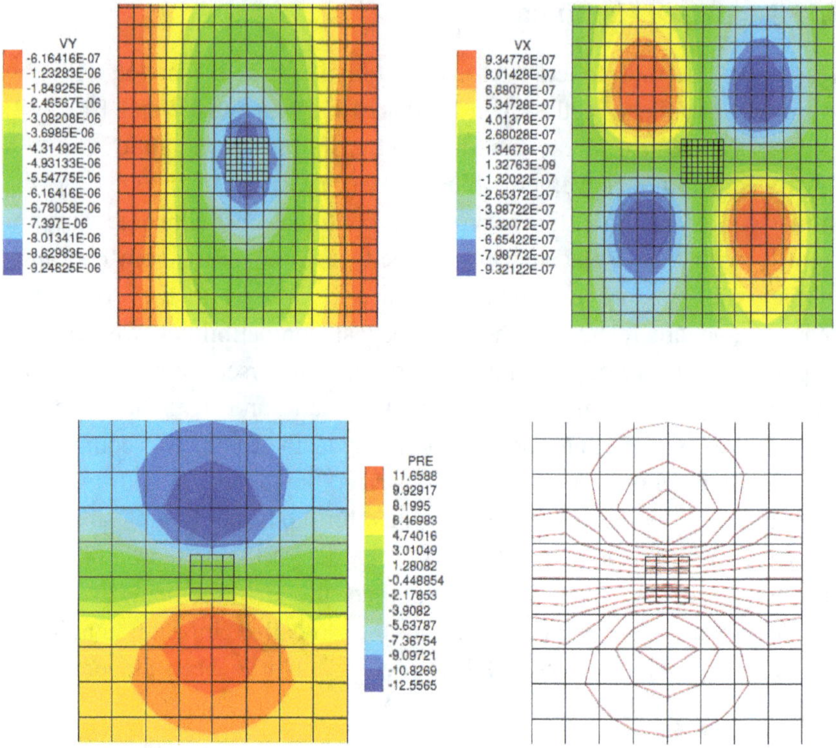

Fig. 4.4. Converged solution for a falling continuum.

Bao *et al.* (2001) and Bathe (1996). Finally, the immersed solid displacements and velocities as well as pressures are clearly consistent with the surrounding viscous fluid as shown in Figs. 4.4 and 4.5.

4.1.1 *Strong and weak forms*

Consider a fluid domain Ω, as depicted in Fig. 4.1, enclosed with a sufficiently smooth boundary $\partial\Omega$, with $\partial\Omega = \Gamma_v \cup \Gamma_f$, where Γ_v and Γ_f stand for the Dirichlet and Neumann boundaries, respectively. Suppose there exists a submerged solid domain Ω_s enclosed by a sufficiently smooth boundary Γ_s, namely, a line for two-dimensional cases and a surface for three-dimensional cases, the entire domain Ω is subdivided into two regions, namely, the solid region Ω_s and the fluid region Ω_f. Therefore,

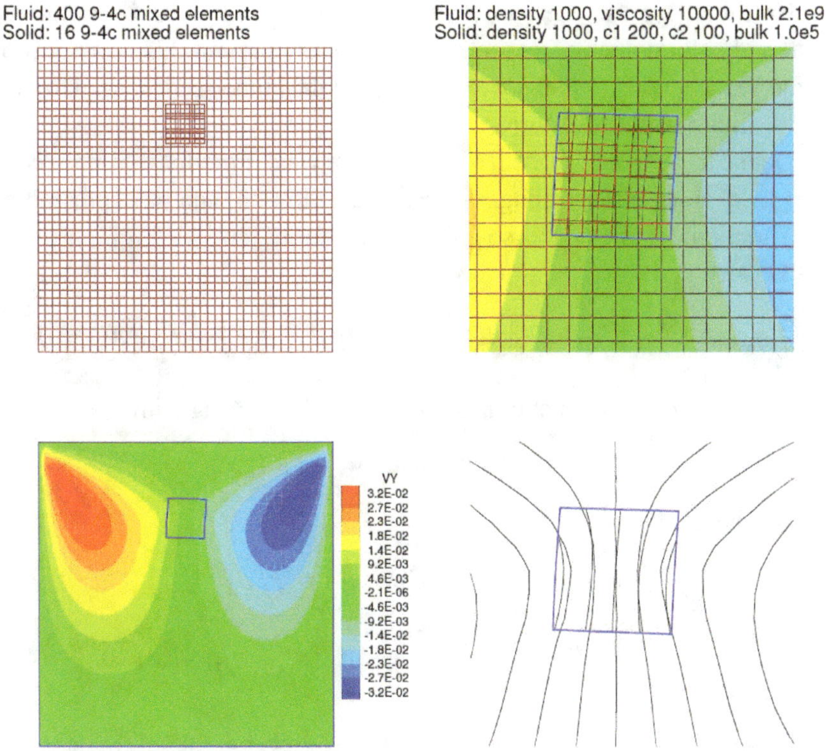

Fig. 4.5. Converged solution for an immersed continuum in a driven cavity.

the boundaries of the solid and fluid regions can be simply expressed as $\partial\Omega_s = \Gamma_s$ and $\partial\Omega_f = \Gamma_s \cup \Gamma_v \cup \Gamma_f$. Denoting $\boldsymbol{\sigma}$ as the stress tensor, and $\mathbf{v}$ as the velocity vector, we establish the following set of governing equations or strong form:

$$\rho_s \dot{v}_i^s = \sigma_{ij,j}^s + \rho_s g_i, \quad \text{in } \Omega_s, \tag{4.1}$$

$$\rho_f \dot{v}_i^f = \sigma_{ij,j}^f + \rho_f g_i, \quad \text{in } \Omega_f, \tag{4.2}$$

$$[v_i] = 0, \quad \text{on } \Gamma_s, \text{ kinematic matching}, \tag{4.3}$$

$$[\sigma_{ij} n_j] = 0, \quad \text{on } \Gamma_s, \text{ dynamic matching}, \tag{4.4}$$

where the surface normal vector $\mathbf{n}$ is aligned with that of the solid domain $\mathbf{n}^s$ and opposite to that of the fluid domain $\mathbf{n}^f$.

Note that we use the subscript or the superscript f and s represent fluid and solid domains, respectively. To clearly present this new form of immersed methods dealing with compressible immersed continuum, we study similar variational forms.

Define the Sobolev space $[H^1_{0,\Gamma_v}(\Omega)]^d$, the weak form for the entire coupled fluid–solid system along with Eqs. (4.1) to (4.4) can be expressed as the variational form or the so-called weak form: $\forall \mathbf{w} \in [H^1_{0,\Gamma_v}(\Omega)]^d$, find $\mathbf{v} \in \Omega$, such that

$$\int_{\Omega_s} w_i \left[\rho_s(\dot{v}_i^s - g_i) - \sigma^s_{ij,j}\right] d\Omega + \int_{\Omega_f} w_i \left[\rho_f(\dot{v}_i^f - g_i) - \sigma^f_{ij,j}\right] d\Omega = 0.$$

Employing the kinematic matching of Eq. (4.3) along with

$$\mathbf{v}^s = \mathbf{v} \quad \text{and} \quad \mathbf{w}^s = \mathbf{w}, \ \forall \mathbf{x} \in \Omega_s,$$

we also have

$$\int_{\Omega} [w_i\rho(\dot{v}_i - g_i) + w_{i,j}\sigma_{ij}]\, d\Omega - \int_{\Gamma^f} w_i^f F_i^p d\Gamma + \int_{\Omega_s} \left[w_i^s(\rho^s - \rho)(\dot{v}_i - g_i) + w^s_{i,j}(\sigma^s_{ij} - \sigma^f_{ij})\right] d\Omega = 0. \quad (4.5)$$

In fact, following the cartoon illustration in Fig. 4.2, $\forall \mathbf{w} \in \left[\mathcal{H}^1_{0,\Gamma_v}(\Omega)\right]^d$,

$$\int_{\Omega_f} w_i(\rho_f\dot{v}_i - \rho_f g_i) d\Omega + \int_{\Omega_f} \frac{\partial w_i}{\partial x_j}\sigma^f_{ij} d\Omega - \int_{\Gamma_f} w_i^f F_i^p d\Gamma - \int_{\Gamma_s} w_i^s \sigma^f_{ij} n^f_j d\Gamma = 0. \quad (4.6)$$

In combination with the following weak form for the fictitious fluid occupying the immersed solid domain, $\forall \mathbf{w} \in \left[\mathcal{H}^1_{0,\Gamma_v}(\Omega)\right]^d$,

$$\int_{\Omega_s} w_i(\rho_f\dot{v}_i - \rho_f g_i) d\Omega + \int_{\Omega_s} \frac{\partial w_i}{\partial x_j}\sigma^f_{ij} d\Omega - \int_{\Gamma_s} w_i^s \sigma^f_{ij} n^s_j d\Gamma = 0. \quad (4.7)$$

To maintain the compatibility of the geometry, we must add and subtract the weak form in Eq. (4.7) with the respective material properties of surrounding viscous fluid and the solid occupying the immersed solid domain. Now, combine with the weak form for the solid, not fictitious fluid, occupying the immersed solid domain, use integration by parts and the divergence theorem, introduce dynamic matching at the interface Γ_s, and combining fluid and solid domains with $\Omega_f \cup \Omega_s = \Omega$, Eqs. (4.5)–(4.7) can be rewritten as: $\forall \mathbf{w} \in \left[\mathcal{H}^1_{0,\Gamma_v}(\Omega)\right]^d$,

$$\int_{\Omega} w_i(\rho_f \dot{v}_i - \rho_f g_i) d\Omega + \int_{\Omega} \frac{\partial w_i}{\partial x_j} \sigma^f_{ij} d\Omega - \int_{\Gamma_f} w^f_i F^p_i d\Gamma - \int_{\Omega_s} w^s_i f^s_i d\Omega = 0, \tag{4.8}$$

where the last term on the left hand side can be elaborated as, $\forall \mathbf{w} \in \left[\mathcal{H}^1_{0,\Gamma_v}(\Omega)\right]^d$,

$$\int_{\Omega_s} w^s_i f^s_i d\Omega \triangleq -\int_{\Omega_s} w_i(\rho_s - \rho_f)(\dot{v}_i - g_i) d\Omega - \int_{\Omega_s} \frac{\partial w_i}{\partial x_j}(\sigma^s_{ij} - \sigma^f_{ij}) d\Omega. \tag{4.9}$$

In Eq. (4.4), the unit surface normal vectors at the immersed solid boundary or FSI interface Γ_s are assigned as $\mathbf{n}^s = -\mathbf{n}^f = \mathbf{n}$; whereas in Eq. (4.9) the term involving the given surface traction $\mathbf{F}^p$ will remain the same if the variational forms are carried out for the entire domain Ω instead of the solid and fluid parts. Thus, the focus will be on the submerged solid Ω_s and its interface with the fluid Γ_s.

Within the same domain Ω^s, we employ the superscripts or subscripts f and s to delineate the stress tensors calculated using the fluid and solid formulations, respectively. Note that the stress tensor and the velocity vector are denoted as $\boldsymbol{\sigma}$ and $\mathbf{v}$, respectively for the entire domain Ω. For convenience, we ignore the superscript or subscript f for the entire domain Ω and retain the superscripts or subscripts s and f for variables within the immersed solid domain Ω_s.

As illustrated in Eq. (4.5), within the entire domain Ω, the external work comes from the external body force $\rho \mathbf{g}$ and the surface traction $\mathbf{F}^p$ at the Neumann boundary Γ_f; whereas the power input within the submerged solid domain Ω_s includes the contribution from the inertial force difference, the buoyancy force, and the internal energy difference. Again, the kinematic matching at the submerged interface Γ_s also implies that the submerged interface will move at the same velocity as that of the fluid-particles in the immediate vicinity. Moreover, in Eq. (4.5), we do not stipulate the material derivative $\dot{\mathbf{v}}$ and the stress $\boldsymbol{\sigma}$, thus the formulation in Eq. (4.5) is applicable to fluid–fluid, fluid–solid, or solid–solid coupled systems. In particular, for fluids, the turbulent and the non-Newtonian models can also be incorporated.

In the case of the incompressible solid immersed in the incompressible fluid, the traditional explicit scheme can be employed, if there is no excessively stiff boundary springs connecting with the tether points Kim *et al.* (2003) and the additional elasticity moduli of the immersed solid are reasonable. This is addressed in the extended immersed boundary method in Chapter 2 or the immersed finite element method in Chapter 3 Wang and Liu (2004); Zhang *et al.* (2004). In this section, we mainly address the issue of implicit formulations for excessively stiff boundary springs and high elastic moduli. We hope to prove some kind of validation and verification for the immersed continuum approaches with this series of driven cavity problems with immersed solids. For the case with one immersed square solid, it is still possible to solve this FSI system with traditional modeling techniques such as the ALE formulation. We will compare the solutions derived from the immersed continuum method with those from the traditional ALE approaches.

The driven cavity problem considered here has the dimension of 1×1 m. A typical immersed solid has the dimension of 0.13×0.13 m. In the immersed continuum method, the background fluid mesh for the entire cavity, which includes the space occupied by the immersed solids, consists of 20×20 $9/4c$ mixed finite elements. A typical immersed solid is represented by 4×4 $9/4c$ mixed finite elements. In the plotting stage, we split each 9-node element into four 4-node elements. As a result, the velocity mesh will be two times denser than the pressure mesh. The mesh construction clearly demonstrates the philosophical difference between the immersed boundary or immersed continuum methods and the traditional ALE approaches. In the immersed continuum method, the solid mesh is right on top of the background fluid mesh, whereas in the ALE formulation, the solid mesh is surrounded by the fluid mesh with a different mesh density.

Preliminary test results of the implicit immersed method for compressible media immersed in another compressible medium have been reported. To compare the dynamical behaviors, the top shear velocity of the cavity is set as $0.1 \sin(2\pi/40t)\,\mathrm{m/s}$. The deformable solid is situated initially in the center of the cavity with zero velocity. The submerged solid is made of a compressible rubber material with the material constant $\kappa_s = 1 \times 10^7\ \mathrm{N/m^2}$ and the density $\rho_s = 1000\ \mathrm{kg/m^3}$. For hard solids we use $C_1 = 400$ and $C_2 = 200$ N/m^2, whereas for soft solids we use $C_1 = 0.4$ and $C_2 = 0.2$ N/m^2. The case with hard solids resembles the driven cavity problem with immersed rigid bodies. In addition, the viscous fluid is represented with compressible model with a constant wave speed.

In this case, we have the dynamic viscosity $\mu = 1$ Pa·s, the bulk modulus $\kappa_f = 2.1 \times 10^6$ dyne/cm^2, and the density $\rho_f = 1000$ kg/m^3. Even with the coarse mesh used, the developed formulation with high-order elements provides reasonable results comparable to a reference solution. In addition, no spatial oscillation and checkerboard pressure bands are observed, as demonstrated. Of course, the proposed description can be used for refined meshes. As one of the main messages, we must point out that in engineering practice, before a large number of finite elements are used, it is always beneficial to employ coarse meshes with high-order elements to obtain a reasonable estimation of complicated problems. As shown in Figs. 4.6 and 4.7, the horizontal velocity distributions predicted with traditional ALE formulation and the immersed continuum method are very close.

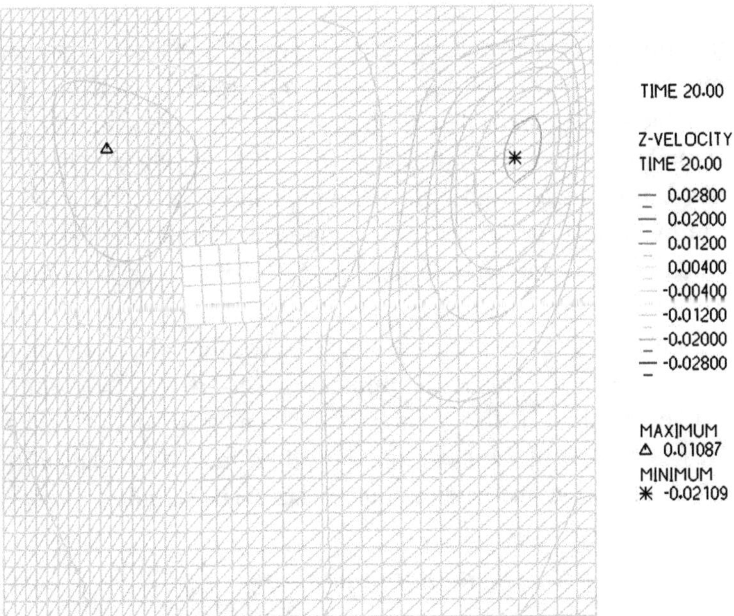

Fig. 4.6. Vertical velocity with ALE formulation.

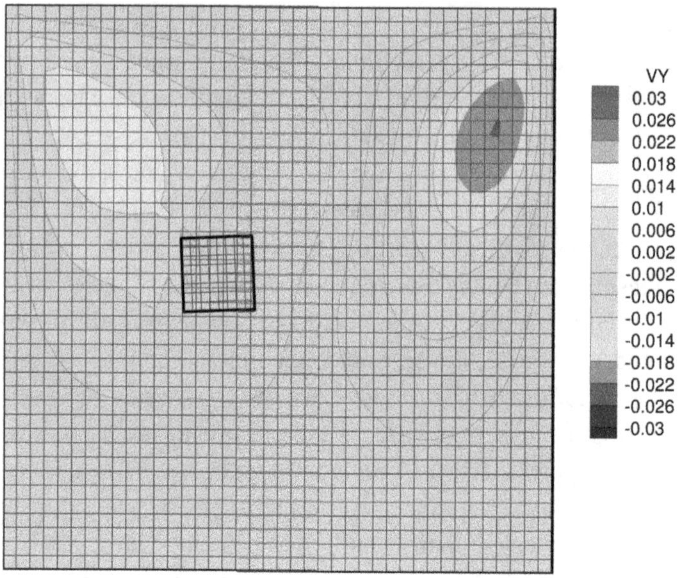

Fig. 4.7. Vertical velocity with the immersed continuum method.

Furthermore, we use a general case of a compressible solid immersed in compressible fluid to cover one seemingly very different scenario, namely, compressible solid immersed in incompressible fluid. In the immersed continuum method, we use a physical argument such that the bulk modulus of the solid can be much higher than the bulk modulus of the fluid. With the assumption that the acoustic wave speed within the fluid is constant, a so-called pseudo compressible fluid model is often used in numerics to mimic incompressible fluid behaviors. Why do we use this type of fluid model coupled with a compressible solid? To use the concept of immersed methods, we need to replace the immersed solid with a fluid similar to the surrounding fluid. If the volume of the immersed solid changes ever slightly, the volume of the corresponding fluid must also change, hence, the fluid model of the surrounding fluid must also be slightly compressible.

4.1.2 *Links to fictitious domain methods*

It is important to point out that the essence of the immersed continuum methods is to introduce the imaginary fluid in the submerged solid domain.

Because such a volume of fluid does not exist physically, to account correctly for the effect of the submerged solid exerted on the surrounding fluid, we must subtract the inertial force, the external body force, and the internal stress effects of such an imaginary fluid volume Ω_s. This is the same idea used in the fictitious domain method in which the fictitious fluid is constraint to have the same behavior of the immersed rigid body.

In the fictitious domain method, the immersed rigid body is governed by

$$\mathbf{v}(\mathbf{x}^s, t) = \bar{\mathbf{v}}(t) + \bar{\boldsymbol{\omega}}(t) \times (\mathbf{x}^s - \bar{\mathbf{x}}(t)), \quad \forall \mathbf{x}^s \text{ in } \Omega_s, \tag{4.10}$$

$$\rho_f \dot{v}_i = \sigma_{ij,j} + \rho_f g_i, \quad \text{in } \Omega_f, \tag{4.11}$$

$$M\dot{\bar{v}}_i = Mg_i + F_i^s, \quad \text{for the rigid body } \Omega_s, \tag{4.12}$$

$$\mathbf{I}\dot{\bar{\boldsymbol{\omega}}} + \bar{\boldsymbol{\omega}} \times \mathbf{I}\bar{\boldsymbol{\omega}} = \mathbf{T}^s, \tag{4.13}$$

where $\mathbf{I}$ and M are the rotational inertia tensor (or matrix) and the mass of the rigid body, respectively; and the resultant torque $\mathbf{T}^s$ and force $\mathbf{F}^s$ due to the fluid traction around the rigid body are expressed as

$$\mathbf{T}^s = -\int_{\Gamma_s} (\mathbf{x}^s - \bar{\mathbf{x}}) \times \boldsymbol{\sigma}\mathbf{n}d\Gamma, \tag{4.14}$$

$$\mathbf{F}^s = -\int_{\Gamma_s} \boldsymbol{\sigma}\mathbf{n}d\Gamma. \tag{4.15}$$

In the fictitious domain method, to synchronize the fictitious fluid with the solid within Ω_s, distributed Lagrangian multipliers are introduced. With the fluid velocity variation $\mathbf{w} \in \left[H^1_{0,\Gamma_v}(\Omega)\right]^d$, the rigid body velocity variation $\bar{\mathbf{w}} \in \mathcal{R}^d$, and the rigid body angular velocity variation $\bar{\boldsymbol{\theta}} \in \mathcal{R}^d$, the governing equations or strong form in Eqs. (4.10) to (4.13) are converted into the variational equations or the so-called weak form,

$$\int_\Omega [\rho_f w_i(\dot{v}_i - g_i) + w_{i,j}\sigma_{ij}]\, d\Omega + r\left[M(\dot{\bar{v}}_i - g_i)\bar{w}_i + (\mathbf{I}\dot{\bar{\boldsymbol{\omega}}} + \bar{\boldsymbol{\omega}} \times \mathbf{I}\bar{\boldsymbol{\omega}}) \cdot \bar{\boldsymbol{\theta}}\right]$$
$$- (\boldsymbol{\lambda}, \mathbf{w} - \bar{\mathbf{w}} - \bar{\boldsymbol{\theta}} \times (\mathbf{x}^s - \bar{\mathbf{x}})) = 0, \tag{4.16}$$

and

$$(\boldsymbol{\mu}, \mathbf{v} - \bar{\mathbf{v}} - \bar{\boldsymbol{\omega}} \times (\mathbf{x}^s - \bar{\mathbf{x}})) = 0, \quad \forall \boldsymbol{\mu} \in \left[H^1(\Omega_s)\right]^d, \tag{4.17}$$

where the inner product is defined as

$$(\boldsymbol{\mu}, \boldsymbol{\lambda}) = \int_{\Omega_s} (\mu_i \lambda_i + l^2 \mu_{i,j} \lambda_{i,j}) d\Omega, \tag{4.18}$$

with a scaling factor l dependent of the characteristic length of Ω_s and $r = 1 - \rho_f/\rho_s$.

Similar to the immersed methods the motion of the imaginary fluid motion is synchronized to the solid motion within the immersed solid domain Ω_s, namely,

$$\mathbf{v}^s = \mathbf{v}^f. \tag{4.19}$$

In fact, the constraint of Eq. (4.19) introduces distributed Lagrangian multipliers which are the equivalent body forces. In the immersed continuum methods, deformable solids instead of rigid bodies are considered. The mixed formulation is similar to the displacement/pressure formulation for acoustic fluids and the Inf-Sup conditions should be considered Bao *et al.* (2001) and Bathe (1996).

By comparing the fictitious domain method with the immersed boundary method, the extended immersed boundary method, the immersed finite element method, or the immersed continuum method, it is not difficult to identify the force density $\mathbf{f}^s + m\ddot{\mathbf{u}}_s$ for the immersed boundary or $\mathbf{f}^s$ for the immersed body (or continuum) in fact represents the Lagrangian multiplier for the constraint of Eq. (4.19). Finally, the same Inf-Sup conditions will be introduced to relax the constraint on volumetric strain and the pressure imposed on the immersed solids and the fictitious fluids.

4.1.3 *Mixed finite element formulations*

It is important to note that here the immersed solid is viewed as impermeable, thus such a volume of fluid does not exist physically. To account for the correct effect of the submerged solid exerting on the surrounding fluid, we must subtract the inertial force, the external body force, and the internal stress effects of such an imaginary fluid volume $\Omega_s(t)$. This idea is very similar to the fictitious domain method, which was proposed to handle immersed rigid bodies (Glowinski *et al.*, 2001). The advantage of the fictitious domain method is the use of a fixed background fluid mesh, whereas the immersed rigid bodies are modeled with the Lagrangian description.

Recently, such an approach has been extended to immersed flexible bodies Van Loon *et al.* (2007); Vos *et al.* (2008); Yu (2005). In full-fledged implicit forms, there must be some hidden mathematical equivalence of these two sets of approaches: namely, the immersed continuum method and the fictitious domain method. Moreover, an Inf-Sup condition similar to the modeling of almost incompressible materials can be derived in the context of the treatment of compressible solids immersed in compressible fluids Bao *et al.* (2001).

As shown in Figs. 4.8–4.9, the pressure distributions predicted with traditional ALE formulation and the immersed continuum method for a driven cavity problem with one immersed soft or hard solid are very close. Similar confirmations with respect to the vertical velocity distributions are also depicted in these figures Wang *et al.* (2009b). Also shown in Figs. 4.10–4.11, for soft immersed solids, the initial shear force of the driven cavity problem causes an obvious shear deformation, whereas for hard immersed solids, rigid body motions such as translations and rotations are observed.

A side benefit of this implicit compressible immersed continuum method is the volume preservation of the immersed solid. In this case, for both soft and hard solids, the material is almost incompressible with a high bulk modulus. Although this implicit matrix-free Newton–Krylov solution scheme is not fast, the procedure is not extremely time consuming or prohibitive to carry out for a typical laptop. In addition, no spatial oscillation and checkerboard pressure bands are observed. Of course, the proper mesh refinement strategies can also be employed.

The benefit of immersed methods is clearly depicted for the case with five immersed solids. For this case, it is no longer feasible to use the ALE formulation, whereas it is a simple task to add a few more deformable solids in the immersed continuum method. Note that in the immersed continuum method the contact between two immersed solids is accomplished for free just as in the material point method and the immersed boundary method.

To quantify the effectiveness of the preconditioner, we compare a typical FSI problem using the matrix-free Newton–Krylov iteration procedures with and without the preconditioner. The total number of time steps is 40. Three criteria are used: the number of Newton–Raphson iterations per time

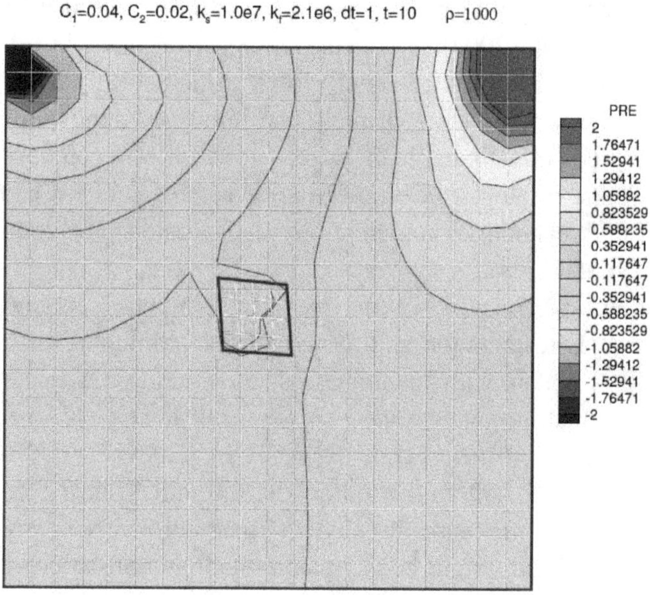

(a)

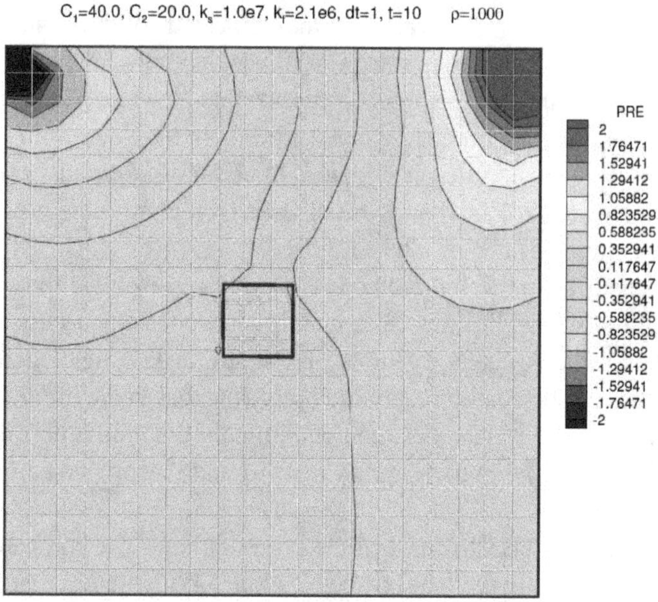

(b)

Fig. 4.8. Pressure snapshot at $t = 10s$ with (a) one soft immersed solid and (b) one hard immersed solid.

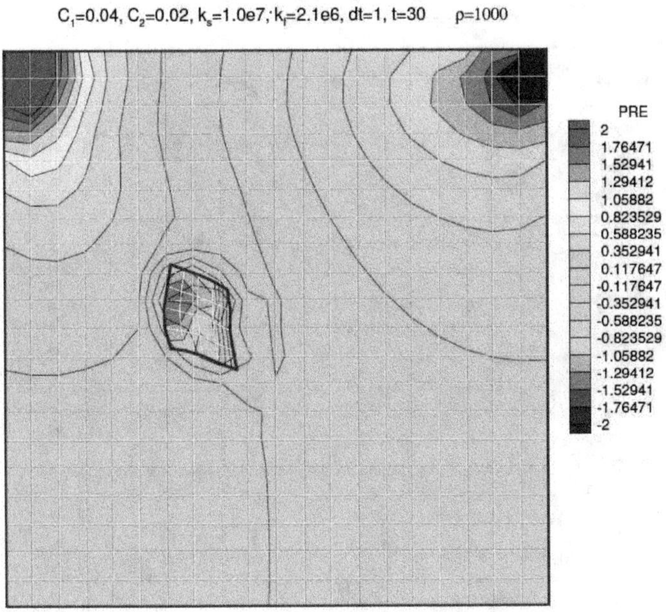

(a)

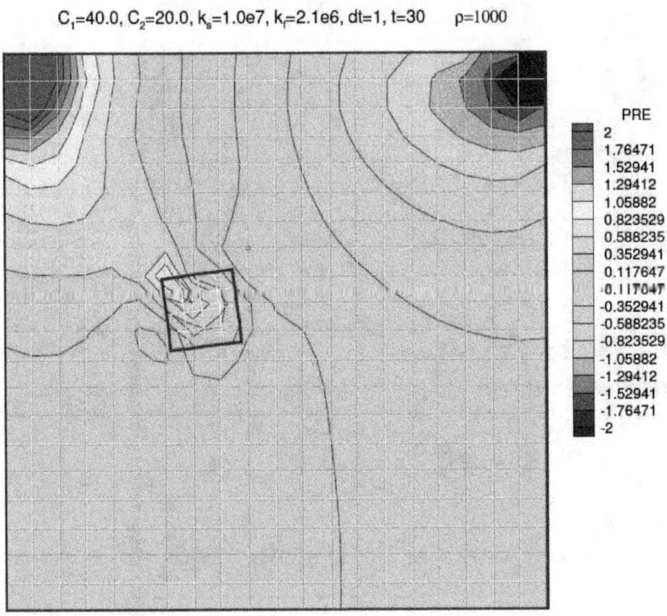

(b)

Fig. 4.9. Pressure snapshot at $t = 30s$ with (a) one soft immersed solid and (b) one hard immersed solid.

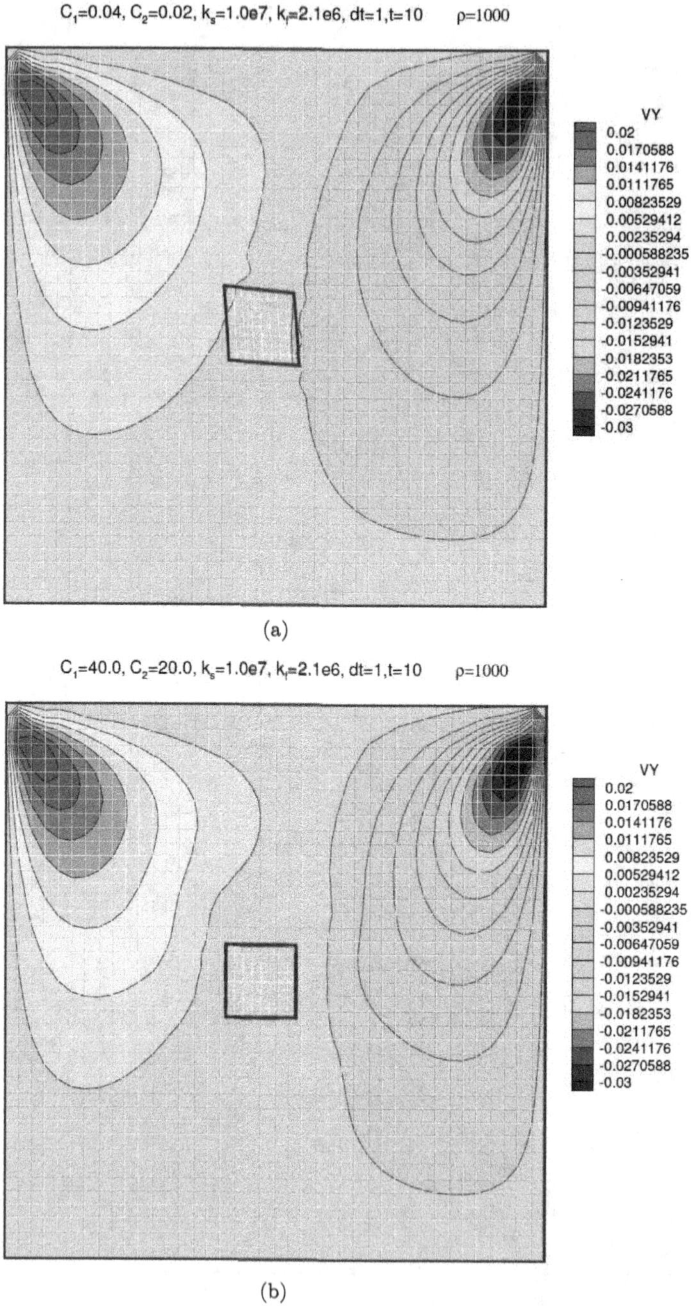

Fig. 4.10. Vertical velocity snapshot at $t = 10s$ with (a) one soft immersed solid and (b) one hard immersed solid.

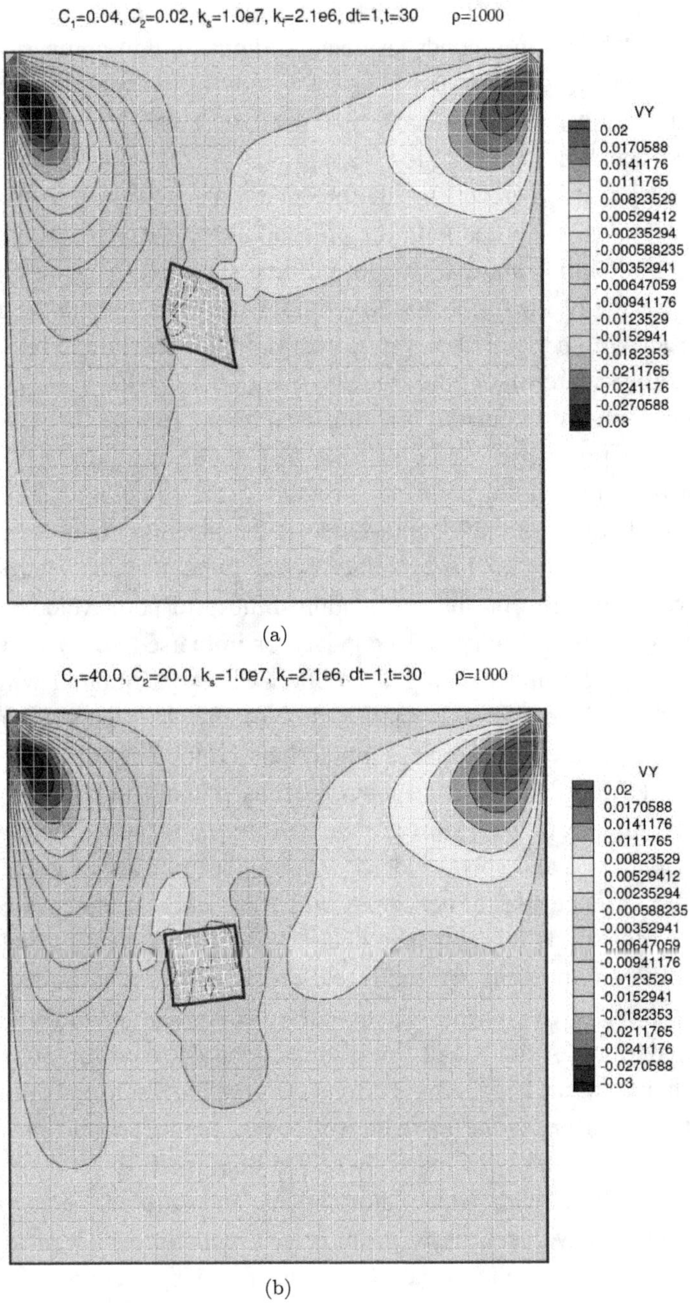

Fig. 4.11. Vertical velocity snapshot at $t = 30s$ with (a) one soft immersed solid and (b) one hard immersed solid.

step, the number of GMRES iterations per Newton–Raphson iteration, and the total CPU time in seconds. A detailed digit by digit comparison also confirms that the velocity and pressure results differ only after the 5 or 6 decimal points, which is consistent with the error criteria set for the Netwon–Krylov iteration as shown in Fig. 4.3. As demonstrated in numerical examples, it is evident that the preconditioner is very effective in all three categories. Note that with the preconditioner the maximum number of GMRES iteration stays at two for all time steps and Newton–Raphson iterations. Finally, the results of the solutions with and without the preconditioner are within the iterative error bounds, in this case, set to be as 10^{-5}, less than the square root of the machine roundoff error 10^{-15}.

In the immersed boundary or immersed continuum methods, in order to satisfy energy conservation, namely, that the energy input to the fluid domain from the elastic boundary or solid is the same as that from the equivalent body force, the same Dirac delta function must be used in both the distribution of the resultant nodal force and the interpolation of the velocity based on the surrounding fluid grid point velocities. The key treatment in the immersed boundary or immersed continuum methods is the synchronization of fluid motion with solid motion within the immersed solid domain Ω_s, namely, Eq. (4.19). It is conceivable that full-fledged augmented Lagrange multipliers, named after Joseph–Louis Lagrange (1736–1813), can be introduced as primary unknowns. In this case, the equivalent body forces can be directly calculated along with independent fluid and solid velocity vectors. Nevertheless, such a procedure will introduce a set of new unknowns equal to the number of velocity unknowns for solids. We conclude that mixed finite element formulations, namely, velocity/pressure for fluids and displacement/pressure for solids, can be successfully adopted to the immersed continuum method.

The coupling of fluids and solids is the central feature in the study of the mechanics of the heart, arteries, veins, microcirculation, and pulmonary blood flow. Currently, the modeling of strong hemodynamic interaction with flexible structures is limited by severe fluid mesh distortions around flexible structures with large deformations and displacements. Recent breakthroughs have been made in immersed methods which prove to be a viable platform for multi-scale and multi-physics modeling of biological systems. These new numerical algorithms for large FSI systems are essential to instability analysis of linear non-autonomous systems; the computation of limit sets, which include stationary points, periodic orbits, quasi-periodic orbits, and strange attractors; and the computation of turning points, which include super- and sub-critical saddle-node, pitchfork, and Hopf bifurcations in parameter space. Of course, it is only the beginning for this type of modeling treatment for complex FSI systems. Further mathematical understanding and study are urgently needed.

In order to handle compressible continuum immersed in compressible fluids, an implicit algorithm must be introduced. Moreover, the velocity/pressure formulation is employed for the compressible viscous fluid and the displacement/pressure formulation for the compressible solid with a hyperelastic material model. For simplicity, in this section, the superscript or subscript f for fluid variables is omitted. For the fluid domain, we adopt an Eulerian kinematic description, therefore, the material derivative of the fluid velocity is expressed as

$$\dot{v}_i = v_{i,t} + v_j v_{i,j}. \tag{4.20}$$

For the solid domain, a Lagrangian kinematic description is used, thus the fluid–solid interface will be tracked automatically by the position of solid particles. Moreover, there is no need for convective terms in the solid domain and the material derivative is the same as the time derivative. Hence, the solid velocity vector $\mathbf{v}^s$ and the acceleration vector $\dot{\mathbf{v}}^s$ can be expressed as

$$\mathbf{v}^s = \dot{\mathbf{u}}^s \quad \text{and} \quad \dot{\mathbf{v}}^s = \ddot{\mathbf{u}}^s, \tag{4.21}$$

with the displacement vector $\mathbf{u}^s(t) = \mathbf{x}^s(t) - \mathbf{x}^s(0)$, where $\mathbf{x}^s(t)$ and $\mathbf{x}^s(0)$ stand for the current and the original material point positions within the solid domain Ω_s.

Of course, the solid domain Ω_s and the material point position $\mathbf{x}^s$ refer to the current solid configurations, and therefore for clarity could be denoted as $\Omega_s(t)$ and $\mathbf{x}^s(t)$, respectively. In order to match with the compressible viscous fluid, the pressure p is subtracted from the stress components σ_{ij} and obtain the deviatoric stress components τ_{ij}, which is illustrated in a Newtonian fluid model,

$$\sigma_{ij} = -p\delta_{ij} + \tau_{ij}, \tag{4.22}$$

with $\tau_{ij} = \mu(v_{j,i} + v_{i,j})$.

Furthermore, to couple with the pressure unknown, the continuity equation of the compressible viscous fluid is expressed as

$$v_{i,i} + \frac{\dot{p}}{\kappa} = 0, \tag{4.23}$$

where κ is the bulk modulus of the fluid; and the material derivative $\dot{p}$ can be simply expressed as $p_{,t} + v_i p_{,i}$.

Note that the bulk modulus κ in the fluid can be simply viewed as a penalty term associated with the divergence of the fluid velocity. Therefore, for convenience, we ignore the convective term of the pressure and obtain $p_{,t} = \dot{p}$. Furthermore, for typical fluids, the change of the fluid density can be easily added through the pressure change or simply ignored. Like the fluid stress tensor, the solid stress tensor is decomposed as a hydrostatic pressure p^s, and a deviatoric stress tensor τ^s_{ij}, and Eq. (1.4).

Unlike the fluid domain, since the Lagrangian description is used for the submerged solid, the treatments of the continuity equation and the Cauchy stress in nonlinear solid mechanics are not as straightforward. As a special case, if the submerged solid is a flexible structure with a linear elastic material law, only the geometrical nonlinearity has to be dealt with.

In this case, suppose the Young's modulus and the Poisson's ratio are E and ν, respectively, the bulk modulus for the solid can be simply expressed as $\kappa^s = E/3(1-2\nu)$. In this work, both the geometrical and material nonlinearities will be included Wang and Liu (2004); Zhang *et al.* (2004). First of all, the solid deformation gradient $F_{ij} = \partial x^s_i(t)/\partial x^s_j(0)$ has to be introduced, from which the Green–Lagrangian strain ε_{ij} can be derived. Consequently the energy conjugate stress S_{ij} and the second Piola–Kirchhoff stress are derived from the elastic energy $\bar{W}$, which is often related to the three invariants of the Cauchy–Green deformation tensor $\mathbf{C}$ defined as $\mathbf{F}^T\mathbf{F}$. Moreover, an elastic energy term $-[p^s + \kappa^s(J_3 - 1)]^2/2\kappa^s$ is added to $\bar{W}$, along with the solid pressure unknown p^s introduced as

$$J_3 - 1 + \frac{p^s}{\kappa^s} = 0, \tag{4.24}$$

where κ^s is the solid bulk modulus and J_3 stands for the determinant of the deformation gradient.

Of course, to match with the expression in Eq. (1.4), the solid Cauchy stress is converted from the second Piola–Kirchhoff stress, in Eq. (1.5).

Finally, since the solid displacement is dependent on the fluid velocity, the primary unknowns for the coupled fluid–solid system are the fluid velocity $\mathbf{v}$, the fluid pressure p, and the solid pressure p^s.

We present here the velocity/pressure formulation for the compressible viscous fluid and the displacement/pressure formulation for the compressible solid with a hyperelastic material model. For simplicity in this section, we omit the superscript or subscript f for fluid variables. Unlike the fluid domain, since we use the Lagrangian description for the submerged solid, the treatments of the continuity equation and the Cauchy stress in nonlinear solid mechanics are not as straightforward. As a special case, if the submerged solid is a flexible structure with a linear elastic material law, we will only have the geometrical nonlinearity to deal with. However, Cauchy stress must still be depicted in the current configuration, which itself is unknown. Define the Sobolev spaces so the weak form of governing equations can be modified as: $\forall q \in \mathcal{L}^2(\Omega)$, $q^s \in \mathcal{L}^2(\Omega_s)$, $\mathbf{w} \in \left[\mathcal{H}^1_{0,\Gamma_v}(\Omega)\right]^d$, which includes $\forall \mathbf{w}^s \in \left[\mathcal{H}^1(\Omega_s)\right]^d$, and we find $\mathbf{v}$ and p in Ω, p^s in Ω_s, such that

$$\begin{aligned}
&\int_\Omega w_i \rho(\dot{v}_i - g_i)d\Omega + \int_\Omega (w_{i,j}\tau_{ij} - p w_{i,i})d\Omega - \int_{\Gamma_f} w_i F_i^p d\Gamma \\
&+ \int_{\Omega_s} [w_i^s(\rho_s - \rho)(\dot{v}_i - g_i) + w_{i,j}^s(\tau_{ij}^s - \tau_{ij}^f) - (p^s - p)w_{i,i}^s]d\Omega \\
&+ \int_\Omega q\left(v_{j,j} + \frac{p_{,t}}{\kappa}\right)d\Omega + \int_{\Omega_s} q^s\left(J_3 - 1 + \frac{p^s}{\kappa^s}\right)d\Omega = 0. \qquad (4.25)
\end{aligned}$$

We recognize that there are two sets of discretizations, namely, one for the Lagrangian solid mesh and one for the Eulerian fluid mesh. The stabilization finite element scheme is similar to what has been published in Wang (2000). Note that the convective terms are hidden in $\dot{v}_i^h$ and the detailed expressions of other stabilized Galerkin formulation for the Navier–Stokes equations can be found in Section 3.3 or (Tezduyar, 1991, 2001; Hughes *et al.*, 1986). In general, the interpolation functions for the velocity vector and the unknown pressures are different. Therefore, we retain the superscripts v and p to denote such differences. For the fluid domain Ω, the following interpolations are used:

$$\mathbf{v}^h = h_i^v \mathbf{v}_i, \quad \mathbf{w}^h = h_i^v \mathbf{w}_i, \quad p^h = h_i^p p_i, \quad q^h = h_i^p q_i, \qquad (4.26)$$

where h_i^v and h_i^p stand for the interpolation functions at node i for the velocity vector and the pressure, and $\mathbf{v}_i$, $\mathbf{w}_i$, p_i, and q_i are the nodal values of the discretized velocity vector, admissible velocity variation, pressure, and pressure variation, respectively.

For the solid domain Ω_s, the discretization is based on the following:

$$\mathbf{u}^{s,h} = h_j^u \mathbf{u}_j^s, \quad \mathbf{w}^{s,h} = h_j^u \mathbf{w}_j^s, \quad p^{s,h} = h_j^{p^s} p_j^s, \quad q^{s,h} = h_j^{p^s} q_j^s, \tag{4.27}$$

where h_j^u and $h_j^{p^s}$ stand for the interpolation functions at node j for the displacement vector and the unknown pressures, and $\mathbf{u}_j^{s,h}$, $\mathbf{w}_j^{s,h}$, $p_j^{s,h}$, and $q_j^{s,h}$ are the nodal values of the discretized displacement vector, admissible velocity variation, pressure, and pressure variation, respectively.

Substituting both discretizations (4.26) and (4.27) into Eq. (1.6), we obtain the following discretization of the weak form: $\forall q^h \in \mathcal{L}^2(\Omega^h)$, $q^{s,h} \in \mathcal{L}^2(\Omega_s^h)$, $\mathbf{w}^h \in \left[\mathcal{H}^{1,h}_{0,\Gamma_v^h}(\Omega^h)\right]^d$, which includes $\mathbf{w}^{s,h} \in \left[\mathcal{H}^{1,h}(\Omega_s^h)\right]^d$,

$$\begin{aligned}
&\int_{\Omega^h} w_{ik} h_i^v \rho \dot{v}_k^h d\Omega - \int_{\Gamma_f^h} w_{ik} h_i^{v,f} F_k^p d\Gamma + \int_{\Omega^h} (w_{ik} h_{i,l}^v \tau_{kl} - p^h w_{ik} h_{i,k}^v) d\Omega \\
&+ \int_{\Omega_s^h} \left[w_{jk}^s h_j^u (\rho_s - \rho)(\dot{v}_k^h - g_k) + w_{jk}^s h_{j,l}^u (\sigma_{kl}^s - \sigma_{kl}^f) \right] d\Omega \\
&- \int_{\Omega^h} w_{ik} h_i^v \rho g_k d\Omega + \int_{\Omega^h} q_i h_i^p \left(v_{l,l}^h + \frac{p_{,t}^h}{\kappa} \right) d\Omega \\
&+ \int_{\Omega_s^h} q_j^s h_j^{p^s} \left(J_3 - 1 + \frac{p^{s,h}}{\kappa^s} \right) d\Omega = 0.
\end{aligned}$$

For clarity, we introduce a displacement nodal unknown vector $\mathbf{U}$, although it is only evaluated in the solid domain Ω_s in which a Lagrangian description is prescribed. In fact, within the solid domain, $\mathbf{U}$ is denoted as $\mathbf{U}^s$ and evolves according to $\dot{\mathbf{U}}^s$ and $\ddot{\mathbf{U}}^s$, which are mapped from the velocity nodal unknown vector $\mathbf{V}$ and acceleration nodal unknown vectors $\dot{\mathbf{V}}$ for the fluid domain. Mathematically, we could say that $\mathbf{v}^s$ is $\mathbf{v}$ directly evaluated at the material point $\mathbf{x}^s$.

Likewise, the pressure nodal unknown vectors $\mathbf{P}$ and $\mathbf{P}^s$ are introduced for fluid and solid domains, respectively. Moreover, in the discussion of numerical procedures, we denote the time derivative of a variable a as $\dot{a}$. For simplicity, we employ the Newton–Raphson iteration, and apply the Newmark time integration schemes as illustrated. The details of this time integration scheme for linear and nonlinear systems have also been elaborated in Bathe and Sonnad (1980), Belytschko *et al.* (2000), and Wang (1997, 1999). Finally, for the entire domain Ω, due to the arbitrariness of the variations w_{ik}, q_i, and q^s_j, we have four equations at each fluid node i and one equation at each solid node j:

$$r^v_{ik} = 0, \quad r^p_i = 0, \quad r^{p^s}_j = 0, \tag{4.28}$$

or

$$\mathbf{r}(\mathbf{V}, \mathbf{P}, \mathbf{P}^s) = \mathbf{0}, \tag{4.29}$$

where the residuals are defined as

$$\begin{aligned}
r^v_{ik} &= \int_{\Omega^h} h^v_i \rho \dot{v}^h_k d\Omega + \int_{\Omega^h} [h^v_{i,l}\tau_{kl} - p^h h^v_{i,k}] d\Omega - \int_{\Gamma^h_f} h^{v,f}_i F^p_k d\Gamma \\
&\quad + \int_{\Omega^h_s} \mathcal{M}\left[h^u_j(\rho_s - \rho)(\dot{v}^h_k - g_k) + h^u_{j,l}(\sigma^s_{kl} - \sigma^f_{kl})\right] d\Omega \\
&\quad - \int_{\Omega^h} h^v_i \rho g_k d\Omega, \\
r^p_i &= \int_{\Omega^h} h^p_i \left(v^h_{l,l} + \frac{p^h_{,t}}{\kappa} \right) d\Omega, \\
r^{p^s}_j &= \int_{\Omega^h_s} h^{p^s}_j \left(J_3 - 1 + \frac{p^{s,h}}{\kappa^s} \right) d\Omega.
\end{aligned}$$

Again, the operator $\mathcal{M}$ stands for a complicated mapping process between the immersed solid nodes with the background fluid nodes within their respective influence domain with typical Reproducing Kernel Particle Methods (RKPM) procedures (Liu *et al.*, 1995; Wang and Liu, 2004).

4.1.4 *Implicit matrix-free Newton–Krylov iteration*

After the temporal discretization or time integration, Newton–Krylov matrix-free iteration procedures will be used to solve the residual equations. In this project, we employ a matrix-free Newton–Krylov iterative procedure for nonlinear sets of equations. In the kth Newton–Raphson iteration at time step $m+1$ of the nonlinear residual equation (4.29), from $\mathcal{R}^N$ to $\mathcal{R}^N$, with N as the number of the total unknowns, we start with a first guess of the incremental unknowns $\Delta\boldsymbol{\Theta}^{k,0}$, namely, $\Delta\mathbf{V}^0$, $\Delta\mathbf{P}^0$, and $\Delta\mathbf{P}^{s,0}$, which often are zero vectors. Then the residual of the linearized systems of equations at the kth Newton–Raphson iteration is evaluated as

$$\mathbf{p} = -\mathbf{r}^{m+1,k-1} - \mathbf{r}_{,v}^{m+1,k-1}\Delta\mathbf{V}^0 - \mathbf{r}_{,p}^{m+1,k-1}\Delta\mathbf{P}^0 - \mathbf{r}_{,p^s}^{m+1,k-1}\Delta\mathbf{P}^{s,0}. \tag{4.30}$$

This error vector is used to construct the n-dimensional Krylov subspace $\mathbf{K}^n$ where $\mathbf{J}$ is the $N \times N$ Jacobian matrix evaluated at time step $m+1$ and the kth Newton–Raphson iteration of the nonlinear residual equation and can be rewritten as

$$\mathbf{J} = \left(\mathbf{r}_{,v}^{m+1,k-1}, \mathbf{r}_{,p}^{m+1,k-1}, \mathbf{r}_{,p^s}^{m+1,k-1}\right). \tag{4.31}$$

The approximate solution $\Delta\boldsymbol{\Theta}$ is written as the combination of the initial guess $\Delta\boldsymbol{\Theta}^{k,0}$ and $\mathbf{z}^n$, with $\mathbf{z}^n \in \mathbf{K}^n$. Note that the dimension of the subspace $\mathbf{K}^n$ is n which is much smaller than the dimension N of the unknown vector $\Delta\boldsymbol{\Theta}$. In matrix-free Newton–Krylov iteration, we do not form the Jacobian matrix. In general, this Jacobian matrix in the immersed boundary or immersed continuum methods has an $O(n^2)$ storage requirement. For large systems with million degrees of freedoms, this Jacobian matrix requires a terabyte storage space, which is beyond the limit of computational facilitates available for most scientific researchers. Based on this understanding, we would also like to design a preconditioning technique without the use of the Jacobian matrix (Knoll and Keyes, 2004; Kay *et al.*, 2002; Tidriri, 1997). First of all, the initial residual vector $\mathbf{p}$ in the kth Newton–Raphson iteration at time step $m+1$ of the nonlinear residual equation is normalized as $\mathbf{v}_1$ with the length $\gamma = \|\mathbf{p}\|_2$. Moreover, we have the corresponding n dimensional residual vector $\mathbf{b} = \gamma\mathbf{e}_1$.

In the generalized minimum residual (GMRES) method, the modified Gram-Schmidt orthogonalization procedures, named after Jorgen Pedersen Gram (1850–1916) and Erhard Schmidt (1876–1959), we then have $\mathbf{q}_i = \boldsymbol{\Lambda}^{-1}\mathbf{v}_i$ and $\mathbf{w} = \mathbf{J}\mathbf{q}_i$, and for $j = 1$ to i, we have $h_{ji} = \mathbf{w}^T\mathbf{v}_j$ and $\mathbf{w}$ is updated with $\mathbf{w} - h_{ji}\mathbf{v}_j$.

As a consequence, we obtain $h_{(i+1)i} = \|\mathbf{w}\|_2$ and $\mathbf{v}_{i+1} = \mathbf{w}/h_{(i+1)i}$ with

$$\mathbf{J}\mathbf{q}_i \simeq \left[\mathbf{r}(\boldsymbol{\Theta}^{m+1,k-1} + e\mathbf{q}_i) - \mathbf{r}(\boldsymbol{\Theta}^{m+1,k-1})\right]/e.$$

If we define $\mathbf{V}_n = (\mathbf{v}_1, \mathbf{v}_2, \ldots, \mathbf{v}_n)$ and $\mathbf{V}_{n+1} = (\mathbf{v}_1, \mathbf{v}_2, \ldots, \mathbf{v}_{n+1})$, a set of orthonormal vectors $\mathbf{v}_i$, with $1 \leq i \leq n$ in the Krylov space $\mathcal{K}^n$ and an $(n+1) \times n$ upper-Hessenberg matrix $\bar{\mathbf{H}}_n$ can be established as

$$\mathbf{J}\mathbf{V}_n = \mathbf{V}_{n+1}\bar{\mathbf{H}}_n,$$

and the remaining process in the GMRES method is to solve the least square problem:

$$\min_{\mathbf{z}\in\mathcal{K}^n} \|\mathbf{p} - \mathbf{J}\mathbf{z}\| \quad \text{or} \quad \min_{\mathbf{y}\in\mathcal{R}^n} \|\mathbf{p} - \mathbf{J}\mathbf{V}_n\mathbf{y}\|. \tag{4.32}$$

Assuming γ is the length of the initial residual vector $\mathbf{p}$ and $\mathbf{e}_1$ is the unit vector, the first column of $(n+1) \times (n+1)$ identity matrix,

$$\min_{\mathbf{y}\in\mathcal{R}^n} \|\gamma\mathbf{e}_1 - \bar{\mathbf{H}}_n\mathbf{y}\|. \tag{4.33}$$

After we establish the elements of an upper $n \times n$ Hessenberg matrix $\mathbf{H}_n$ as well as an upper $(n+1) \times n$ Hessenberg matrix $\bar{\mathbf{H}}_n$, for $j = 1$ to n, and $i = 1$ to $j-1$, a factorization of $\mathbf{H}_n$ is carried out

$$h_{ij} = c_i h_{ij} + s_i h_{(i+1)j} \quad \text{and} \quad h_{(i+1)j} = -s_i h_{ij} + c_i h_{(i+1)j}, \tag{4.34}$$

where the entities of the rotation processes are calculated as

$$r = \sqrt{h_{jj}^2 + h_{(j+1)j}^2}, \quad c_j = h_{jj}/r, \quad \text{and} \quad s_j = h_{(j+1)j}/r.$$

Through this rotation process, the upper Hessenberg matrix is converted to a diagonal matrix with the coefficients defined as: for $j = 1$ to n

$$h_{jj} = r, \quad p_j = c_j b_j, \quad \text{and} \quad p_{j+1} = -s_j b_j. \tag{4.35}$$

Finally, the termination criteria: if $|b_{n+1}| < \varepsilon$, the solution vector $\Delta\boldsymbol{\Theta}$, or rather $\Delta\mathbf{V}$, $\Delta\mathbf{P}$, and $\Delta\mathbf{P}^s$ is expressed as

$$\Delta\boldsymbol{\Theta}^{k,n} = \Delta\boldsymbol{\Theta}^{k,0} + \sum_{i=1}^{n} y_i \mathbf{q}^i,$$

or

$$\langle \Delta\mathbf{V}, \Delta\mathbf{P}, \Delta\mathbf{P}^s \rangle = \Delta\mathbf{\Theta}^{k,0} + \sum_{i=1}^{n} y_i \mathbf{q}^i.$$

As a final remark, if the initial guess $\Delta\mathbf{\Theta}^{k,0}$ does not produce a good estimate within a sufficiently small Krylov subspace $\mathbf{K}^n$. $\Delta\mathbf{\Theta}^{k,n}$ will be introduced as an updated initial guess and the GMRES iteration procedure will continue until a solution with the desired accuracy is obtained. By choosing the proper preconditioner, the inner GMRES iteration can be mininized to two to three steps while the outer Newton iteration still depends on how close to the solution the initial guess is.

In the matrix-free Newton–Krylov iteration, the target is the solution vector $\Delta\mathbf{\Theta}$, or rather $\Delta\mathbf{V}$ for fluid velocity unknowns, $\Delta\mathbf{P}$ for fluid pressure unknowns, and $\Delta\mathbf{P}^s$ for independent solid pressure unknowns. In this monolithic implicit immersed method with iterative solutions based on Newton–Krylov and generalized minimal residual method (GMRES) along with the numerical construction of the Jacobian matrix in the Newton–Raphson iteration, no specific analytical equation can be established other than an abstract operator $\tilde{N}$ which is implemented in recursive, iterative, and nested subroutines.

Note here for time increments, we use the Newmark time integration, named after Nathan Mortimore Newmark (1910–1981) with two constants α and β,

$$\begin{aligned} \mathbf{b}(t+\Delta t) &= \mathbf{b}(t) + \dot{\mathbf{b}}(t)\Delta t + \left[\left(\frac{1}{2} - \beta\right)\dot{\mathbf{b}}(t) + \beta\dot{\mathbf{b}}(t+\Delta t)\right]\Delta t^2, \\ \dot{\mathbf{b}}(t+\Delta t) &= \dot{\mathbf{b}}(t) + \left[(1-\alpha)\ddot{\mathbf{b}}(t) + \alpha\ddot{\mathbf{b}}(t+\Delta t)\right]\Delta t, \end{aligned} \tag{4.36}$$

along with

$$\begin{aligned} \Delta\mathbf{b}(t+\Delta t) &= \frac{\beta}{\alpha}\Delta t \Delta\dot{\mathbf{b}}(t+\Delta t), \\ \Delta\dot{\mathbf{b}}(t+\Delta t) &= \alpha\Delta t \Delta\ddot{\mathbf{b}}(t+\Delta t). \end{aligned} \tag{4.37}$$

In engineering practice (Bathe, 1996), it is often that we choose $\alpha = 1/2$ and $\beta = 1/6$, respectively. As a special case, when $\alpha = 1/2$ and

$\beta = 1/4$, we have the trapezoidal temporal discretization is recovered as the following:

$$\begin{aligned} \mathbf{b}(t+\Delta t) &= \mathbf{b}(t) + \frac{\Delta t}{2}\left[\dot{\mathbf{b}}(t) + \dot{\mathbf{b}}(t+\Delta t)\right], \\ \dot{\mathbf{b}}(t+\Delta t) &= \dot{\mathbf{b}}(t) + \frac{\Delta t}{2}\left[\ddot{\mathbf{b}}(t) + \ddot{\mathbf{b}}(t+\Delta t)\right], \end{aligned} \tag{4.38}$$

and consequently,

$$\mathbf{b}(t+\Delta t) = \mathbf{b}(t) + \dot{\mathbf{b}}(t)\Delta t + \frac{\Delta t^2}{4}\left[\ddot{\mathbf{b}}(t) + \ddot{\mathbf{b}}(t+\Delta t)\right],$$

and more importantly,

$$\begin{aligned} \Delta\mathbf{b}(t+\Delta t) &= \frac{\Delta t}{2}\Delta\dot{\mathbf{b}}(t+\Delta t), \\ \Delta\dot{\mathbf{b}}(t+\Delta t) &= \frac{\Delta t}{2}\Delta\ddot{\mathbf{b}}(t+\Delta t), \end{aligned} \tag{4.39}$$

where $\mathbf{b}$ stands for the fluid velocity unknowns $\mathbf{V}$, the fluid pressure unknowns $\mathbf{P}$, and the solid pressure unknowns $\mathbf{P}^s$.

Furthermore, the solid velocities are based on the projections of the background fluid velocities within the finite support region. Thus, the solid displacements at the FSI interface and the interior are integrated with the time integration scheme as stipulated in Eqs. (4.38) and (4.39). To be more specific, the update processes occur in very incremental iterations within one time step, which means, the positions of the immersed structures or immersed solids are updated in all iterations within the matrix-free Newton–Krylov iteration.

For specific elaboration of the combination of Newton–Raphson iteration with Newton–Krylov iteration, we introduce

$$\mathbf{A} = \begin{bmatrix} -1 & 1 & 0 & 0 & 0 & 0 \\ 0 & -1 & 1 & 0 & 0 & 0 \\ 0 & -1 & 0 & 1 & 0 & 0 \\ 0 & 0 & -1 & 1 & 0 & 0 \\ 0 & 0 & -1 & 0 & 1 & 0 \\ 0 & 0 & 0 & 0 & -1 & 1 \\ 0 & 0 & 0 & -1 & 0 & 1 \end{bmatrix}, \quad \text{and}$$

$$\mathbf{H} = \begin{bmatrix} h_{11} & h_{12} & h_{13} & h_{14} & \cdots & h_{1k} \\ h_{21} & h_{22} & h_{23} & h_{24} & \cdots & h_{2k} \\ 0 & h_{32} & h_{33} & h_{34} & \cdots & h_{3k} \\ 0 & 0 & h_{43} & h_{44} & \cdots & h_{4k} \\ 0 & 0 & 0 & \cdots & \cdots & \cdots \\ 0 & 0 & 0 & 0 & \cdots & h_{(k+1)k} \end{bmatrix};$$

therefore, for $k = 1$

$$\begin{bmatrix} c_1 & s_1 \\ -s_1 & c_1 \end{bmatrix} \begin{bmatrix} h_{11}^{(1)} \\ h_{21}^{(1)} \end{bmatrix} y_1 = \begin{bmatrix} c_1 & s_1 \\ -s_1 & c_1 \end{bmatrix} \begin{pmatrix} p_1^{(1)} \\ 0 \end{pmatrix}, \quad \text{and}$$

$$\begin{bmatrix} h_{11}^{(2)} \\ 0 \end{bmatrix} \quad y_1 = \begin{pmatrix} p_1^{(2)} \\ p_2^{(2)} \end{pmatrix};$$

with $p_1^{(1)} = \beta$, $p_2^{(1)} = 0$, $p_1^{(2)} = c_1\beta$, $p_2^{(2)} = -s_1\beta$. Naturally, if $p_2^{(2)}$ or rather $-s_1\beta$ is sufficiently small, the second equation will be approximately satisfied, and the iteration will stop at the level when $k = 1$.
Moreover, for $k = 2$

$$\begin{bmatrix} c_1 & s_1 & 0 \\ -s_1 & c_1 & 0 \\ 0 & 0 & 1 \end{bmatrix} \begin{bmatrix} h_{11}^{(1)} & h_{12}^{(1)} \\ h_{21}^{(1)} & h_{22}^{(1)} \\ 0 & h_{32}^{(1)} \end{bmatrix} \begin{bmatrix} y_1 \\ y_2 \end{bmatrix} = \begin{bmatrix} c_1 & s_1 & 0 \\ -s_1 & c_1 & 0 \\ 0 & 0 & 1 \end{bmatrix} \begin{pmatrix} p_1^{(1)} \\ 0 \\ 0 \end{pmatrix},$$

$$\begin{bmatrix} 1 & 0 & 0 \\ 0 & c_2 & s_2 \\ 0 & -s_2 & c_2 \end{bmatrix} \begin{bmatrix} h_{11}^{(2)} & h_{12}^{(2)} \\ 0 & h_{22}^{(2)} \\ 0 & h_{32}^{(2)} \end{bmatrix} \begin{bmatrix} y_1 \\ y_2 \end{bmatrix} = \begin{bmatrix} 1 & 0 & 0 \\ 0 & c_2 & s_2 \\ 0 & -s_2 & c_2 \end{bmatrix} \begin{pmatrix} p_1^{(2)} \\ p_2^{(2)} \\ 0 \end{pmatrix},$$

$$\begin{bmatrix} h_{11}^{(2)} & h_{12}^{(2)} \\ 0 & h_{22}^{(3)} \\ 0 & 0 \end{bmatrix} \begin{bmatrix} y_1 \\ y_2 \end{bmatrix} = \begin{pmatrix} p_1^{(2)} \\ p_2^{(3)} \\ p_3^{(3)} \end{pmatrix};$$

with $p_1^{(1)} = \beta$, $p_2^{(1)} = 0$, $p_3^{(1)} = 0$; $p_1^{(2)} = c_1\beta$, $p_2^{(2)} = -s_1\beta$, $p_3^{(2)} = 0$; $p_1^{(3)} = p_1^{(2)}$, $p_2^{(3)} = c_2 p_2^{(2)}$, $p_3^{(3)} = -s_2 p_2^{(2)}$.

Naturally, if $p_3^{(3)}$ or rather $-s_2 p_2^{(2)}$ is sufficiently small, the third equation will be approximately satisfied, and the iteration will stop at the level when $k = 2$. Finally, for $k = 3$

$$\begin{bmatrix} c_1 & s_1 & 0 & 0 \\ -s_1 & c_1 & 0 & 0 \\ 0 & 0 & 1 & 0 \\ 0 & 0 & 0 & 1 \end{bmatrix} \begin{bmatrix} h_{11}^{(1)} & h_{12}^{(1)} & h_{13}^{(1)} \\ h_{21}^{(1)} & h_{22}^{(1)} & h_{23}^{(1)} \\ 0 & h_{32}^{(1)} & h_{33}^{(1)} \\ 0 & 0 & h_{43}^{(1)} \end{bmatrix} \begin{bmatrix} y_1 \\ y_2 \\ y_3 \end{bmatrix}$$

$$= \begin{bmatrix} c_1 & s_1 & 0 & 0 \\ -s_1 & c_1 & 0 & 0 \\ 0 & 0 & 1 & 0 \\ 0 & 0 & 0 & 1 \end{bmatrix} \begin{pmatrix} p_1^{(1)} \\ 0 \\ 0 \\ 0 \end{pmatrix},$$

$$\begin{bmatrix} 1 & 0 & 0 & 0 \\ 0 & c_2 & s_2 & 0 \\ 0 & -s_2 & c_2 & 0 \\ 0 & 0 & 0 & 1 \end{bmatrix} \begin{bmatrix} h_{11}^{(2)} & h_{12}^{(2)} & h_{13}^{(2)} \\ 0 & h_{22}^{(2)} & h_{23}^{(2)} \\ 0 & h_{32}^{(1)} & h_{33}^{(1)} \\ 0 & 0 & h_{43}^{(1)} \end{bmatrix} \begin{bmatrix} y_1 \\ y_2 \\ y_3 \end{bmatrix}$$

$$= \begin{bmatrix} 1 & 0 & 0 & 0 \\ 0 & c_2 & s_2 & 0 \\ 0 & -s_2 & c_2 & 0 \\ 0 & 0 & 0 & 1 \end{bmatrix} \begin{pmatrix} p_1^{(2)} \\ p_2^{(2)} \\ 0 \\ 0 \end{pmatrix},$$

$$\begin{bmatrix} 1 & 0 & 0 & 0 \\ 0 & 1 & 0 & 0 \\ 0 & 0 & c_3 & s_3 \\ 0 & 0 & -s_3 & c_3 \end{bmatrix} \begin{bmatrix} h_{11}^{(2)} & h_{12}^{(2)} & h_{13}^{(2)} \\ 0 & h_{22}^{(3)} & h_{23}^{(3)} \\ 0 & 0 & h_{33}^{(3)} \\ 0 & 0 & h_{43}^{(1)} \end{bmatrix} \begin{bmatrix} y_1 \\ y_2 \\ y_3 \end{bmatrix}$$

$$= \begin{bmatrix} 1 & 0 & 0 & 0 \\ 0 & 1 & 0 & 0 \\ 0 & 0 & c_3 & s_3 \\ 0 & 0 & -s_3 & c_3 \end{bmatrix} \begin{pmatrix} p_1^{(2)} \\ p_2^{(3)} \\ p_3^{(3)} \\ 0 \end{pmatrix},$$

$$\begin{bmatrix} h_{11}^{(2)} & h_{12}^{(2)} & h_{13}^{(2)} \\ 0 & h_{22}^{(2)} & h_{23}^{(2)} \\ 0 & 0 & h_{33}^{(4)} \\ 0 & 0 & 0 \end{bmatrix} \begin{bmatrix} y_1 \\ y_2 \\ y_3 \end{bmatrix} = \begin{pmatrix} p_1^{(2)} \\ p_2^{(3)} \\ p_3^{(4)} \\ p_4^{(4)} \end{pmatrix};$$

with $p_1^{(1)} = \beta$, $p_2^{(1)} = 0$, $p_3^{(1)} = 0$, $p_4^{(1)} = 0$; $p_1^{(2)} = c_1\beta$, $p_2^{(2)} = -s_1\beta$, $p_3^{(2)} = 0$, $p_4^{(2)} = 0$; $p_1^{(3)} = p_1^{(2)}$, $p_2^{(3)} = c_2 p_2^{(2)}$, $p_3^{(3)} = -s_2 p_2^{(2)}$, $p_4^{(3)} = 0$; $p_1^{(4)} = p_1^{(2)}$, $p_2^{(4)} = p_2^{(3)}$, $p_3^{(4)} = c_3 p_3^{(3)}$, $p_4^{(4)} = -s_3 p_3^{(3)}$.

Naturally, if $p_4^{(4)}$ or rather $-s_2 p_3^{(3)}$ is sufficiently small, the third equation will be approximately satisfied, and the iteration will stop at the level when $k = 3$. In Eq. (4.5), within the entire domain Ω, the external work comes from the external body force $\rho_f \mathbf{g}$, and the surface traction $\mathbf{F}^p$ at the Neumann boundary Γ_f; whereas the power input within the submerged solid domain Ω_s includes the contribution from the inertial force difference, buoyancy force, and internal energy difference. Again, the kinematic matching at the submerged interface Γ_s also implies that the submerged interface will move at the same velocity as that of the fluid particles in the immediate vicinity. In Eq. (4.5), we again do not stipulate the material derivative $d\mathbf{v}/dt$ and the stress $\boldsymbol{\sigma}$. Hence the turbulent and the non-Newtonian fluid models can eventually be incorporated.

In the explicit implementation, just as in the immersed boundary method, in recent finite element extensions (Wang and Liu, 2004; Zhang *et al.*, 2004), we introduce the following two key equations to synchronize the fluid occupying the submerged solid domain Ω_s with the solid and distribute the solid force $\mathbf{f}^s$:

$$f_i^{fsi} = \int_{\Omega_s} f_i^s \delta(\mathbf{x} - \mathbf{x}^s) d\Omega, \tag{4.40}$$

$$v_i^s = \int_{\Omega} v_i \delta(\mathbf{x} - \mathbf{x}^s) d\Omega, \tag{4.41}$$

where $\mathbf{f}^{fsi}$ represents the same equivalent body force as in the immersed boundary method.

It is very important to recognize that $\mathbf{f}^s$ is the force density within the solid domain Ω_s, whereas $\mathbf{f}^{fsi}$ is the equivalent body force over the entire domain Ω. The physical significance of $\mathbf{f}^s$ and $\mathbf{f}^{fsi}$ is quite different. As a consequence, the entire FSI system is represented with the same set of governing equations (2.15) or strong form and the corresponding variational form (2.16), or the so-called weak form.

Equations (4.40) and (4.41) are comparable to Eqs. (2.13) and (2.14) in the IB method. In Eq. (4.9), $\mathbf{f}^s$ can be considered as the equivalent force density within the submerged solid domain Ω_s. In fact, this force density directly corresponds to the rigid link between the fluid occupying the submerged solid domain Ω_s and the solid. In other words, the force density $\mathbf{f}^s$ stands for the Lagrange multiplier corresponding to the constraint in Eq. (4.41). Of course, the definition in Eq. (4.9) also matches the virtual power input from the submerged solid domain Ω_s.

As described in Fig. 4.12, the solutions derived from the implicit immersed methods compare well with traditional ALE based FSI modeling techniques. Moreover, as shown in Figs. 4.13 and 4.14, with one immersed solid, it is still possible to have a mesh density in ADINA FSI modeling in comparison with the immersed continuum method. Although the simulation approaches are completely different, with comparable meshes, the results are very much similar. For the immersed continuum, the rubber material is modeled with $C_1 = 2000$ Pa, $C_2 = 1000$ Pa, bulk modulus for the immersed continuum and the background fluid are $k_s = 1.0 \times 10^8$ Pa and $k_f = 2.1 \times 10^7$ Pa, respectively. The kinematic viscosity ν is 1 m^2/s. Both the immersed continuum and the fluid have the same density, namely, $\rho_f = \rho_s = 1000$ kg/m^3. However, with multiple immersed solids, it is no longer possible to have a comparable mesh density in ADINA FSI modeling or

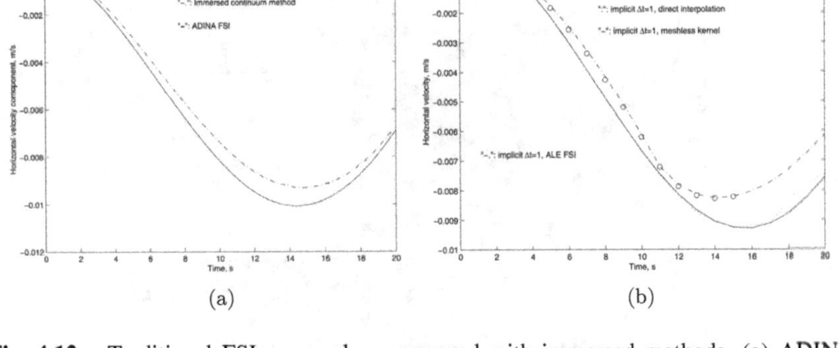

(a) (b)

Fig. 4.12. Traditional FSI approaches compared with immersed methods. (a) ADINA comparison and (b) Explicit and implicit.

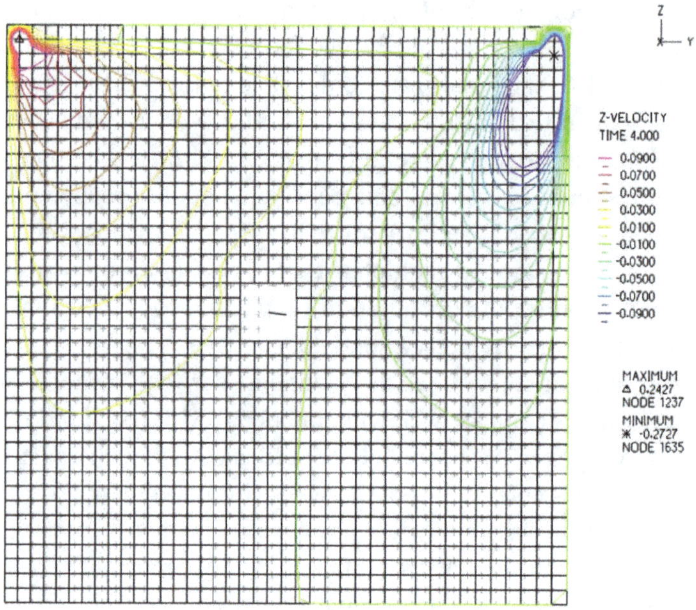

Fig. 4.13. Snapshots of results from the ADINA FSI solver.

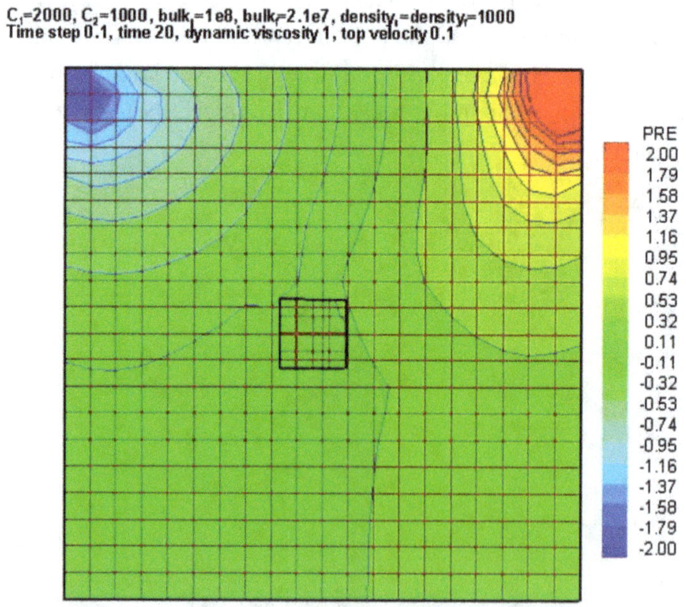

Fig. 4.14. Snapshots of results from the immersed continuum method.

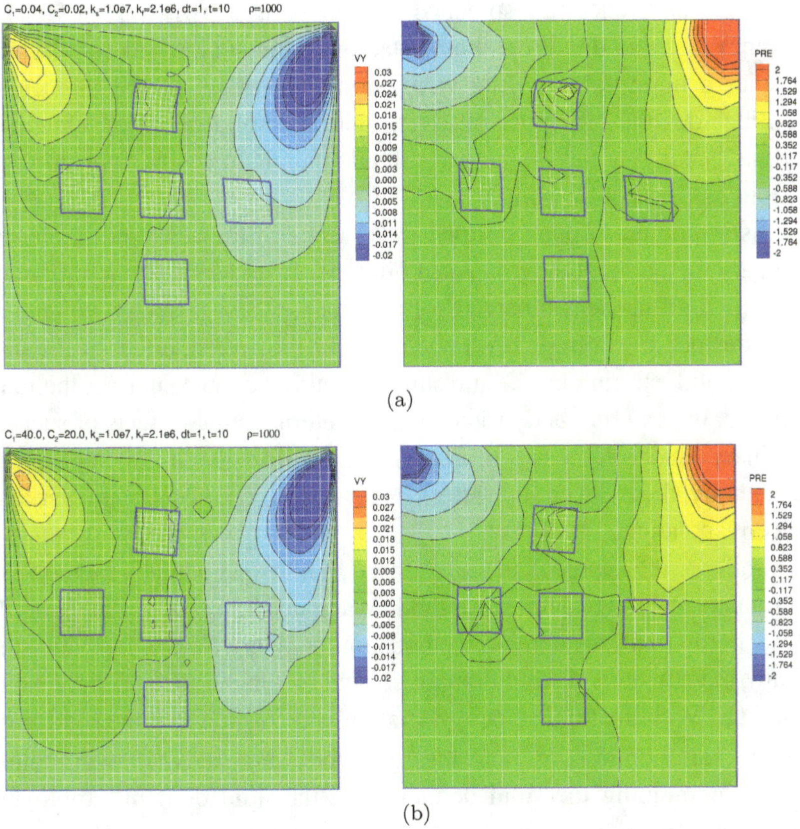

Fig. 4.15. Pressure snapshot with (a) five soft immersed solid at $t = 10s$ and (b) five hard immersed solid at $t = 10s$.

any other traditional ALE based FSI modelling. Nevertheless, with the immersed continuum method, such an expansion from one immersed solid to multiple solids is straightforward. As depicted in Figs. 4.14 and 4.15, with five compressible solids immersed in the same driven cavity problem with a compressible viscous fluid, the results are encouraging.

Note that the inertial force difference $-\int_{\Omega_s} w_i(\rho_s - \rho_f)\dot{v}_i d\Omega$ of the submerged solid continuum corresponds to the inertial force $-\int_{\Gamma_s} w_i^s m\ddot{u}_i^s d\Gamma$ of the submerged elastic boundary, whereas the internal energy difference $-\int_{\Omega_s} w_{i,j}(\sigma_{ij}^s - \sigma_{ij}^f)d\Omega$ of the submerged solid continuum corresponds to the elastic force $-\int_{\Gamma_s} w_i^s f_i^s d\Gamma$ of the submerged elastic boundary. In comparison with the submerged elastic boundary, the contribution of the submerged solid includes an additional term to account for the external body force difference (the so-called buoyancy) $\int_{\Omega_s} w_i(\rho_s - \rho_f)g_i d\Omega$. This buoyancy force is the direct manifestation of the submerged solid which occupies a finite volume. Likewise, the inertial effect of the submerged solid includes the difference between fluid and solid densities, assuming the fluid occupying the submerged solid domain is forced to have the same motions as those of the solid.

It is important to point out that the essence of the extended immersed boundary method and the Immersed Finite Element Method is to introduce the same fluid in the submerged solid domain. Because such a volume of fluid does not exist physically, to account for the correct effect of the submerged solid exerting on the surrounding fluid, we must subtract the inertial force, the external body force, and the internal stress effects of such an imaginary fluid volume Ω_s. Moreover, for a fluid volume moving with an elastic body, the induced fluid stress in general is significantly smaller than the corresponding solid stress.

Similar to the approach in the extended immersed boundary method and the Immersed Finite Element Method (Wang and Liu, 2004; Zhang *et al.*, 2004), in the fictitious domain method, an imaginary fluid is introduced to occupy the submerged solid domain Ω_s and to synchronize with the solid within Ω_s. With the fluid velocity variation $\mathbf{w} \in [H^1_{0,\Gamma_v}(\Omega)]^d$, the rigid body velocity variation $\bar{\mathbf{w}} \in R^d$, and the rigid body angular velocity variation $\bar{\boldsymbol{\theta}} \in R^d$, combining the solid domain with the fluid domain, employing the integration by parts, the divergence theorem, and Eqs. (4.10), (4.14), and (4.15), we can convert the governing equations or strong form in Eqs. (4.11)–(4.13) into the variational equations or the so-called weak form,

$$\int_{\Omega} [\rho_f w_i(\dot{v}_i - g_i) + w_{i,j}\sigma_{ij}] d\Omega + r[M(\dot{\bar{v}}_i - g_i)\bar{w}_i + (\mathbf{I}\dot{\bar{\boldsymbol{\omega}}} + \bar{\boldsymbol{\omega}} \times \mathbf{I}\bar{\boldsymbol{\omega}}) \cdot \bar{\boldsymbol{\theta}}] = 0,$$

with $r = 1 - \rho_f/\rho_s$.

Since the solid occupying Ω_s is a rigid body, Eq. (4.42) can be rewritten as Eq. (4.10). Of course, on the fluid–solid interface Γ_s, Eq. (4.10) is manifested as Eq. (4.42). The key treatment in the immersed boundary method, the extended immersed boundary method, and the Immersed Finite Element Method, is to introduce the delta function to synchronize the fluid motion with the solid motion within the immersed solid domain Ω_s, namely, Eq. (4.19).

The no-slip boundary condition on the interface Γ_s yields,

$$\mathbf{v}(\mathbf{x},t) = \bar{\mathbf{v}}(t) + \bar{\boldsymbol{\omega}}(t) \times (\mathbf{x} - \bar{\mathbf{x}}(t)), \quad \forall \mathbf{x} \in \Gamma_s, \tag{4.42}$$

where $\bar{\mathbf{x}}$, $\bar{\mathbf{v}}$, $\bar{\boldsymbol{\omega}}$, and $\mathbf{x}$ stand for the current position of the mass center, the velocity, and the angular velocity, and the position on the interface of the rigid body.

In fact, the constraint of Eq. (4.19) introduces the (distributed) Lagrangian multiplier as the equivalent body force. In the fictitious domain method, a similar (distributed) Lagrangian multiplier $\boldsymbol{\lambda}$ is introduced, along with the following traditional mixed formulation, we obtain, $\forall \mathbf{w} \in [H^1_{0,\Gamma_v}(\Omega)]^d$ and $\boldsymbol{\lambda} \in [H^1(\Omega_s)]^d$

$$\int_\Omega [\rho_f w_i(\dot{v}_i - g_i) + w_{i,j}\sigma_{ij}]d\Omega + r[M(\dot{\bar{v}}_i - g_i)\bar{w}_i + (\mathbf{I}\dot{\bar{\boldsymbol{\omega}}} + \bar{\boldsymbol{\omega}} \times \mathbf{I}\bar{\boldsymbol{\omega}}) \cdot \bar{\boldsymbol{\theta}}]$$
$$-(\boldsymbol{\lambda}, \mathbf{w} - \bar{\mathbf{w}} - \bar{\boldsymbol{\theta}} \times (\mathbf{x}^s - \bar{\mathbf{x}})) = 0,$$

and

$$(\boldsymbol{\mu}, \mathbf{v} - \bar{\mathbf{v}} - \bar{\boldsymbol{\omega}} \times (\mathbf{x}^s - \bar{\mathbf{x}})) = 0, \quad \forall \boldsymbol{\mu} \in [H^1(\Omega_s)]^d,$$

where the inner product is defined as

$$(\boldsymbol{\mu}, \boldsymbol{\lambda}) = \int_{\Omega_s} (\mu_i \lambda_i + l^2 \mu_{i,j} \lambda_{i,j}) d\Omega,$$

with a scaling factor l dependent of the characteristic length of Ω_s.

The in-house developed non-symmetric column solver is based on the symmetric column solver published in Bathe (1996). The source codes along with those of the implicit matrix-free Newton–Raphson iterative immersed continuum methods will be released with this book. It is interesting to point out that in some cases, for example, ALE-based FSI system in which a full Jacobian could be formulated, the direct solver does provide a comparable solution with much less computational time as shown in Fig. 4.16.

In order to better illustrate the coupling of the immersed continuum method with the traditional ALE FSI model, in Fig. 4.17 a set of preliminary results of a collapsible tube experiencing large displacements and large strains are also depicted. Note that there is no external fluid and around the non-slip fluid–structure interface or the collapsible boundary, the displacements are continuous. The established method will also include

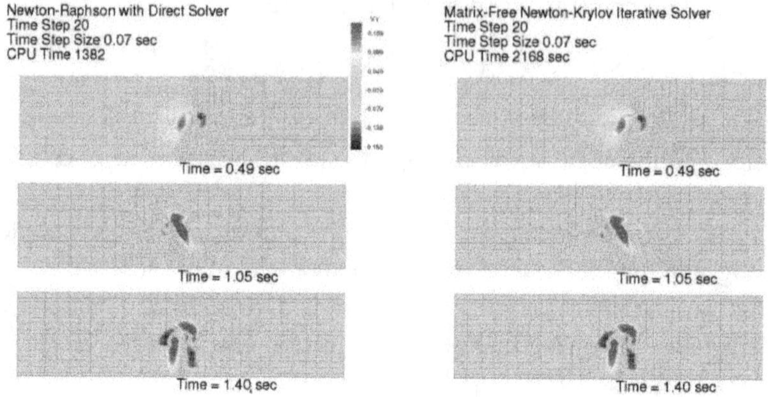

Fig. 4.16. Non-symmetric direct and matrix-free Newton–Krylov iterative solvers.

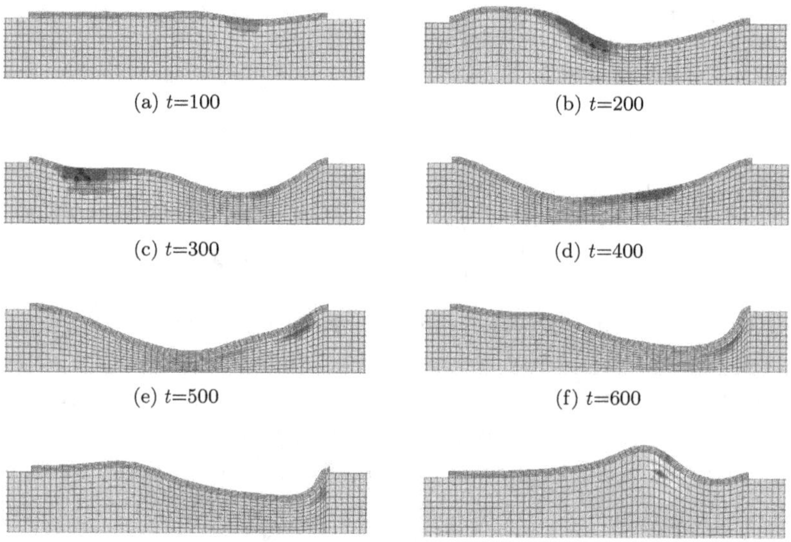

(a) t=100 (b) t=200

(c) t=300 (d) t=400

(e) t=500 (f) t=600

(g) t=700 (h) t=800

Fig. 4.17. Vertical velocity snapshots of the collapsible tube problem.

the scenario with deformable cells within collapsible tubes. Again, for a typical converged solution as shown in Figs. 4.4 and 4.5, the immersed solid displacement, the surrounding and background fluid velocity, the pressure unknowns for both fluid and solid domains are simultaneously converged. Moreover, because of the restrictions of the Inf-Sup conditions, the rate of convergence for the pressure is in general one order of magnitude lower than that of the displacement of the immersed solid and the velocity of the fluid.

4.2 Compact Schemes

In this section, we will try to adhere to the convention in this book that boldfaced lower case letters represent vectors and boldfaced upper case letters matrices. In addition, regular lower case letters will denote magnitudes of the corresponding vectors or just scalar; regular upper case letters will denote determinant of the corresponding matrices or just scalar. However, we would like to circumvent the tradition of a mixture of superscripts and subscripts in the expressions related to covariant and contravariant vectors as stipulated in Section 1.1.2.

Consider a position vector $\mathbf{x}$ in a Cartesian coordinate system. If the starting point of the vector is the origin, this position vector corresponds to a unique point (x, y, z). In vector notations, this vector is often represented as $\langle x, y, z\rangle$. In order to illustrate clearly with indicial notations, we often interchangeably adopt $\langle x_1, x_2, x_3\rangle$ to represent $\mathbf{x}$. Likewise, the corresponding position vector in a general or curvilinear coordinate system is denoted as $\boldsymbol{\xi}$ or $\langle \xi, \eta, \zeta\rangle$ or $\langle \xi_1, \xi_2, \xi_3\rangle$. In many engineering practices, it is essential to transform the governing equations from the Cartesian coordinate system to curvilinear coordinate systems (Kreiss, 1983). Many vector quantities involved in the transformation between these two systems are related to the so-called covariant and contravariant vectors. Traditionally, covariant and contravariant vectors are represented with a combination of superscripts and subscripts. In this section, we will try to avoid such complications and adhere to straightforward notation commonly accepted in applied mathematics and engineering fields.

4.2.1 *Curvilinear transformations*

According to multi-variable calculus and differential geometry, a smooth surface in three-dimensional space can be represented with a single equation $F(x, y, z) = 0$, relating all three coordinates, or a parametric representation with two independent parameters, namely,

$$x = x(\xi_1, \xi_2);$$

$$y = y(\xi_1, \xi_2);$$
$$z = z(\xi_1, \xi_2).$$

Likewise, a smooth curve in three-dimensional space can be represented with two equations $F_1(x,y,z) = 0$ and $F_2(x,y,z) = 0$, relating all three coordinates, or a parametric representation with one parameter, namely,

$$x = x(\xi_1);$$
$$y = y(\xi_1);$$
$$z = z(\xi_1).$$

According to the convention, tangent vectors with respect to general coordinate ξ_i are defined as covariant vectors $\mathbf{t}_i$, i.e.,

$$\mathbf{t}_i = \frac{\partial \mathbf{x}}{\partial \xi_i}. \tag{4.43}$$

Moreover, the cross-product of any two of them will point to a so-called normal or gradient direction, which is aligned with the so-called contravariant vector direction. To facilitate a concise representation, we introduce the cyclic order of indices (i, j, k) and (l, m, n).

For instance, we choose $\mathbf{t}_1$ and $\mathbf{t}_2$ as the tangential directions of a surface represented with two parameters ξ_1 and ξ_2. Then the surface normal direction is aligned with $\mathbf{t}_1 \times \mathbf{t}_2$. In fact, we often call this normal $\mathbf{n}_3$ and the surface ξ_3-surface.

In order to clarify the issue related to the magnitude of these vectors and the conventional metric treatments of covariant and contravariant vectors, let's take a look at some basic properties of 3×3 matrices. First of all, let's look at the transformations between the differential forms $d\mathbf{x}$ and $d\boldsymbol{\xi}$ and the gradient forms $\boldsymbol{\nabla}$ and $\bar{\boldsymbol{\nabla}}$. In general, we define a transformation matrix $\mathbf{T}$ as the Jacobian matrix, namely,

$$\frac{\partial(x_1, x_2, x_3)}{\partial(\xi_1, \xi_2, \xi_3)} = \begin{bmatrix} \frac{\partial x_1}{\partial \xi_1} & \frac{\partial x_1}{\partial \xi_2} & \frac{\partial x_1}{\partial \xi_3} \\ \frac{\partial x_2}{\partial \xi_1} & \frac{\partial x_2}{\partial \xi_2} & \frac{\partial x_2}{\partial \xi_3} \\ \frac{\partial x_3}{\partial \xi_1} & \frac{\partial x_3}{\partial \xi_2} & \frac{\partial x_3}{\partial \xi_3} \end{bmatrix}.$$

In differential transformation from $\boldsymbol{\xi}$ to $\mathbf{x}$, we have $d\mathbf{x} = \mathbf{T}d\boldsymbol{\xi}$, or

$$\begin{pmatrix} dx \\ dy \\ dz \end{pmatrix} = \begin{bmatrix} \frac{\partial x}{\partial \xi} & \frac{\partial x}{\partial \eta} & \frac{\partial x}{\partial \zeta} \\ \frac{\partial y}{\partial \xi} & \frac{\partial y}{\partial \eta} & \frac{\partial y}{\partial \zeta} \\ \frac{\partial z}{\partial \xi} & \frac{\partial z}{\partial \eta} & \frac{\partial z}{\partial \zeta} \end{bmatrix} \begin{pmatrix} d\xi \\ d\eta \\ d\zeta \end{pmatrix}. \tag{4.44}$$

Likewise, the gradient form of the transformation, $\mathbf{\nabla} = \mathbf{C}\bar{\mathbf{\nabla}}$, can also be expressed as

$$\begin{pmatrix} \frac{\partial}{\partial x} \\ \frac{\partial}{\partial y} \\ \frac{\partial}{\partial z} \end{pmatrix} = \begin{bmatrix} \frac{\partial \xi}{\partial x} & \frac{\partial \eta}{\partial x} & \frac{\partial \zeta}{\partial x} \\ \frac{\partial \xi}{\partial y} & \frac{\partial \eta}{\partial y} & \frac{\partial \zeta}{\partial y} \\ \frac{\partial \xi}{\partial z} & \frac{\partial \eta}{\partial z} & \frac{\partial \zeta}{\partial z} \end{bmatrix} \begin{pmatrix} \frac{\partial}{\partial \xi} \\ \frac{\partial}{\partial \eta} \\ \frac{\partial}{\partial \zeta} \end{pmatrix}. \tag{4.45}$$

At this point, we also would like to explicitly express the dual transformation from curvilinear coordinate systems to the Cartesian coordinate system. Namely, instead of transformation from (ξ, η, ζ) to (x, y, z), we switch the transformation order, namely, from (x, y, z) to (ξ, η, ζ). This illustration is not redundant or superfluous. In fact, by detailed depiction of those transformations, we will have a better understanding of the intriguing relationship between contravariant and covariant vectors.

In differential transformation from $d\mathbf{x}$ to $\boldsymbol{\xi}$, we have $d\boldsymbol{\xi} = \mathbf{T}^{-1}d\mathbf{x}$, or

$$\begin{pmatrix} d\xi \\ d\eta \\ d\zeta \end{pmatrix} = \begin{bmatrix} \frac{\partial \xi}{\partial x} & \frac{\partial \xi}{\partial y} & \frac{\partial \xi}{\partial z} \\ \frac{\partial \eta}{\partial x} & \frac{\partial \eta}{\partial y} & \frac{\partial \eta}{\partial z} \\ \frac{\partial \zeta}{\partial x} & \frac{\partial \zeta}{\partial y} & \frac{\partial \zeta}{\partial z} \end{bmatrix} \begin{pmatrix} dx \\ dy \\ dz \end{pmatrix}. \tag{4.46}$$

Likewise, the gradient form of the transformation, $\bar{\boldsymbol{\nabla}} = \mathbf{C}^{-1}\boldsymbol{\nabla}$, can also be expressed as

$$\begin{pmatrix} \frac{\partial}{\partial \xi} \\ \frac{\partial}{\partial \eta} \\ \frac{\partial}{\partial \zeta} \end{pmatrix} = \begin{bmatrix} \frac{\partial x}{\partial \xi} & \frac{\partial y}{\partial \xi} & \frac{\partial z}{\partial \xi} \\ \frac{\partial x}{\partial \eta} & \frac{\partial y}{\partial \eta} & \frac{\partial z}{\partial \eta} \\ \frac{\partial x}{\partial \zeta} & \frac{\partial y}{\partial \zeta} & \frac{\partial z}{\partial \zeta} \end{bmatrix} \begin{pmatrix} \frac{\partial}{\partial x} \\ \frac{\partial}{\partial y} \\ \frac{\partial}{\partial z} \end{pmatrix}. \tag{4.47}$$

Note that the ith column of the transformation matrix $\mathbf{T}$ or $\langle T_{1i}, T_{2i}, T_{3i} \rangle$ can be viewed as the tangential vector $\mathbf{t}_i$ and the Jacobian refers to the determinant of the matrix $\mathbf{T}$ or its transpose. Of course, it is very helpful to note that the determinant of the transpose of a square matrix remains the same. Furthermore, based on the definition of the cross product, for instance, $\mathbf{t}_2 \times \mathbf{t}_3$,

$$\mathbf{t}_2 \times \mathbf{t}_3 = \begin{vmatrix} \mathbf{i} & \mathbf{j} & \mathbf{k} \\ \frac{\partial x_1}{\partial \xi_2} & \frac{\partial x_2}{\partial \xi_2} & \frac{\partial x_3}{\partial \xi_2} \\ \frac{\partial x_1}{\partial \xi_3} & \frac{\partial x_2}{\partial \xi_3} & \frac{\partial x_3}{\partial \xi_3} \end{vmatrix}, \tag{4.48}$$

based on which it is not difficult to prove that $\mathbf{t}_1 \cdot (\mathbf{t}_2 \times \mathbf{t}_3) = \det(\mathbf{T}^T) = J$.

In fact, similar relationships hold for $\mathbf{t}_i \cdot (\mathbf{t}_j \times \mathbf{t}_k) = J$. Clearly, compare Eqs. (4.47) and (4.48), we have $\mathbf{C}^T = \mathbf{T}^{-1}$. Moreover, according to linear algebra definition, $J\mathbf{C}$ is in fact the adjoint matrix of $\mathbf{T}$, denoted as $\mathbf{B}$, namely, $T_{ik}B_{jk} = J\delta_{ij}$. We can easily identify that the definition of the i column of the matrix $\mathbf{B}$ also written as $\langle B_{1i}, B_{2i}, B_{3i} \rangle$ is exactly the same as the cross production $\mathbf{t}_j \times \mathbf{t}_k$. For simplicity of the presentation, we also denote the ith column of the matrix $\mathbf{B}$ as $\mathbf{b}_i$. If we connect the matrix inverse concept, the cross-product definition, and cyclic orders of indices (i, j, k) and (l, m, n), we are now ready to connect the traditional definitions of covariant and contravariant vectors.

The traditional definition of contravariant vectors represented as

$$\mathbf{c}_i = \frac{\partial \xi_i}{\partial \mathbf{x}},$$

which is in fact the ith column of the transformation matrix $\mathbf{C}$, or $\langle C_{1i}, C_{2,i}, C_{3,i}\rangle$, and more importantly we also have $\mathbf{c}_i = \mathbf{b}_i/J$.

Furthermore, $\mathbf{b}_i$ is aligned with the unit surface normal vector $\mathbf{n}_i$ of the ξ_i-surface with ξ_j and ξ_k as independent parameters. In order to represent the unit tangent and normal vectors, we need to denote the magnitudes of covariant vector $\mathbf{t}_i$ and contravariant vector $\mathbf{c}_i$. Again, if we compare with the transformation matrices in Eqs. (4.45) and (4.47), we can identify that covariant vectors correspond to the columns of the transformation matrix in Eq. (4.43), denoted as $\mathbf{T}$; whereas contravariant vectors correspond to the columns of the transformation matrix in Eq. (4.46), denoted as $\mathbf{C}$. In general, we have $\mathbf{b}_i = \mathbf{t}_j \times \mathbf{t}_k$, and the lth component of the vector $\mathbf{b}_i$ is written as

$$B_{li} = JC_{li} = \frac{\partial x_m}{\partial \xi_j}\frac{\partial x_n}{\partial \xi_k} - \frac{\partial x_n}{\partial \xi_j}\frac{\partial x_m}{\partial \xi_k}.$$

During the coordinate transformation, the differential arc length ds_i along the ξ_i line is defined as $ds_i = |\mathbf{t}_i| = \sqrt{\mathbf{t}_i \cdot \mathbf{t}_i}$. Therefore, the unit tangent vector along the ξ_i line is denoted as $\boldsymbol{\tau}_i = \mathbf{t}_i/ds_i$. Similarly, the differential area dS_i along the surface governed by two independent parameters ξ_j and ξ_k is defined as

$$dS_i = |\mathbf{t}_j \times \mathbf{t}_k| d\xi_j d\xi_k.$$

Define the magnitude of the vector $\mathbf{b}_i$ as

$$b_i = |\mathbf{b}_i| = \sqrt{\mathbf{b}_i \cdot \mathbf{b}_i}$$

and the magnitude of the vector $\mathbf{c}_i$ as

$$c_i = |\mathbf{c}_i| = \sqrt{\mathbf{c}_i \cdot \mathbf{c}_i}.$$

Obviously, we have $b_i = Jc_i$. Moreover, the unit surface normal vector of the ξ_i surface is denoted as

$$\mathbf{n}_i = \mathbf{b}_i/b_i = \mathbf{c}_i/c_i = J\mathbf{c}_i/b_i$$

and the differential area dS_i is expressed as $b_i d\xi_j d\xi_k$.

Finally, the differential volume dV is expressed as the volume of a differential parallel pipe, i.e.,

$$dV = \mathbf{t}_1 \cdot (\mathbf{t}_2 \times \mathbf{t}_3) d\xi_1 d\xi_2 d\xi_3 = J d\xi_1 d\xi_2 d\xi_3.$$

At this point, divergence, gradient, curl, and Laplacian operators are ready to be disseminated with respect to curvilinear coordinate systems. First of all, let's start from the divergence theorem

$$\int_V \nabla\cdot\mathbf{a}dV = \oint_S \mathbf{a}\cdot\mathbf{n}dS.$$

Note that for simplicity, we denote the general multi-dimensional integration with one integration sign. However, the true details of the integration is defined by V as a volume, S as a surface, or C as a curve (or contour). In curvilinear coordinate systems, we have

$$\begin{aligned}\int_V \nabla\cdot\mathbf{a}Jd\xi_i d\xi_j d\xi_k &= \sum_i\left(\int_{\xi_i^+}\mathbf{a}\cdot\mathbf{n}_i dS_i - \int_{\xi_i^-}\mathbf{a}\cdot\mathbf{n}_i dS_i\right)\\ &= \sum_i \frac{\partial}{\partial\xi_i}(\mathbf{a}\cdot\mathbf{b})d\xi_i d\xi_j d\xi_k,\end{aligned}$$

which holds for all selections of the vector $\mathbf{a}$.

Therefore, by choosing $\mathbf{a}=\mathbf{e}_j$, where $\mathbf{e}_j$ is the jth unit vector in the Cartesian coordinate system, we can easily verify that for $j=1,2,3$, we have

$$\sum_i \frac{\partial}{\partial\xi_i}B_{ji} = \sum_i \frac{\partial}{\partial\xi_i}(JC_{ji}) = 0.$$

This important geometrical conservation relation will be used to connect non-conservative forms and conservative forms. In general, for any vector $\mathbf{a}$, we have the conservative form of the divergence

$$\nabla\cdot\mathbf{a} = \frac{1}{J}\sum_i \frac{\partial}{\partial\xi_i}(J\mathbf{a}\cdot\mathbf{c}_i),$$

and the non-conservative form of the divergence

$$\nabla\cdot\mathbf{a} = \frac{1}{J}\sum_i \mathbf{c}_i\cdot\frac{\partial\mathbf{a}}{\partial\xi_i}.$$

Furthermore, consider $a_m = a_n = 0$ and $a_l = f$, we have the conservative form of the gradient

$$\frac{\partial f}{\partial x_l} = \frac{1}{J}\sum_i \frac{\partial}{\partial\xi_i}(B_{li}f),$$

and the non-conservative form of the divergence

$$\frac{\partial f}{\partial x_l} = \frac{1}{J}\sum_i B_{li}\frac{\partial f}{\partial \xi_i}.$$

It is also interesting to substitute $f = \xi_j$ and verify the following transformation relationship

$$\frac{\partial \xi_j}{\partial x_l} = C_{lj} = \frac{B_{lj}}{J}.$$

Now, utilize the indicial form of the curl operator and the cyclic order of (i, j, k) and (l, m, n), we can easily verify the conservative form of the curl

$$(\nabla \times \mathbf{a})_l = \frac{\partial a_n}{\partial x_m} - \frac{\partial a_m}{\partial x_n} = \frac{1}{J}\sum_i \frac{\partial}{\partial \xi_i}(B_{mi}a_n - B_{ni}a_m),$$

and the non-conservative form of the curl

$$(\nabla \times \mathbf{a})_l = \frac{1}{J}\sum_i \left(B_{mi}\frac{\partial a_n}{\partial \xi_i} - B_{ni}\frac{\partial a_m}{\partial \xi_i}\right).$$

Denote

$$d_{ik} = \mathbf{c}_i \cdot \mathbf{c}_k = \frac{1}{J^2}\sum_j B_{ji}B_{jk} = \frac{\mathbf{b}_i \cdot \mathbf{b}_j}{J^2},$$

and replace $\mathbf{a}$ with ∇f, we can easily have the conservative form of the Laplacian

$$\nabla^2 f = \frac{1}{J}\sum_i\sum_j\sum_k \frac{\partial}{\partial \xi_i}\left[B_{ji}\frac{1}{J}\frac{\partial}{\partial \xi_k}(B_{jk}f)\right],$$

and the non-conservative form of the Laplacian

$$\nabla^2 f = \sum_i\sum_k d_{ik}\frac{\partial^2 f}{\partial \xi_i \partial \xi_k} + \sum_k \left(\nabla^2 \xi_k \frac{\partial f}{\partial \xi_k}\right),$$

with

$$\nabla^2 \xi_k = \frac{1}{J}\sum_i\sum_j B_{ji}\frac{\partial}{\partial \xi_i}\left(\frac{1}{J}B_{jk}\right).$$

Hence, the geometrically conservative form of the normal derivative on the ξ_i surface is expressed as

$$\left.\frac{\partial f}{\partial n}\right|_{\xi_i} = \mathbf{n}_i \cdot \nabla f = \frac{1}{Jb_i}\sum_k\sum_l \frac{\partial}{\partial \xi_l}(B_{kl}f)B_{ki},$$

and the corresponding non-conservative form

$$\left.\frac{\partial f}{\partial n}\right|_{\xi_i} = \frac{J}{b_i}\sum_l d_{li}\frac{\partial f}{\partial \xi_l}.$$

Similarly, the tangential derivative on the ξ_i line in geometrically conservative form is written as

$$\left.\frac{\partial f}{\partial \tau}\right|_{\xi_i} = \boldsymbol{\tau}_i \cdot \nabla f = \frac{1}{Jt_i}\sum_k\sum_l \frac{\partial}{\partial \xi_l}(B_{kl}f)T_{ki},$$

and the corresponding non-conservative form

$$\left.\frac{\partial f}{\partial n}\right|_{\xi_i} = \frac{1}{t_i}\frac{\partial f}{\partial \xi_i}.$$

Likewise, integrations can also be set up following the conservative and non-conservative forms. In computational fluid dynamics, it is often required to have a moving mesh capability. In this case, we need to expand the coordinate transformation between (x,y,z) and (ξ,η,ζ) to the transformation between (t,x,y,z) and (τ,ξ,η,ζ). In general, we have

$$\begin{aligned} t &= \tau; \\ x &= x(\tau,\xi,\eta,\zeta); \\ y &= y(\tau,\xi,\eta,\zeta); \\ z &= z(\tau,\xi,\eta,\zeta). \end{aligned}$$

The corresponding transformation in gradient form can be written as

$$\begin{pmatrix} \dfrac{\partial}{\partial t} \\ \dfrac{\partial}{\partial x} \\ \dfrac{\partial}{\partial y} \\ \dfrac{\partial}{\partial z} \end{pmatrix} = \begin{bmatrix} 1 & \dfrac{\partial \xi}{\partial t} & \dfrac{\partial \eta}{\partial t} & \dfrac{\partial \zeta}{\partial t} \\ 0 & \dfrac{\partial \xi}{\partial x} & \dfrac{\partial \eta}{\partial x} & \dfrac{\partial \zeta}{\partial x} \\ 0 & \dfrac{\partial \xi}{\partial y} & \dfrac{\partial \eta}{\partial y} & \dfrac{\partial \zeta}{\partial y} \\ 0 & \dfrac{\partial \xi}{\partial z} & \dfrac{\partial \eta}{\partial z} & \dfrac{\partial \zeta}{\partial z} \end{bmatrix} \begin{pmatrix} \dfrac{\partial}{\partial \tau} \\ \dfrac{\partial}{\partial \xi} \\ \dfrac{\partial}{\partial \eta} \\ \dfrac{\partial}{\partial \zeta} \end{pmatrix}; \tag{4.49}$$

and the corresponding transformation in differential form can be written as

$$\begin{pmatrix} dt \\ dx \\ dy \\ dz \end{pmatrix} = \begin{bmatrix} 1 & 0 & 0 & 0 \\ \frac{\partial x}{\partial \tau} & \frac{\partial x}{\partial \xi} & \frac{\partial x}{\partial \eta} & \frac{\partial x}{\partial \zeta} \\ \frac{\partial y}{\partial \tau} & \frac{\partial y}{\partial \xi} & \frac{\partial y}{\partial \eta} & \frac{\partial y}{\partial \zeta} \\ \frac{\partial z}{\partial \tau} & \frac{\partial z}{\partial \xi} & \frac{\partial z}{\partial \eta} & \frac{\partial z}{\partial \zeta} \end{bmatrix} \begin{pmatrix} d\tau \\ d\xi \\ d\eta \\ d\zeta \end{pmatrix}. \tag{4.50}$$

Consequently, it is straightforward to show that

$$\frac{\partial J}{\partial t} + \frac{\partial}{\partial \xi_j}(JC_{ij}) = 0, \tag{4.51}$$

or

$$\frac{\partial J}{\partial t} + \frac{\partial}{\partial \xi_j}(B_{ij}) = 0. \tag{4.52}$$

4.2.2 *Conservation forms for compressible flow*

The governing equations are derived from the mass conservation, the linear momentum conservation, and the energy conservation laws. In addition, one state equation is introduced to related pressure p, specific volume v, and temperature T. The specific internal energy u and specific enthalpy h are also expressed as a function of temperature, namely, $du = c_v dT$ and $dh = c_p dT$, where c_v and c_p are the specific heats for constant volume and constant pressure, respectively. Note that in general, for liquids and solids, we normally have $du = c_v dT \simeq c_p dT$. However, for gas, due to the compressibility, the adiabatic exponent γ is defined as the ratio between two specific heats, i.e., c_p/c_v.

Moreover, for ideal gases, in which the inter-molecule potentials are not significant except during the collisions, we commonly used equation of state as stipulated below

$$p = \rho RT, \tag{4.53}$$

where the density ρ is the reciprocal of the specific volume v, and the gas constant R often expressed as $R_u/\bar{M}$, with the universal gas constant R_u and the molar mass $\bar{M}$.

Define the total mass m and the total volume V, we have the specific volume $v = V/m$ and the mole number defined as $n = m/\bar{M}$, essentially a number unit equal to 6.022×10^{23} units of any substance such as atoms or molecules or ions, a so-called Avogadro' number, named after Lorenzo Romano Amedeo Carlo Avogadro (1776–1856). Equation (4.53) can be expressed in the familiar form in physics,

$$pV = nR_uT. \tag{4.54}$$

Furthermore, it is not difficult to identify that the gas constant R can be also expressed as $c_p - c_v$, and as a consequence, the specific internal energy u and the enthalpy h can also be written as $\frac{RT}{\gamma-1}$ and $\frac{\gamma RT}{\gamma-1}$, respectively. Traditionally, tensor notations are introduced for these governing equations. In this book, we choose the familiar indicial notations. Define the density or specific volume as ρ, the velocity at a specific location as v_i, for $i = 1,2,3$, representing all three directions.

Select a typical differential block (dx, dy, dz), implementing the conservation laws, the continuity equation is expressed as

$$\frac{\partial \rho}{\partial t} + \frac{\partial}{\partial x_i}(\rho v_i) = 0.$$

Likewise, the linear momentum equation in x_i direction can also be expressed as

$$\frac{\partial(\rho v_i)}{\partial t} + \frac{\partial}{\partial x_j}(\rho v_i v_j + p) = \frac{\partial \tau_{ij}}{\partial x_j} + f_i,$$

where $\mathbf{f}$ stands for the body force vector.

In the Newtonian fluid, the stress tensor $\boldsymbol{\sigma}$ minus the mechanical pressure term is denoted as $\boldsymbol{\tau}$ and expressed as

$$\tau_{ij} = -\lambda \frac{\partial v_k}{\partial x_k}\delta_{ij} + \mu\left(\frac{\partial v_i}{\partial x_j} + \frac{\partial v_j}{\partial x_i}\right),$$

with the Lame constants or parameters μ and λ, named after Gabriel Lame (1795–1870), more specifically with λ defined as $2\mu/3$.

To introduce the stress tensor properly, we must recognize two different pressure definitions. In mechanics, the pressure p refers to the average of all normal stress, i.e., $p = -\tau_{ii}/3$. In this section, let's use p_m to represent such a mechanical pressure. In thermodynamics, the pressure p

often refers to the average linear momentum change in statistical thermodynamical equilibrium of a fluid, liquid or gas. The following experimental relationship often relates these two pressures,

$$p_m = p - \alpha \frac{\partial v_i}{\partial x_i}, \tag{4.55}$$

where α is an empirical coefficient called expansion coefficient, or bulk viscosity.

According to the second law of thermodynamics, the constant α must be positive. This suggests that the thermodynamic pressure tends to be higher than the mechanical pressure when the volume is expanding namely the mechanical pressure is decreasing and vice versa. In other words, the thermodynamic pressure always tends to lag behind the mechanical pressure. For diluted monatomic gasses, the constant α is equal to 0. However, for water α is about three times larger than μ, and for complex liquids like benzene, α can be over 100 times larger. In general, the effects on the flow of the term $\nabla \cdot \mathbf{v}$ are small. However, inside the shock wave over a molecular scale or over very short time in high-intensity ultrasonic situations, the expansion viscosity becomes significant.

Combining with Eq. (4.55), we have

$$\sigma_{ij} = -\left[p + (\lambda - \alpha)\frac{\partial v_k}{\partial x_k}\right]\delta_{ij} + \mu\left(\frac{\partial v_i}{\partial x_j} + \frac{\partial v_j}{\partial x_i}\right).$$

For most flows including most compressible flows, we can approximate the stress tensor by

$$\sigma_{ij} = -p\delta_{ij} + \mu\left(\frac{\partial v_i}{\partial x_j} + \frac{\partial v_j}{\partial x_i}\right),$$

which is the conventional form for Newtonian fluids.

Furthermore, define the total specific mechanical energy as e, we have

$$e = u + \frac{1}{2}v_i v_i.$$

Substitute into Eq. (4.53), the pressure p can also be written as

$$p = \rho(\gamma - 1)u = \rho(\gamma - 1)\left(e - \frac{1}{2}v_i v_i\right).$$

In addition, for isentropic (quasistatic, adiabatic, and reversible) gas, according to the first law of thermodynamics, namely,

$$de = -pd\left(\frac{1}{\rho}\right) + dq,$$

where the reciprocal of the density ρ is in fact the specific volume v, and for isentropic gas we have $dq = 0$.

Therefore, we have

$$c_v dT = -pd\left(\frac{1}{\rho}\right),$$

and combine with the ideal gas law in Eq. (4.53), we also have

$$\frac{c_v}{R}d\left(\frac{p}{\rho}\right) = -pd\left(\frac{1}{\rho}\right). \tag{4.56}$$

Considering the gas constant R is also written as $c_p - c_v = (\gamma - 1)c_v$, we can rewrite Eq. (4.56) as

$$\gamma\frac{d\rho}{\rho} = \frac{dp}{p}.$$

Consequently, for isentropic gas, the pressure p is also governed by the following polytropic relation between the pressure p and the density ρ

$$p = \rho^\gamma C,$$

where C is a constant.

Hence, the speed of sound c defined as

$$c^2 = \frac{dp}{d\rho} = \gamma\frac{p}{\rho}.$$

Moreover, the specific enthalpy form in engineering fields can be expressed as

$$h + \frac{1}{2}v_i v_i = u + \frac{1}{2}v_i v_i + \frac{p}{\rho} = \frac{c^2}{(\gamma - 1)} + \frac{1}{2}v_i v_i. \tag{4.57}$$

Note that for cases where the altitude or elevation change is significant, for example, the hydroelectrical system, the gravitational potential is also of importance in the total specific energy representation, and Eq. (4.57) can be simply expressed as

$$h+\frac{1}{2}v_iv_i+gz=u+\frac{1}{2}v_iv_i+gz+\frac{p}{\rho}=\frac{c^2}{(\gamma-1)}+gz+\frac{1}{2}v_iv_i, \tag{4.58}$$

in which the total specific mechanical energy e can be expressed as

$$e=u+\frac{1}{2}v_iv_i+gz, \tag{4.59}$$

where z represents the elevation of the system and g is the gravitational constant.

Finally, the conservation of energy with the introduction of the heat conduction with the isotropic conductivity k is written as

$$\frac{\partial(\rho e)}{\partial t}+\frac{\partial}{\partial x_i}[(\rho e+p)v_i]-\frac{\partial}{\partial x_i}\left(k\frac{\partial T}{\partial x_i}\right)=\frac{\partial}{\partial x_i}(\tau_{ij}v_j)+f_iv_i+q,$$

where q stands for the heat source per unit volume and time and is different from the state variables or the primary unknown vector is defined as a column vector expressed as

$$\mathbf{q}=\langle\rho,\rho v_j,\rho e\rangle,$$

with $j=1,2,3$.

Furthermore, the governing equations are expressed in a unified conservative form

$$\frac{\partial\mathbf{q}}{\partial t}+\frac{\partial\mathbf{F}_i}{\partial x_i}=\frac{\partial\mathbf{P}_i}{\partial x_i}, \tag{4.60}$$

where the flux vectors are defined as a combination of two vectors

$$\mathbf{F}_i=\langle\rho v_i,\rho v_iv_j,(\rho e+p)v_i\rangle+\langle 0,p\delta_{ij},0\rangle$$

for $i=1,2,3$ and $j=1,2,3$; whereas the ith power vectors $\mathbf{P}_i$ for $i=1,2,3$ are also defined as

$$\mathbf{P}_i=\left\langle 0,\tau_{ij},\tau_{ik}v_k-k\frac{\partial T}{\partial x_i}\right\rangle.$$

Using the geometrical conservation identity in Eq. (4.51) along with the transformation, the conservative form of the governing equations in the Cartesian coordinate system can be written as

$$\frac{\partial(J\mathbf{q})}{\partial\tau}+\frac{\partial(B_{ji}\mathbf{F}_j)}{\partial\xi_i}=\frac{\partial(B_{ji}\mathbf{P}_j)}{\partial x_i}. \tag{4.61}$$

In engineering practice, we often use the non-dimensional forms of these governing equations. Define L along with u_∞, ρ_∞, T_∞, p_∞, k_∞, and

μ_∞ as the characteristic length (for example, the chord length of an airfoil) along with the far field velocity magnitude, density, temperature, pressure, thermal conductivity, and dynamic viscosity. A typical procedure of non-dimensionalization can be expressed as

$$\begin{aligned}
\bar{x} &= \frac{x}{L}, \quad \bar{y} = \frac{y}{L}, \quad \bar{z} = \frac{z}{L}, \quad \bar{t} = \frac{tu_\infty}{L}; \\
\bar{v}_i &= \frac{v_i}{u_\infty}, \quad \bar{a} = \frac{a}{u_\infty}, \quad \bar{\rho} = \frac{\rho}{\rho_\infty}, \quad \bar{T} = \frac{T}{T_\infty}; \\
\bar{p} &= \frac{p}{\rho_\infty u_\infty^2}, \quad \bar{e} = \frac{e}{u_\infty^2}, \quad \bar{\mu} = \frac{\mu}{\mu_\infty}, \quad \bar{k} = \frac{k}{k_\infty}.
\end{aligned} \tag{4.62}$$

In Eq. (4.62), the bar indicates non-dimensional quantities. However, once we have converted all variables, both dependent and independent, to non-dimensional forms. According to Sutherland's law, named after William Sutherland (1859–1911), the dynamic viscosity μ for an ideal gas is given as

$$\mu(T) = \mu_\infty \left(\frac{T}{T_\infty}\right)^{3/2} \frac{T_\infty + C}{T + C}, \tag{4.63}$$

where the temperature constant for air C is 110 K.

Consequently, with $\bar{C} = C/T_\infty$, we have the following non-dimensional form of Eq. (4.63),

$$\bar{\mu}(\bar{T}) = \bar{T}^{3/2} \frac{1 + \bar{C}}{\bar{T} + \bar{C}}. \tag{4.64}$$

Revert to regular notations for simplicity, we have the non-dimensional forms of the following two conservation equations

$$\frac{\partial(\rho v_i)}{\partial t} + \frac{\partial}{\partial x_j}(\rho v_i v_j + p) = \frac{1}{Re}\left(\frac{\partial \tau_{ij}}{\partial x_j} + f_i\right),$$

$$\frac{\partial(\rho e)}{\partial t} + \frac{\partial}{\partial x_i}[(\rho e + p)v_i] - \frac{1}{Sc}\frac{\partial}{\partial x_i}(k\frac{\partial T}{\partial x_i}) = \frac{1}{Re}\left[\frac{\partial}{\partial x_i}(\tau_{ij}v_j) + f_i v_i + q\right],$$

where the scaling factor Sc is defined as $(\gamma - 1)RePrM_\infty^2$ with the Reynolds number $Re = u_\infty L/\mu_\infty$, named after Osborne Reynolds (1842–1912), the Prandtl number $Pr = c_p\mu_\infty/k_\infty$, named after Ludwig Prandtl (1875–1953), the far field Mach number $M_\infty = u_\infty/\sqrt{\gamma RT_\infty}$, named after Ernst Waldfried Josef Wenzel Mach (1838–1916).

Consequently, we should rewrite the power vector as

$$\mathbf{P}_i = \left\langle 0, \tau_{ij}, \tau_{ik}v_k - \frac{1}{Sc}k\frac{\partial T}{\partial x_i} \right\rangle,$$

and the final conservative form of the governing equation in the general coordinate system is written as

$$\frac{\partial(J\mathbf{q})}{\partial \tau} + \frac{\partial(B_{ji}\mathbf{F}_j)}{\partial \xi_i} = \frac{1}{Re}\frac{\partial(B_{ji}\mathbf{P}_j)}{\partial x_i}. \tag{4.65}$$

4.2.3 *Multigrid preconditioner*

In this section, we will also explore the use of V-cycle multigrid solver for three-dimensional FSI problems. In multigrid methods, the sweep or iteration is based on the conventional iterative methods, the coarsening of the fine grid, often by a factor of 2 in all directions, transforms the low wave number error into a high wave number error at all subsequent grid levels (McCormick, 1987; Axelsson, 1996). There are a number of issues that have effects on the performance of multigrid accelerations. The key issue is the grid coarsening. In general, geometric coarsening is much more effective than algebraic coarsening and carries less than two times the work of a regular fine grid generation (Wang, 2008).

At each grid level, we perform m (often 1 or 2) Gauss–Seidel or Successive Overrelaxation sweeps. The coarsening starts with a factor of 2 in all directions until the first dimension is exhausted. In the implementation, we store the right-hand side source or boundary terms in the working array **f**.

If the initial guess $\Delta\mathbf{\Theta}^{k,0}$ does not produce a good estimate within a sufficiently small Krylov subspace $\mathbf{K}^n$. $\Delta\mathbf{\Theta}^{k,n}$ will be introduced as an updated initial guess and the GMRES iteration procedure will continue until a solution with the desired accuracy is obtained. Moreover, different preconditioners for implicit matrix-free Newton–Krylov solution scheme can be introduced.

In the implementation, we store the right-hand side source or boundary terms in the working array **f**. In order to use various iteration or relaxation schemes, we use the array **v** to store the iteration results **u** based on $\mathbf{Au} = \mathbf{f}$, and the buffer array **w** to calculate the error $\mathbf{r} = \mathbf{f} - \mathbf{Au}$ and to store the intermediate result during the iteration process. Convention dictates that we store contiguously the solutions and the right-hand side vectors on the nested grids in single arrays. The data structures of **f**, **w**, and **v**, as depicted in Wang (2008), are identical with the cascade of hierarchical meshes stacked on top of each other. The buffer array **w** is particularly needed for iterative procedures like the Successive Over Relaxation (SOR) method.

If the computational domain is a rectangular box, say

$$(2^{n_1}+1)\times(2^{n_2}+1)\times(2^{n_3}+1),$$

with $n_1 \leq n_2 \leq n_3$, we coarsen the mesh uniformly in three dimensions from the level 1 to n_1.

Then, we solve exactly the n_1 level mesh

$$(2^1+1)\times(2^{n_2-n_1+1}+1)\times(2^{n_3-n_1+1}+1),$$

using either the tri-diagonal solver if $n_1 = n_2$, or the nonsymmetric column solver if $n_1 < n_2$.

With respect to the data structures, the three-dimensional sweep at the level k, with $1 \leq k \leq n_1$, the size of the working array is

$$(2^{n_1-k+1}+1)(2^{n_2-k+1}+1)(2^{n_3-k+1}+1),$$

which is stored from the position n onward.

Consider a typical V-cycle multigrid scheme as depicted in Wang (2008), and assign ν_1 and ν_2 as the iteration numbers of the down and up cycle sweeps. The computational flow can then be illustrated as $\mathbf{v}^h \leftarrow MV^h(\mathbf{v}^h, \mathbf{f}^h)$:

(1) Relax ν_1 times on $\mathbf{K}^h\mathbf{b}^h = \mathbf{f}^h$ with given initial guess $\mathbf{v}^h = \mathbf{0}$;
(2) If Ω^h coarsest grid, then go to 4), else $\mathbf{f}^{2h} \leftarrow \mathbf{I}_h^{2h}(\mathbf{f}^h - \mathbf{K}^h\mathbf{v}^h)$, $\mathbf{v}^{2h} \leftarrow \mathbf{0}$, $\mathbf{v}^{2h} \leftarrow MV^{2h}(\mathbf{v}^{2h}, \mathbf{f}^{2h})$;
(3) Convert $\mathbf{v}^h \leftarrow \mathbf{v}^h + \mathbf{I}_{2h}^h\mathbf{v}^{2h}$;
(4) Relax ν_2 times on $\mathbf{K}^h\mathbf{b}^h = \mathbf{f}^h$ with an initial guess $\mathbf{v}^h$.

For simplicity of the notation, we introduce $\bar{i} = 2i$, $\bar{j} = 2j$ and $\bar{k} = 2k$. Results from a three-dimensional implementation are presented in Wang (2008). There are several popular iterative methods for the solution of linear systems of equations, among which Jacobi, weighted or damped

Jacobi, Gauss-Seidel, and Successive Overrelaxation (SOR) methods are prevalently used in engineering practice. An overview of various iterative solution techniques and their variations are available in Axelsson (1996). In essence, all iterative solution methods for the solution of an elliptic pressure equation suffer from poor convergence rates for low spatial wave numbers. Since in engineering practice, subsonic pressure fields are almost always smooth, there is a need to introduce the multigrid solution technique which has been proven to be effective in eliminating those low wave number errors (McCormick, 1987) and significantly speed up the solution process.

Although in multigrid methods, the sweep or iteration is based on the conventional iterative methods, the coarsening of the fine grid, often by a factor of 2 in all directions, transforms the low wave number error into a high wave number error at all subsequent grid levels. Another benefit of the multigrid method is that the solution time is of order $N\log(N)$ where N represents the number of grid unknowns. On the other hand, the Gauss-Seidel solution time is of order N^2. Since most iterative methods are effective in reducing high wave number errors, the coarsening of the fine grid results in a tremendous speed up for the solver.

There are a number of issues that have effects on the performance of multigrid accelerations. The key issue is the grid coarsening. In general, geometric coarsening is much more effective than algebraic coarsening and only carries less than two times the work of a regular fine grid generation. Nevertheless, the geometric multigrid solver is not a strictly black box solver and special treatment of the mesh coarsening and the generation of the coefficients must be considered. This is of particular importance if the computational domain has different mesh densities in three directions.

In this section, we employ a V-cycle multigrid method with various mesh coarsening schemes. At each grid level, we perform m (often 1 or 2) Gauss–Seidel or Successive Overrelaxation sweeps. The coarsening starts with a factor of 2 in all directions until the first dimension is exhausted. Subsequently, we adopt three different coarsening procedures for the cubic and rectangular boxes. A typical 19-point stencil based on moving mesh discretizations is used for the treatment of the boundary points. However, if needed, different discretization schemes can be directly employed by simply replacing the iteration sweep procedures.

Let's look at the finite difference operator with 19-point stencil for the following governing equation,

$$\overline{\nabla} \cdot \nabla p = f, \tag{4.66}$$

with

$$\nabla = \begin{pmatrix} \dfrac{\partial}{\partial x_1} \\ \dfrac{\partial}{\partial x_2} \\ \dfrac{\partial}{\partial x_3} \end{pmatrix} \quad \text{and} \quad \overline{\nabla} = \begin{pmatrix} \dfrac{\partial}{\partial \overline{x}_1} \\ \dfrac{\partial}{\partial \overline{x}_2} \\ \dfrac{\partial}{\partial \overline{x}_3} \end{pmatrix},$$

with

$$\frac{\partial}{\partial x_i} = \frac{\partial}{\partial \xi}\frac{\partial \xi}{\partial x_i} + \frac{\partial}{\partial \eta}\frac{\partial \eta}{\partial x_i} + \frac{\partial}{\partial \zeta}\frac{\partial \zeta}{\partial x_i} = \xi_{x_i}\frac{\partial}{\partial \xi} + \eta_{x_i}\frac{\partial}{\partial \eta} + \zeta_{x_i}\frac{\partial}{\partial \zeta},$$

$$\frac{\partial}{\partial \overline{x}_i} = \frac{\partial}{\partial \xi}\frac{\partial \xi}{\partial \overline{x}_i} + \frac{\partial}{\partial \eta}\frac{\partial \eta}{\partial \overline{x}_i} + \frac{\partial}{\partial \zeta}\frac{\partial \zeta}{\partial \overline{x}_i} = \xi_{\overline{x}_i}\frac{\partial}{\partial \xi} + \eta_{\overline{x}_i}\frac{\partial}{\partial \eta} + \zeta_{\overline{x}_i}\frac{\partial}{\partial \zeta}.$$

For a typical grid point (i, j, k), the pressure unknown p is denoted as $P_{i,j,k}$, and the left-hand side of Eq. (4.66) can be discretized as

$$\left(\frac{\partial}{\partial \overline{x}_m}\frac{\partial}{\partial x_m}p\right)_{i,j,k} = (\xi_{\overline{x}_m})_{i,j,k}A_{\xi m} + (\eta_{\overline{x}_m})_{i,j,k}A_{\eta m} + (\zeta_{\overline{x}_m})_{i,j,k}A_{\zeta m}, \tag{4.67}$$

with

$$\begin{aligned} A_{\xi m} = {} & \frac{(\xi_{x_m})_{i+1/2,j,k}}{\Delta\xi^2}(P_{i+1,j,k} - P_{i,j,k}) - \frac{(\xi_{x_m})_{i-1/2,j,k}}{\Delta\xi^2}(P_{i,j,k} - P_{i-1,j,k}) \\ & + \frac{(\eta_{x_m})_{i+1,j,k}}{4\Delta\xi\Delta\eta}(P_{i+1,j+1,k} - P_{i+1,j-1,k}) \\ & - \frac{(\eta_{x_m})_{i-1,j,k}}{4\Delta\xi\Delta\eta}(P_{i-1,j+1,k} - P_{i-1,j-1,k}) \\ & + \frac{(\zeta_{x_m})_{i+1,j,k}}{4\Delta\xi\Delta\zeta}(P_{i+1,j,k+1} - P_{i+1,j,k-1}) \end{aligned}$$

$$
\begin{aligned}
&-\frac{(\zeta_{x_m})_{i-1,j,k}}{4\Delta\xi\Delta\zeta}(P_{i-1,j,k+1}-P_{i-1,j,k-1}),\\
A_{\eta m} = &\frac{(\xi_{x_m})_{i,j+1,k}}{4\Delta\xi\Delta\eta}(P_{i+1,j+1,k}-P_{i-1,j+1,k})\\
&-\frac{(\xi_{x_m})_{i,j-1,k}}{4\Delta\xi\Delta\eta}(P_{i+1,j-1,k}-P_{i-1,j-1,k})\\
&+\frac{(\eta_{x_m})_{i,j+1/2,k}}{\Delta\eta^2}(P_{i,j+1,k}-P_{i,j,k})\\
&-\frac{(\eta_{x_m})_{i,j-1/2,k}}{\Delta\eta^2}(P_{i,j,k}-P_{i,j-1,k})\\
&+\frac{(\zeta_{x_m})_{i,j+1,k}}{4\Delta\eta\Delta\zeta}(P_{i,j+1,k+1}-P_{i,j+1,k-1})\\
&-\frac{(\zeta_{x_m})_{i,j-1,k}}{4\Delta\eta\Delta\zeta}(P_{i,j-1,k+1}-P_{i,j-1,k+1}),\\
A_{\zeta m} = &\frac{(\xi_{x_m})_{i,j,k+1}}{4\Delta\xi\Delta\zeta}(P_{i+1,j,k+1}-P_{i-1,j,k+1})\\
&-\frac{(\xi_{x_m})_{i,j,k-1}}{4\Delta\xi\Delta\zeta}(P_{i+1,j,k-1}-P_{i-1,j,k-1})\\
&+\frac{(\eta_{x_m})_{i,j,k+1}}{4\Delta\eta\Delta\zeta}(P_{i,j+1,k+1}-P_{i,j-1,k+1})\\
&-\frac{(\eta_{x_m})_{i,j,k-1}}{4\Delta\eta\Delta\zeta}(P_{i,j+1,k-1}-P_{i,j-1,k-1})\\
&+\frac{(\zeta_{x_m})_{i,j,k+1/2}}{\Delta\zeta^2}(P_{i,j,k+1}-P_{i,j,k})\\
&-\frac{(\zeta_{x_m})_{i,j,k-1/2}}{\Delta\zeta^2}(P_{i,j,k}-P_{i,j,k-1}).
\end{aligned}
$$

Thus, derived from Eq. (4.67), the coefficients of the 19-stencil can be expressed as

$$
P_{i,j,k}: -\frac{(\xi_{\bar{x}_m})_{i,j,k}}{\Delta\xi^2}\left[(\xi_{x_m})_{i+1/2,j,k}+(\xi_{x_m})_{i-1/2,j,k}\right]
$$

$$-\frac{(\eta_{\bar{x}_m})_{i,j,k}}{\Delta\eta^2}\left[(\eta_{x_m})_{i,j+1/2,k}+(\eta_{x_m})_{i,j-1/2,k}\right]$$

$$-\frac{(\zeta_{\bar{x}_m})_{i,j,k}}{\Delta\zeta^2}\left[(\zeta_{x_m})_{i,j,k+1/2}+(\zeta_{x_m})_{i,j,k-1/2}\right],$$

$$P_{i\pm1,j,k}:\frac{(\xi_{\bar{x}_m})_{i,j,k}}{\Delta\xi^2}(\xi_{x_m})_{i\pm1/2,j,k},$$

$$P_{i,j\pm1,k}:\frac{(\eta_{\bar{x}_m})_{i,j,k}}{\Delta\eta^2}(\eta_{x_m})_{i,j\pm1/2,k},$$

$$P_{i,j,k\pm1}:\frac{(\zeta_{\bar{x}_m})_{i,j,k}}{\Delta\zeta^2}(\zeta_{x_m})_{i,j,k\pm1/2},$$

$$P_{i\pm1,j+1,k}:\left[\pm(\xi_{\bar{x}_m})_{i,j,k}(\eta_{x_m})_{i\pm1,j,k}\pm(\eta_{\bar{x}_m})_{i,j,k}(\xi_{x_m})_{i,j+1,k}\right]\frac{1}{4\Delta\xi\Delta\eta},$$

$$P_{i\pm1,j-1,k}:\left[\mp(\xi_{\bar{x}_m})_{i,j,k}(\eta_{x_m})_{i\pm1,j,k}\mp(\eta_{\bar{x}_m})_{i,j,k}(\xi_{x_m})_{i,j-1,k}\right]\frac{1}{4\Delta\xi\Delta\eta};$$

$$P_{i+1,j,k\pm1}:\left[\pm(\zeta_{\bar{x}_m})_{i,j,k}(\xi_{x_m})_{i,j,k+1}\pm(\xi_{\bar{x}_m})_{i,j,k}(\zeta_{x_m})_{i+1,j,k}\right]\frac{1}{4\Delta\xi\Delta\zeta},$$

$$P_{i-1,j,k\pm1}:\left[\mp(\zeta_{\bar{x}_m})_{i,j,k}(\xi_{x_m})_{i,j,k+1}\mp(\xi_{\bar{x}_m})_{i,j,k}(\zeta_{x_m})_{i-1,j,k}\right]\frac{1}{4\Delta\xi\Delta\zeta},$$

$$P_{i,j\pm1,k+1}:\left[\pm(\eta_{\bar{x}_m})_{i,j,k}(\zeta_{x_m})_{i,j\pm1,k}\pm(\eta_{x_m})_{i,j,k+1}(\zeta_{\bar{x}_m})_{i,j,k}\right]\frac{1}{4\Delta\eta\Delta\zeta},$$

$$P_{i,j\pm1,k-1}:\left[\mp(\eta_{\bar{x}_m})_{i,j,k}(\zeta_{x_m})_{i,j+1,k}\mp(\zeta_{x_m})_{i,j,k-1}(\zeta_{\bar{x}_m})_{i,j,k}\right]\frac{1}{4\Delta\eta\Delta\zeta}.$$

In Table 4.1, we have the general three-dimensional (3D) mapping of the discretized coefficients of our Laplacian operator. Naturally, C_i with $1 \le i \le 19$ must follow the definition of the 19-point stencil. Moreover, to facilitate the use of multigrid iteration, the 3D parametric space, is represented with a grid $N_1(2^{n_1})$, $N_2(2^{n_2})$, and $N_3(2^{n_3})$ with respect to ξ, η, and ζ.

The detailed implementation aspects of treating the boundary conditions are illustrated in the next section, which is slightly different in the array numbering. In the following discussion, for simplicity, we define the

Table 4.1. Coefficient mapping.

Coefficient	Stencil Points	Coefficient	Stencil Points
C_1	i,j,k		
C_2	i+1,j,k	C_3	i-1,j,k
C_4	i,j+1,k	C_5	i,j-1,k
C_6	i,j,k+1	C_7	i,j,k-1
C_8	i+1,j+1,k	C_9	i+1,j-1,k
C_{10}	i-1,j+1,k	C_{11}	i-1,j-1,k
C_{12}	i+1,j,k+1	C_{13}	i+1,j,k-1
C_{14}	i-1,j,k+1	C_{15}	i-1,j,k-1
C_{16}	i,j+1,k+1	C_{17}	i,j+1,k-1
C_{18}	i,j-1,k+1	C_{19}	i,j-1,k-1

interior station values at

$$\begin{cases} 1 \le i \le N_1 - 1, \\ 1 \le j \le N_2 - 1, \\ 1 \le k \le N_3 - 1; \end{cases}$$

with boundary conditions at

$$\begin{cases} i = 0, \quad N_1, \\ j = 0, \quad N_2, \\ k = 0, \quad N_3. \end{cases}$$

In general, the grid points at

$$\begin{cases} 0 \le i \le N_1, \\ 0 \le j \le N_2, \\ 0 \le k \le N_3; \end{cases}$$

are called the domain proper, whereas the ghost region refers to the grid points at

$$\begin{cases} i = -1, \quad N_1 + 1, \\ j = -1, \quad N_2 + 1, \\ k = -1, \quad N_3 + 1. \end{cases}$$

It is interesting to point out that if the periodic boundary conditions are used, stations

$$\begin{cases} i = -1, \\ j = -1, \\ k = -1; \end{cases}$$

are identical to stations

$$\begin{cases} i = N_1, \\ j = N_2, \\ k = N_3. \end{cases}$$

Likewise, stations depected as

$$\begin{cases} i = 0, \\ j = 0, \\ k = 0; \end{cases}$$

are in fact the same as

$$\begin{cases} i = N_1 + 1, \\ j = N_2 + 1, \\ k = N_3 + 1. \end{cases}$$

Now, the important step to devise a strategy for iterative solver is to properly construct the matrix, if we were to have such an assembled matrix. First of all, we introduce the following lexicographic ordering,

$$\mathbf{a}_i = \begin{bmatrix} \mathbf{a}_{i,0} \\ \mathbf{a}_{i,1} \\ \vdots \\ \mathbf{a}_{i,N_2} \end{bmatrix}, \quad \mathbf{a}_{i,j} = \begin{bmatrix} P_{i,j,0} \\ P_{i,j,1} \\ \vdots \\ P_{i,j,N_3} \end{bmatrix}, \quad \mathbf{f}_i = \begin{bmatrix} \mathbf{f}_{i,0} \\ \mathbf{f}_{i,1} \\ \vdots \\ \mathbf{f}_{i,N_2} \end{bmatrix} \quad \text{and} \quad \mathbf{f}_{i,j} = \begin{bmatrix} f_{i,j,0} \\ f_{i,j,1} \\ \vdots \\ f_{i,j,N_3} \end{bmatrix},$$

with $0 \leq i \leq N_1$ and $0 \leq j \leq N_2$.

Thus, the final assembly of the discretized governing equation $\mathbf{Ka} = \mathbf{f}$, can be expressed as

$$\begin{bmatrix} \mathbf{A} & \mathbf{B}_1 & \mathbf{0} & \dots & \mathbf{0} & \mathbf{0} & \mathbf{0} \\ \mathbf{B}_2 & \mathbf{A} & \mathbf{B}_1 & \dots & \mathbf{0} & \mathbf{0} & \mathbf{0} \\ \mathbf{0} & \mathbf{B}_2 & \mathbf{A} & \dots & \mathbf{0} & \mathbf{0} & \mathbf{0} \\ \vdots & \vdots & \vdots & \ddots & \vdots & \vdots & \vdots \\ \mathbf{0} & \mathbf{0} & \mathbf{0} & \dots & \mathbf{A} & \mathbf{B}_1 & \mathbf{0} \\ \mathbf{0} & \mathbf{0} & \mathbf{0} & \dots & \mathbf{B}_2 & \mathbf{A} & \mathbf{B}_1 \\ \mathbf{0} & \mathbf{0} & \mathbf{0} & \dots & \mathbf{0} & \mathbf{B}_2 & \mathbf{A} \end{bmatrix} \begin{bmatrix} \mathbf{a}_0 \\ \mathbf{a}_1 \\ \mathbf{a}_2 \\ \vdots \\ \mathbf{a}_{N_1-2} \\ \mathbf{a}_{N_1-1} \\ \mathbf{a}_{N_1} \end{bmatrix} = \begin{bmatrix} \mathbf{f}_0 \\ \mathbf{f}_1 \\ \mathbf{f}_2 \\ \vdots \\ \mathbf{f}_{N_1-2} \\ \mathbf{f}_{N_1-1} \\ \mathbf{f}_{N_1} \end{bmatrix}, \tag{4.68}$$

where $\mathbf{f}_0, \dots, \mathbf{f}_{N_1}$ are coefficients derived from the right-hand side of Eq. (4.66) and boundary conditions, and the block matrices are defined as follows:

$$\mathbf{A} = \begin{bmatrix} \mathbf{C} & \mathbf{D}_1 & \mathbf{0} & \dots & \mathbf{0} & \mathbf{0} & \mathbf{0} \\ \mathbf{D}_2 & \mathbf{C} & \mathbf{D}_1 & \dots & \mathbf{0} & \mathbf{0} & \mathbf{0} \\ \mathbf{0} & \mathbf{D}_2 & \mathbf{C} & \dots & \mathbf{0} & \mathbf{0} & \mathbf{0} \\ \vdots & \vdots & \vdots & \ddots & \vdots & \vdots & \vdots \\ \mathbf{0} & \mathbf{0} & \mathbf{0} & \dots & \mathbf{C} & \mathbf{D}_1 & \mathbf{0} \\ \mathbf{0} & \mathbf{0} & \mathbf{0} & \dots & \mathbf{D}_2 & \mathbf{C} & \mathbf{D}_1 \\ \mathbf{0} & \mathbf{0} & \mathbf{0} & \dots & \mathbf{0} & \mathbf{D}_2 & \mathbf{C} \end{bmatrix} \quad \text{and}$$

$$\mathbf{C} = \begin{bmatrix} C_1 & C_6 & 0 & \dots & 0 & 0 & 0 \\ C_7 & C_1 & C_6 & \dots & 0 & 0 & 0 \\ 0 & C_7 & C_1 & \dots & 0 & 0 & 0 \\ \vdots & \vdots & \vdots & \ddots & \vdots & \vdots & \vdots \\ 0 & 0 & 0 & \dots & C_1 & C_6 & 0 \\ 0 & 0 & 0 & \dots & C_7 & C_1 & C_6 \\ 0 & 0 & 0 & \dots & 0 & C_7 & C_1 \end{bmatrix};$$

$$\mathbf{D}_1 = \begin{bmatrix} C_4 & C_{16} & 0 & \dots & 0 & 0 & 0 \\ C_{17} & C_4 & C_{16} & \dots & 0 & 0 & 0 \\ 0 & C_{17} & C_4 & \dots & 0 & 0 & 0 \\ \vdots & \vdots & \vdots & \ddots & \vdots & \vdots & \vdots \\ 0 & 0 & 0 & \dots & C_4 & C_{16} & 0 \\ 0 & 0 & 0 & \dots & C_{17} & C_4 & C_{16} \\ 0 & 0 & 0 & \dots & 0 & C_{17} & C_4 \end{bmatrix} \quad \text{and}$$

$$\mathbf{D}_2 = \begin{bmatrix} C_5 & C_{18} & 0 & \dots & 0 & 0 & 0 \\ C_{19} & C_5 & C_{18} & \dots & 0 & 0 & 0 \\ 0 & C_{19} & C_5 & \dots & 0 & 0 & 0 \\ \vdots & \vdots & \vdots & \ddots & \vdots & \vdots & \vdots \\ 0 & 0 & 0 & \dots & C_5 & C_{18} & 0 \\ 0 & 0 & 0 & \dots & C_{19} & C_5 & C_{18} \\ 0 & 0 & 0 & \dots & 0 & C_{19} & C_5 \end{bmatrix};$$

$$\mathbf{B}_1 = \begin{bmatrix} \mathbf{E} & \mathbf{F}_1 & \mathbf{0} & \dots & \mathbf{0} & \mathbf{0} & \mathbf{0} \\ \mathbf{F}_2 & \mathbf{E} & \mathbf{F}_1 & \dots & \mathbf{0} & \mathbf{0} & \mathbf{0} \\ \mathbf{0} & \mathbf{F}_2 & \mathbf{E} & \dots & \mathbf{0} & \mathbf{0} & \mathbf{0} \\ \vdots & \vdots & \vdots & \ddots & \vdots & \vdots & \vdots \\ \mathbf{0} & \mathbf{0} & \mathbf{0} & \dots & \mathbf{E} & \mathbf{F}_1 & \mathbf{0} \\ \mathbf{0} & \mathbf{0} & \mathbf{0} & \dots & \mathbf{F}_2 & \mathbf{E} & \mathbf{F}_1 \\ \mathbf{0} & \mathbf{0} & \mathbf{0} & \dots & \mathbf{0} & \mathbf{F}_2 & \mathbf{E} \end{bmatrix} \quad \text{and}$$

$$\mathbf{E} = \begin{bmatrix} C_2 & C_{12} & 0 & \dots & 0 & 0 & 0 \\ C_{13} & C_2 & C_{12} & \dots & 0 & 0 & 0 \\ 0 & C_{13} & C_2 & \dots & 0 & 0 & 0 \\ \vdots & \vdots & \vdots & \ddots & \vdots & \vdots & \vdots \\ 0 & 0 & 0 & \dots & C_2 & C_{12} & 0 \\ 0 & 0 & 0 & \dots & C_{13} & C_2 & C_{12} \\ 0 & 0 & 0 & \dots & 0 & C_{13} & C_2 \end{bmatrix},$$

$$\mathbf{F}_1 = \begin{bmatrix} C_8 & 0 & \dots & 0 & 0 \\ 0 & C_8 & \dots & 0 & 0 \\ \vdots & \vdots & \ddots & \vdots & \vdots \\ 0 & 0 & \dots & C_8 & 0 \\ 0 & 0 & \dots & 0 & C_8 \end{bmatrix} \quad \text{and}$$

$$\mathbf{F}_2 = \begin{bmatrix} C_9 & 0 & \dots & 0 & 0 \\ 0 & C_9 & \dots & 0 & 0 \\ \vdots & \vdots & \ddots & \vdots & \vdots \\ 0 & 0 & \dots & C_9 & 0 \\ 0 & 0 & \dots & 0 & C_9 \end{bmatrix},$$

$$\mathbf{B}_2 = \begin{bmatrix} \mathbf{G} & \mathbf{H}_1 & \mathbf{0} & \dots & \mathbf{0} & \mathbf{0} & \mathbf{0} \\ \mathbf{H}_2 & \mathbf{G} & \mathbf{H}_1 & \dots & \mathbf{0} & \mathbf{0} & \mathbf{0} \\ \mathbf{0} & \mathbf{H}_2 & \mathbf{G} & \dots & \mathbf{0} & \mathbf{0} & \mathbf{0} \\ \vdots & \vdots & \vdots & \ddots & \vdots & \vdots & \vdots \\ \mathbf{0} & \mathbf{0} & \mathbf{0} & \dots & \mathbf{G} & \mathbf{H}_1 & \mathbf{0} \\ \mathbf{0} & \mathbf{0} & \mathbf{0} & \dots & \mathbf{H}_2 & \mathbf{G} & \mathbf{H}_1 \\ \mathbf{0} & \mathbf{0} & \mathbf{0} & \dots & \mathbf{0} & \mathbf{H}_2 & \mathbf{G} \end{bmatrix} \quad \text{and}$$

$$\mathbf{G} = \begin{bmatrix} C_3 & C_{14} & 0 & \dots & 0 & 0 & 0 \\ C_{15} & C_3 & C_{14} & \dots & 0 & 0 & 0 \\ 0 & C_{15} & C_3 & \dots & 0 & 0 & 0 \\ \vdots & \vdots & \vdots & \ddots & \vdots & \vdots & \vdots \\ 0 & 0 & 0 & \dots & C_3 & C_{14} & 0 \\ 0 & 0 & 0 & \dots & C_{15} & C_3 & C_{14} \\ 0 & 0 & 0 & \dots & 0 & C_{15} & C_3 \end{bmatrix};$$

$$\mathbf{H}_1 = \begin{bmatrix} C_{10} & 0 & \dots & 0 & 0 \\ 0 & C_{10} & \dots & 0 & 0 \\ \vdots & \vdots & \ddots & \vdots & \vdots \\ 0 & 0 & \dots & C_{10} & 0 \\ 0 & 0 & \dots & 0 & C_{10} \end{bmatrix} \quad \text{and}$$

$$\mathbf{H}_2 = \begin{bmatrix} C_{11} & 0 & \dots & 0 & 0 \\ 0 & C_{11} & \dots & 0 & 0 \\ \vdots & \vdots & \ddots & \vdots & \vdots \\ 0 & 0 & \dots & C_{11} & 0 \\ 0 & 0 & \dots & 0 & C_{11} \end{bmatrix}.$$

Obviously, for boundary stations, the assembled matrix must be modified with respect to the ghost cell values. To derive the ghost cell values, we must employ boundary conditions. In this case, we have

$$\text{at } i = 0,\ \frac{\partial p}{\partial x_2} = 0 \text{ (Neumann) and at } i = N_1, \frac{\partial p}{\partial x_2} = 0 \text{ (Neumann)}$$

$$\text{at } j = 0,\ \frac{\partial p}{\partial n} = g_o \text{ (Neumann) and at } j = N_2, \frac{\partial p}{\partial n} = g_{N_2} \text{ (Neumann)}$$

$$\text{at } k = 0,\ \frac{\partial p}{\partial x_3} = 0 \text{ (Neumann) and at } k = N_3, p = h \text{ (Dirichlet)}$$

$$\frac{\partial p}{\partial x_m} = \xi_{x_m}\frac{\partial p}{\partial \xi} + \eta_{x_m}\frac{\partial p}{\partial \eta} + \zeta_{x_m}\frac{\partial p}{\partial \zeta},$$

$$\frac{\partial n}{\partial x_m} = \xi_{x_m}\frac{\partial n}{\partial \xi} + \eta_{x_m}\frac{\partial n}{\partial \eta} + \zeta_{x_m}\frac{\partial n}{\partial \zeta},$$

$$\frac{\partial p}{\partial n} = n_{x_m}\frac{\partial p}{\partial x_m} = R_1\frac{\partial p}{\partial \xi} + R_2\frac{\partial p}{\partial \eta} + R_3\frac{\partial p}{\partial \zeta},$$

with $1 \le m \le 3$, and

$$R_1 = \xi_{x_m} n_{x_m}, R_2 = \eta_{x_m} n_{x_m}, \quad \text{and} \quad R_3 = \zeta_{x_m} n_{x_m}.$$

Therefore, we have

at $i = 0$:

$$(\xi_{x_2})_{0,j,k}\frac{P_{1,j,k} - P_{-1,j,k}}{2\Delta\xi} + (\eta_{x_2})_{0,j,k}\frac{P_{0,j+1,k} - P_{0,j-1,k}}{2\Delta\eta} + (\zeta_{x_2})_{0,j,k}\frac{P_{0,j,k+1} - P_{0,j,k-1}}{2\Delta\zeta} = 0, \tag{4.69}$$

at $i = N_1$:

$$(\xi_{x_2})_{N_1,j,k}\frac{P_{N_1+1,j,k} - P_{N_1-1,j,k}}{2\Delta\xi} + (\eta_{x_2})_{N_1,j,k}\frac{P_{N_1,j+1,k} - P_{N_1,j-1,k}}{2\Delta\eta} + (\zeta_{x_2})_{N_1,j,k}\frac{P_{N_1,j,k+1} - P_{N_1,j,k-1}}{2\Delta\zeta} = 0; \tag{4.70}$$

at $j = 0$:

$$(R_1)_{i,0,k}\frac{P_{i+1,0,k} - P_{i-1,0,k}}{2\Delta\xi} + (R_2)_{i,0,k}\frac{P_{i,1,k} - P_{i,-1,k}}{2\Delta\eta} + (R_3)_{i,0,k}\frac{P_{i,0,k+1} - P_{i,0,k-1}}{2\Delta\zeta} = g_o, \tag{4.71}$$

at $j = N_2$:

$$(R_1)_{i,N_2,k}\frac{P_{i+1,N_2,k} - P_{i-1,N_2,k}}{2\Delta\xi} + (R_2)_{i,N_2,k}\frac{P_{i,N_2+1,k} - P_{i,N_2-1,k}}{2\Delta\eta} + (R_3)_{i,N_2,k}\frac{P_{i,N_2,k+1} - P_{i,N_2,k-1}}{2\Delta\zeta} = g_{N_2}; \tag{4.72}$$

at $k = 0$:

$$(\xi_{x_3})_{i,j,0}\frac{P_{i+1,j,0} - P_{i-1,j,0}}{2\Delta\xi} + (\eta_{x_3})_{i,j,0}\frac{P_{i,j+1,0} - P_{i,j-1,0}}{2\Delta\eta} + (\zeta_{x_3})_{i,j,0}\frac{P_{i,j,1} - P_{i,j,-1}}{2\Delta\zeta} = 0, \tag{4.73}$$

at $k = N_3$:

$$P_{i,j,N_3} = h. \tag{4.74}$$

Note that from Eqs. (4.69) and (4.70), for $1 \le j \le N_2 - 1$ and $1 \le k \le N_3 - 1$, we can directly modified the structure of the assembled matrices. For simplicity, we introduce the modification matrices $\mathbf{A}^{1a}$, $\mathbf{B}_1^{1a}$, $\mathbf{C}^{1a}$, $\mathbf{D}_1^{1a}$, $\mathbf{D}_2^{1a}$, and $\mathbf{E}^{1a}$; and $\mathbf{A}^{1b}$, $\mathbf{B}_2^{1b}$, $\mathbf{C}^{1b}$, $\mathbf{D}_1^{1b}$, $\mathbf{D}_2^{1b}$, and $\mathbf{G}^{1b}$ in corresponding to $\mathbf{A}$, $\mathbf{B}_1$, $\mathbf{C}$, $\mathbf{D}_1$, $\mathbf{D}_2$, $\mathbf{E}$, and $\mathbf{G}$, respectively. Likewise, from Eqs. (4.71) and (4.72), for $1 \le i \le N_1 - 1$ and $1 \le k \le N_3 - 1$, we also introduce the corresponding modification matrices $\mathbf{A}^{2a}$, $\mathbf{B}_1^{2a}$, $\mathbf{B}_2^{2a}$, $\mathbf{C}^{2a}$, $\mathbf{D}_1^{2a}$, $\mathbf{E}^{2a}$, and $\mathbf{G}^{2a}$; and $\mathbf{A}^{2b}$, $\mathbf{B}_1^{2b}$, $\mathbf{B}_2^{2b}$, $\mathbf{C}^{2b}$, $\mathbf{D}_2^{2b}$, $\mathbf{E}^{2b}$, and $\mathbf{G}^{2b}$. Moreover, from Eq. (4.73), for $1 \le i \le N_1 - 1$ and $1 \le j \le N_2 - 1$, we have the modification matrices $\mathbf{A}^{3a}$, $\mathbf{B}_1^{3a}$, $\mathbf{B}_2^{3a}$, $\mathbf{C}^{3a}$, $\mathbf{D}_1^{3a}$, $\mathbf{D}_2^{3a}$, $\mathbf{E}^{3a}$, and $\mathbf{G}^{3a}$.

Now, according to the structure of the matrix, we are ready to incorporate the modifications of the assembled matrix. Thus, the final assembly of the geometry matrix before employing the ghost cell values can be written as

$$\begin{bmatrix} \tilde{\mathbf{A}}_a & \tilde{\mathbf{B}}_1 & \mathbf{0} & \dots & \mathbf{0} & \mathbf{0} & \mathbf{0} \\ \hat{\mathbf{B}}_2 & \hat{\mathbf{A}} & \hat{\mathbf{B}}_1 & \dots & \mathbf{0} & \mathbf{0} & \mathbf{0} \\ \mathbf{0} & \hat{\mathbf{B}}_2 & \hat{\mathbf{A}} & \dots & \mathbf{0} & \mathbf{0} & \mathbf{0} \\ \vdots & \vdots & \vdots & \ddots & \vdots & \vdots & \vdots \\ \mathbf{0} & \mathbf{0} & \mathbf{0} & \dots & \hat{\mathbf{A}} & \hat{\mathbf{B}}_1 & \mathbf{0} \\ \mathbf{0} & \mathbf{0} & \mathbf{0} & \dots & \hat{\mathbf{B}}_2 & \hat{\mathbf{A}} & \hat{\mathbf{B}}_1 \\ \mathbf{0} & \mathbf{0} & \mathbf{0} & \dots & \mathbf{0} & \tilde{\mathbf{B}}_2 & \tilde{\mathbf{A}}_b \end{bmatrix} \begin{bmatrix} \mathbf{a}_0 \\ \mathbf{a}_1 \\ \mathbf{a}_2 \\ \vdots \\ \mathbf{a}_{N_1-2} \\ \mathbf{a}_{N_1-1} \\ \mathbf{a}_{N_1} \end{bmatrix} = \begin{bmatrix} \hat{\mathbf{f}}_0 \\ \hat{\mathbf{f}}_1 \\ \hat{\mathbf{f}}_2 \\ \vdots \\ \hat{\mathbf{f}}_{N_1-2} \\ \hat{\mathbf{f}}_{N_1-1} \\ \hat{\mathbf{f}}_{N_1} \end{bmatrix}. \tag{4.75}$$

And the modifications of submatrices are derived as

$$\begin{aligned} \tilde{\mathbf{A}}_a &= \mathbf{A} + \mathbf{A}^{1a} + \mathbf{A}^{2a} + \mathbf{A}^{2b} + \mathbf{A}^{3a}, \\ \tilde{\mathbf{A}}_b &= \mathbf{A} + \mathbf{A}^{1b} + \mathbf{A}^{2a} + \mathbf{A}^{2b} + \mathbf{A}^{3a}, \\ \hat{\mathbf{A}} &= \mathbf{A} + \mathbf{A}^{2a} + \mathbf{A}^{2b} + \mathbf{A}^{3a}; \\ \tilde{\mathbf{B}}_1 &= \mathbf{B}_1 + \mathbf{B}_1^{1a} + \mathbf{B}_1^{2a} + \mathbf{B}_1^{2b} + \mathbf{B}_1^{3a}, \\ \tilde{\mathbf{B}}_2 &= \mathbf{B}_2 + \mathbf{B}_2^{1b} + \mathbf{B}_2^{2a} + \mathbf{B}_2^{2b} + \mathbf{B}_2^{3a}, \\ \hat{\mathbf{B}}_1 &= \mathbf{B}_1 + \mathbf{B}_1^{2a} + \mathbf{B}_1^{2b} + \mathbf{B}_1^{3a}; \\ \hat{\mathbf{B}}_2 &= \mathbf{B}_2 + \mathbf{B}_2^{2a} + \mathbf{B}_2^{2b} + \mathbf{B}_2^{3a}, \\ \hat{\mathbf{f}}_{i,0} &= \mathbf{f}_{i,0} + \mathbf{f}_{i,0}^{2a}, \\ \hat{\mathbf{f}}_{i,N_2} &= \mathbf{f}_{i,N_2} + \mathbf{f}_{i,N_2}^{2b}; \end{aligned}$$

with

$$\mathbf{A}^{1a} = \mathbf{B}_2 \begin{bmatrix} \mathbf{C}^{1a} & \mathbf{D}_1^{1a} & \mathbf{0} & \dots & \mathbf{0} & \mathbf{0} & \mathbf{0} \\ \mathbf{D}_2^{1a} & \mathbf{C}^{1a} & \mathbf{D}_1^{1a} & \dots & \mathbf{0} & \mathbf{0} & \mathbf{0} \\ \mathbf{0} & \mathbf{D}_2^{1a} & \mathbf{C}^{1a} & \dots & \mathbf{0} & \mathbf{0} & \mathbf{0} \\ \vdots & \vdots & \vdots & \ddots & \vdots & \vdots & \vdots \\ \mathbf{0} & \mathbf{0} & \mathbf{0} & \dots & \mathbf{C}^{1a} & \mathbf{D}_1^{1a} & \mathbf{0} \\ \mathbf{0} & \mathbf{0} & \mathbf{0} & \dots & \mathbf{D}_2^{1a} & \mathbf{C}^{1a} & \mathbf{D}_1^{1a} \\ \mathbf{0} & \mathbf{0} & \mathbf{0} & \dots & \mathbf{0} & \mathbf{D}_2^{1a} & \mathbf{C}^{1a} \end{bmatrix},$$

$$\mathbf{C}^{1a} = \begin{bmatrix} 0 & \frac{\Delta\xi(\zeta_{x_2})_{0,j,1}}{\Delta\zeta(\xi_{x_2})_{0,j,1}} & \dots & 0 & 0 \\ -\frac{\Delta\xi(\zeta_{x_2})_{0,j,0}}{\Delta\zeta(\xi_{x_2})_{0,j,0}} & 0 & \dots & 0 & 0 \\ \vdots & \vdots & \ddots & \vdots & \vdots \\ 0 & 0 & \dots & 0 & \frac{\Delta\xi(\zeta_{x_2})_{0,j,N_3}}{\Delta\zeta(\xi_{x_2})_{0,j,N_3}} \\ 0 & 0 & \dots & -\frac{\Delta\xi(\zeta_{x_2})_{0,j,N_3-1}}{\Delta\zeta(\xi_{x_2})_{0,j,N_3-1}} & 0 \end{bmatrix},$$

$$\mathbf{D}_1^{1a} = \begin{bmatrix} \frac{\Delta\xi(\eta_{x_2})_{0,j+1,0}}{\Delta\eta(\xi_{x_2})_{0,j+1,0}} & 0 & \dots & 0 & 0 \\ 0 & \frac{\Delta\xi(\eta_{x_2})_{0,j+1,1}}{\Delta\eta(\xi_{x_2})_{0,j+1,1}} & \dots & 0 & 0 \\ \vdots & \vdots & \ddots & \vdots & \vdots \\ 0 & 0 & \dots & \frac{\Delta\xi(\eta_{x_2})_{0,j+1,N_3-1}}{\Delta\eta(\xi_{x_2})_{0,j+1,N_3-1}} & 0 \\ 0 & 0 & \dots & 0 & \frac{\Delta\xi(\eta_{x_2})_{0,j+1,N_3}}{\Delta\eta(\xi_{x_2})_{0,j+1,N_3}} \end{bmatrix},$$

$$\mathbf{D}_2^{1a} = \begin{bmatrix} -\frac{\Delta\xi(\eta_{x_2})_{0,j-1,0}}{\Delta\eta(\xi_{x_2})_{0,j-1,0}} & 0 & \dots & 0 & 0 \\ 0 & -\frac{\Delta\xi(\eta_{x_2})_{0,j-1,1}}{\Delta\eta(\xi_{x_2})_{0,j-1,1}} & \dots & 0 & 0 \\ \vdots & \vdots & \ddots & \vdots & \vdots \\ 0 & 0 & \dots & -\frac{\Delta\xi(\eta_{x_2})_{0,j-1,N_3-1}}{\Delta\eta(\xi_{x_2})_{0,j-1,N_3-1}} & 0 \\ 0 & 0 & \dots & 0 & -\frac{\Delta\xi(\eta_{x_2})_{0,j-1,N_3}}{\Delta\eta(\xi_{x_2})_{0,j-1,N_3}} \end{bmatrix},$$

$$\mathbf{E}^{1a} = \begin{bmatrix} 1 & 0 & \dots & 0 & 0 \\ 0 & 1 & \dots & 0 & 0 \\ \vdots & \vdots & \ddots & \vdots & \vdots \\ 0 & 0 & \dots & 1 & 0 \\ 0 & 0 & \dots & 0 & 1 \end{bmatrix} \quad \text{and} \quad \mathbf{B}_1^{1a} = \mathbf{B}_2 \begin{bmatrix} \mathbf{E}^{1a} & \mathbf{0} & \dots & \mathbf{0} & \mathbf{0} \\ \mathbf{0} & \mathbf{E}^{1a} & \dots & \mathbf{0} & \mathbf{0} \\ \vdots & \vdots & \ddots & \vdots & \vdots \\ \mathbf{0} & \mathbf{0} & \dots & \mathbf{E}^{1a} & \mathbf{0} \\ \mathbf{0} & \mathbf{0} & \dots & \mathbf{0} & \mathbf{E}^{1a} \end{bmatrix};$$

$$\mathbf{A}^{1b} = \mathbf{B}_1 \begin{bmatrix} \mathbf{C}^{1b} & \mathbf{D}_1^{1b} & \mathbf{0} & \dots & \mathbf{0} & \mathbf{0} & \mathbf{0} \\ \mathbf{D}_2^{1b} & \mathbf{C}^{1b} & \mathbf{D}_1^{1b} & \dots & \mathbf{0} & \mathbf{0} & \mathbf{0} \\ \mathbf{0} & \mathbf{D}_2^{1b} & \mathbf{C}^{1b} & \dots & \mathbf{0} & \mathbf{0} & \mathbf{0} \\ \vdots & \vdots & \vdots & \ddots & \vdots & \vdots & \vdots \\ \mathbf{0} & \mathbf{0} & \mathbf{0} & \dots & \mathbf{C}^{1b} & \mathbf{D}_1^{1b} & \mathbf{0} \\ \mathbf{0} & \mathbf{0} & \mathbf{0} & \dots & \mathbf{D}_2^{1b} & \mathbf{C}^{1b} & \mathbf{D}_1^{1b} \\ \mathbf{0} & \mathbf{0} & \mathbf{0} & \dots & \mathbf{0} & \mathbf{D}_2^{1b} & \mathbf{C}^{1b} \end{bmatrix};$$

$$\mathbf{B}_2^{1b} = \mathbf{B}_1 \begin{bmatrix} \mathbf{G}^{1b} & \mathbf{0} & \dots & \mathbf{0} & \mathbf{0} \\ \mathbf{0} & \mathbf{G}^{1b} & \dots & \mathbf{0} & \mathbf{0} \\ \vdots & \vdots & \ddots & \vdots & \vdots \\ \mathbf{0} & \mathbf{0} & \dots & \mathbf{G}^{1b} & \mathbf{0} \\ \mathbf{0} & \mathbf{0} & \dots & \mathbf{0} & \mathbf{G}^{1b} \end{bmatrix} \quad \text{and} \quad \mathbf{E}^{1b} = \begin{bmatrix} 1 & 0 & \dots & 0 & 0 \\ 0 & 1 & \dots & 0 & 0 \\ \vdots & \vdots & \ddots & \vdots & \vdots \\ 0 & 0 & \dots & 1 & 0 \\ 0 & 0 & \dots & 0 & 1 \end{bmatrix},$$

$$\mathbf{C}^{1b} = \begin{bmatrix} 0 & -\frac{\Delta\xi(\zeta_{x_2})_{N_1,j,1}}{\Delta\zeta(\xi_{x_2})_{N_1,j,1}} & \dots & 0 & 0 \\ \frac{\Delta\xi(\zeta_{x_2})_{N_1,j,0}}{\Delta\zeta(\xi_{x_2})_{N_1,j,0}} & 0 & \dots & 0 & 0 \\ \vdots & \vdots & \ddots & \vdots & \vdots \\ 0 & 0 & \dots & 0 & -\frac{\Delta\xi(\zeta_{x_2})_{N_1,j,N_3}}{\Delta\zeta(\xi_{x_2})_{N_1,j,N_3}} \\ 0 & 0 & \dots & \frac{\Delta\xi(\zeta_{x_2})_{N_1,j,N_3-1}}{\Delta\zeta(\xi_{x_2})_{N_1,j,N_3-1}} & 0 \end{bmatrix},$$

$$\mathbf{D}_1^{1b} = \begin{bmatrix} -\frac{\Delta\xi(\eta_{x_2})_{N_1,j+1,0}}{\Delta\eta(\xi_{x_2})_{N_1,j+1,0}} & 0 & \dots & 0 & 0 \\ 0 & -\frac{\Delta\xi(\eta_{x_2})_{N_1,j+1,1}}{\Delta\eta(\xi_{x_2})_{N_1,j+1,1}} & \dots & 0 & 0 \\ \vdots & \vdots & \ddots & \vdots & \vdots \\ 0 & 0 & \dots & -\frac{\Delta\xi(\eta_{x_2})_{N_1,j+1,N_3-1}}{\Delta\eta(\xi_{x_2})_{N_1,j+1,N_3-1}} & 0 \\ 0 & 0 & \dots & 0 & -\frac{\Delta\xi(\eta_{x_2})_{N_1,j+1,N_3}}{\Delta\eta(\xi_{x_2})_{N_1,j+1,N_3}} \end{bmatrix},$$

$$\mathbf{D}_2^{1b} = \begin{bmatrix} \frac{\Delta\xi(\eta_{x_2})_{N_1,j-1,0}}{\Delta\eta(\xi_{x_2})_{N_1,j-1,0}} & 0 & \dots & 0 & 0 \\ 0 & \frac{\Delta\xi(\eta_{x_2})_{N_1,j-1,1}}{\Delta\eta(\xi_{x_2})_{N_1,j-1,1}} & \dots & 0 & 0 \\ \vdots & \vdots & \ddots & \vdots & \vdots \\ 0 & 0 & \dots & \frac{\Delta\xi(\eta_{x_2})_{N_1,j-1,N_3-1}}{\Delta\eta(\xi_{x_2})_{N_1,j-1,N_3-1}} & 0 \\ 0 & 0 & \dots & 0 & \frac{\Delta\xi(\eta_{x_2})_{N_1,j-1,N_3}}{\Delta\eta(\xi_{x_2})_{N_1,j-1,N_3}} \end{bmatrix},$$

$$\mathbf{A}^{2a} = \begin{bmatrix} \mathbf{C}^{2a} & \mathbf{D}_1^{2a} & \dots & \mathbf{0} & \mathbf{0} \\ \mathbf{0} & \mathbf{0} & \dots & \mathbf{0} & \mathbf{0} \\ \vdots & \vdots & \ddots & \vdots & \vdots \\ \mathbf{0} & \mathbf{0} & \dots & \mathbf{0} & \mathbf{0} \\ \mathbf{0} & \mathbf{0} & \dots & \mathbf{0} & \mathbf{0} \end{bmatrix}, \quad \mathbf{B}_1^{2a} = \begin{bmatrix} \mathbf{E}^{2a} & \mathbf{0} & \dots & \mathbf{0} & \mathbf{0} \\ \mathbf{0} & \mathbf{0} & \dots & \mathbf{0} & \mathbf{0} \\ \vdots & \vdots & \ddots & \vdots & \vdots \\ \mathbf{0} & \mathbf{0} & \dots & \mathbf{0} & \mathbf{0} \\ \mathbf{0} & \mathbf{0} & \dots & \mathbf{0} & \mathbf{0} \end{bmatrix}, \quad \text{and}$$

$$\mathbf{B}_2^{2a} = \begin{bmatrix} \mathbf{G}^{2a} & \mathbf{0} & \dots & \mathbf{0} & \mathbf{0} \\ \mathbf{0} & \mathbf{0} & \dots & \mathbf{0} & \mathbf{0} \\ \vdots & \vdots & \ddots & \vdots & \vdots \\ \mathbf{0} & \mathbf{0} & \dots & \mathbf{0} & \mathbf{0} \\ \mathbf{0} & \mathbf{0} & \dots & \mathbf{0} & \mathbf{0} \end{bmatrix};$$

$$\mathbf{C}^{2a} = \begin{bmatrix} 0 & \frac{\Delta\eta(R_3)_{i,0,1}}{\Delta\zeta(R_2)_{i,0,1}} & \dots & 0 & 0 \\ -\frac{\Delta\eta(R_3)_{i,0,0}}{\Delta\zeta(R_2)_{i,0,0}} & 0 & \dots & 0 & 0 \\ \vdots & \vdots & \ddots & \vdots & \vdots \\ 0 & 0 & \dots & 0 & \frac{\Delta\eta(R_3)_{i,0,N_3}}{\Delta\zeta(R_2)_{i,0,N_3}} \\ 0 & 0 & \dots & -\frac{\Delta\eta(R_3)_{i,0,N_3-1}}{\Delta\zeta(R_2)_{i,0,N_3-1}} & 0 \end{bmatrix},$$

$$\mathbf{E}^{2a} = \begin{bmatrix} \frac{\Delta\eta(R_1)_{i,0,0}}{\Delta\xi(R_2)_{i,0,0}} & 0 & \dots & 0 & 0 \\ 0 & \frac{\Delta\eta(R_1)_{i,0,1}}{\Delta\xi(R_2)_{i,0,1}} & \dots & 0 & 0 \\ \vdots & \vdots & \ddots & \vdots & \vdots \\ 0 & 0 & \dots & \frac{\Delta\eta(R_1)_{i,0,N_3-1}}{\Delta\xi(R_2)_{i,0,N_3-1}} & 0 \\ 0 & 0 & \dots & 0 & \frac{\Delta\eta(R_1)_{i,0,N_3}}{\Delta\xi(R_2)_{i,0,N_3}} \end{bmatrix},$$

$$\mathbf{G}^{2a} = \begin{bmatrix} -\dfrac{\Delta\eta(R_1)_{i,0,0}}{\Delta\xi(R_2)_{i,0,0}} & 0 & \dots & 0 & 0 \\ 0 & -\dfrac{\Delta\eta(R_1)_{i,0,1}}{\Delta\xi(R_2)_{i,0,1}} & \dots & 0 & 0 \\ \vdots & \vdots & \ddots & \vdots & \vdots \\ 0 & 0 & \dots & -\dfrac{\Delta\eta(R_1)_{i,0,N_3-1}}{\Delta\xi(R_2)_{i,0,N_3-1}} & 0 \\ 0 & 0 & \dots & 0 & -\dfrac{\Delta\eta(R_1)_{i,0,N_3}}{\Delta\xi(R_2)_{i,0,N_3}} \end{bmatrix},$$

$$\mathbf{A}^{2b} = \begin{bmatrix} \mathbf{0} & \dots & \mathbf{0} & \mathbf{0} & \\ \mathbf{0} & \dots & \mathbf{0} & \mathbf{0} & \\ \vdots & \ddots & \vdots & \vdots & \\ \mathbf{0} & \mathbf{0} & \dots & \mathbf{0} & \mathbf{0} \\ \mathbf{0} & \mathbf{0} & \dots & \mathbf{D}_2^{2b} & \mathbf{C}^{2b} \end{bmatrix}, \quad \mathbf{D}_1^{2a} = \begin{bmatrix} 1 & 0 & \dots & 0 & 0 \\ 0 & 1 & \dots & 0 & 0 \\ \vdots & \vdots & \ddots & \vdots & \vdots \\ 0 & 0 & \dots & 1 & 0 \\ 0 & 0 & \dots & 0 & 1 \end{bmatrix}, \quad \text{and}$$

$$\mathbf{C}^{3a} = \begin{bmatrix} 0 & 1 & \dots & 0 & 0 \\ 0 & 0 & \dots & 0 & 0 \\ \vdots & \vdots & \ddots & \vdots & \vdots \\ 0 & 0 & \dots & 0 & 0 \\ 0 & 0 & \dots & 0 & 0 \end{bmatrix},$$

$$\mathbf{B}_1^{2b} = \begin{bmatrix} \mathbf{0} & \mathbf{0} & \dots & \mathbf{0} & \mathbf{0} \\ \mathbf{0} & \mathbf{0} & \dots & \mathbf{0} & \mathbf{0} \\ \vdots & \vdots & \ddots & \vdots & \vdots \\ \mathbf{0} & \mathbf{0} & \dots & \mathbf{0} & \mathbf{0} \\ \mathbf{0} & \mathbf{0} & \dots & \mathbf{0} & \mathbf{E}^{2b} \end{bmatrix}, \quad \mathbf{B}_2^{2b} = \begin{bmatrix} \mathbf{0} & \mathbf{0} & \dots & \mathbf{0} & \mathbf{0} \\ \mathbf{0} & \mathbf{0} & \dots & \mathbf{0} & \mathbf{0} \\ \vdots & \vdots & \ddots & \vdots & \vdots \\ \mathbf{0} & \mathbf{0} & \dots & \mathbf{0} & \mathbf{0} \\ \mathbf{0} & \mathbf{0} & \dots & \mathbf{0} & \mathbf{G}^{2b} \end{bmatrix}, \quad \text{and}$$

$$\mathbf{D}_2^{2b} = \begin{bmatrix} 1 & 0 & \dots & 0 & 0 \\ 0 & 1 & \dots & 0 & 0 \\ \vdots & \vdots & \ddots & \vdots & \vdots \\ 0 & 0 & \dots & 1 & 0 \\ 0 & 0 & \dots & 0 & 1 \end{bmatrix},$$

$$
\mathbf{C}^{2b} = \begin{bmatrix}
0 & -\dfrac{\Delta\eta(R_3)_{i,N_2,1}}{\Delta\zeta(R_2)_{i,N_2,1}} & \cdots & 0 & 0 \\
\dfrac{\Delta\eta(R_3)_{i,N_2,0}}{\Delta\zeta(R_2)_{i,N_2,0}} & 0 & \cdots & 0 & 0 \\
\vdots & \vdots & \ddots & \vdots & \vdots \\
0 & 0 & \cdots & 0 & -\dfrac{\Delta\eta(R_3)_{i,N_2,N_3}}{\Delta\zeta(R_2)_{i,N_2,N_3}} \\
0 & 0 & \cdots & \dfrac{\Delta\eta(R_3)_{i,N_2,N_3-1}}{\Delta\zeta(R_2)_{i,N_2,N_3-1}} & 0
\end{bmatrix},
$$

$$
\mathbf{E}^{2b} = \begin{bmatrix}
-\dfrac{\Delta\eta(R_1)_{i,N_2,0}}{\Delta\xi(R_2)_{i,N_2,0}} & 0 & \cdots & 0 & 0 \\
0 & -\dfrac{\Delta\eta(R_1)_{i,N_2,1}}{\Delta\xi(R_2)_{i,N_2,1}} & \cdots & 0 & 0 \\
\vdots & \vdots & \ddots & \vdots & \vdots \\
0 & 0 & \cdots & -\dfrac{\Delta\eta(R_1)_{i,N_2,N_3-1}}{\Delta\xi(R_2)_{i,N_2,N_3-1}} & 0 \\
0 & 0 & \cdots & 0 & -\dfrac{\Delta\eta(R_1)_{i,N_2,N_3}}{\Delta\xi(R_2)_{i,N_2,N_3}}
\end{bmatrix},
$$

$$
\mathbf{G}^{2b} = \begin{bmatrix}
\dfrac{\Delta\eta(R_1)_{i,N_2,0}}{\Delta\xi(R_2)_{i,N_2,0}} & 0 & \cdots & 0 & 0 \\
0 & \dfrac{\Delta\eta(R_1)_{i,N_2,1}}{\Delta\xi(R_2)_{i,N_2,1}} & \cdots & 0 & 0 \\
\vdots & \vdots & \ddots & \vdots & \vdots \\
0 & 0 & \cdots & \dfrac{\Delta\eta(R_1)_{i,N_2,N_3-1}}{\Delta\xi(R_2)_{i,N_2,N_3-1}} & 0 \\
0 & 0 & \cdots & 0 & \dfrac{\Delta\eta(R_1)_{i,N_2,N_3}}{\Delta\xi(R_2)_{i,N_2,N_3}}
\end{bmatrix},
$$

$$
\mathbf{f}^{2a}_{i,0} = \begin{pmatrix}
-\dfrac{2\Delta\eta}{(R_2)_{i,0,0}} g_o \\
-\dfrac{2\Delta\eta}{(R_2)_{i,0,1}} g_o \\
\vdots \\
-\dfrac{2\Delta\eta}{(R_2)_{i,0,N_3-1}} g_o \\
-\dfrac{2\Delta\eta}{(R_2)_{i,0,N_3}} g_o
\end{pmatrix}
\quad \text{and} \quad
\mathbf{f}^{2a}_{iN_2} = \begin{pmatrix}
\dfrac{2\Delta\eta}{(R_2)_{i,N_2,0}} g_{N_2} \\
\dfrac{2\Delta\eta}{(R_2)_{i,N_2,1}} g_{N_2} \\
\vdots \\
\dfrac{2\Delta\eta}{(R_2)_{i,N_2,N_3-1}} g_{N_2} \\
\dfrac{2\Delta\eta}{(R_2)_{i,N_2,N_3}} g_{N_2}
\end{pmatrix},
$$

$$\mathbf{A}^{3a} = \begin{bmatrix} \mathbf{C}^{3a} & \mathbf{D}_1^{3a} & \mathbf{0} & \dots & \mathbf{0} & \mathbf{0} & \mathbf{0} \\ \mathbf{D}_2^{3a} & \mathbf{C}^{3a} & \mathbf{D}_1^{3a} & \dots & \mathbf{0} & \mathbf{0} & \mathbf{0} \\ \mathbf{0} & \mathbf{D}_2^{3a} & \mathbf{C}^{3a} & \dots & \mathbf{0} & \mathbf{0} & \mathbf{0} \\ \vdots & \vdots & \vdots & \ddots & \vdots & \vdots & \vdots \\ \mathbf{0} & \mathbf{0} & \mathbf{0} & \dots & \mathbf{C}^{3a} & \mathbf{D}_1^{3a} & \mathbf{0} \\ \mathbf{0} & \mathbf{0} & \mathbf{0} & \dots & \mathbf{D}_2^{3a} & \mathbf{C}^{3a} & \mathbf{D}_1^{3a} \\ \mathbf{0} & \mathbf{0} & \mathbf{0} & \dots & \mathbf{0} & \mathbf{D}_2^{3a} & \mathbf{C}^{3a} \end{bmatrix} \quad \text{and}$$

$$\mathbf{B}_1^{3a} = \begin{bmatrix} \mathbf{E}^{3a} & \mathbf{F}_1^{3a} & \mathbf{0} & \dots & \mathbf{0} & \mathbf{0} & \mathbf{0} \\ \mathbf{F}_2^{3a} & \mathbf{E}^{3a} & \mathbf{F}_1^{3a} & \dots & \mathbf{0} & \mathbf{0} & \mathbf{0} \\ \mathbf{0} & \mathbf{F}_2^{3a} & \mathbf{E}^{3a} & \dots & \mathbf{0} & \mathbf{0} & \mathbf{0} \\ \vdots & \vdots & \vdots & \ddots & \vdots & \vdots & \vdots \\ \mathbf{0} & \mathbf{0} & \mathbf{0} & \dots & \mathbf{E}^{3a} & \mathbf{F}_1^{3a} & \mathbf{0} \\ \mathbf{0} & \mathbf{0} & \mathbf{0} & \dots & \mathbf{F}_2^{3a} & \mathbf{E}^{3a} & \mathbf{F}_1^{3a} \\ \mathbf{0} & \mathbf{0} & \mathbf{0} & \dots & \mathbf{0} & \mathbf{F}_2^{3a} & \mathbf{E}^{3a} \end{bmatrix};$$

$$\mathbf{B}_2^{3a} = \begin{bmatrix} \mathbf{G}^{3a} & \mathbf{H}_1^{3a} & \mathbf{0} & \dots & \mathbf{0} & \mathbf{0} & \mathbf{0} \\ \mathbf{H}_2^{3a} & \mathbf{G}^{3a} & \mathbf{H}_1^{3a} & \dots & \mathbf{0} & \mathbf{0} & \mathbf{0} \\ \mathbf{0} & \mathbf{H}_2^{3a} & \mathbf{G}^{3a} & \dots & \mathbf{0} & \mathbf{0} & \mathbf{0} \\ \vdots & \vdots & \vdots & \ddots & \vdots & \vdots & \vdots \\ \mathbf{0} & \mathbf{0} & \mathbf{0} & \dots & \mathbf{G}^{3a} & \mathbf{H}_1^{3a} & \mathbf{0} \\ \mathbf{0} & \mathbf{0} & \mathbf{0} & \dots & \mathbf{H}_2^{3a} & \mathbf{G}^{3a} & \mathbf{H}_1^{3a} \\ \mathbf{0} & \mathbf{0} & \mathbf{0} & \dots & \mathbf{0} & \mathbf{H}_2^{3a} & \mathbf{G}^{3a} \end{bmatrix},$$

$$\mathbf{D}_1^{3a} = \begin{bmatrix} \frac{\Delta\zeta(\eta_{x_3})_{i,j,0}}{\Delta\eta(\zeta_{x_3})_{i,j,0}} & 0 & \dots & 0 & 0 \\ 0 & 0 & \dots & 0 & 0 \\ \vdots & \vdots & \ddots & \vdots & \vdots \\ 0 & 0 & \dots & 0 & 0 \\ 0 & 0 & \dots & 0 & 0 \end{bmatrix} \quad \text{and} \quad \mathbf{D}_2^{3a} = \begin{bmatrix} -\frac{\Delta\zeta(\eta_{x_3})_{i,j,0}}{\Delta\eta(\zeta_{x_3})_{i,j,0}} & 0 & \dots & 0 & 0 \\ 0 & 0 & \dots & 0 & 0 \\ \vdots & \vdots & \ddots & \vdots & \vdots \\ 0 & 0 & \dots & 0 & 0 \\ 0 & 0 & \dots & 0 & 0 \end{bmatrix};$$

$$\mathbf{E}^{3a} = \mathbf{B}_1 \begin{bmatrix} \frac{\Delta\zeta(\xi_{x_3})_{i,j,0}}{\Delta\xi(\zeta_{x_3})_{i,j,0}} & 0 & \dots & 0 & 0 \\ 0 & 0 & \dots & 0 & 0 \\ \vdots & \vdots & \ddots & \vdots & \vdots \\ 0 & 0 & \dots & 0 & 0 \\ 0 & 0 & \dots & 0 & 0 \end{bmatrix} \quad \text{and} \quad \mathbf{G}^{3a} = \begin{bmatrix} -\frac{\Delta\zeta(\xi_{x_3})_{i,j,0}}{\Delta\xi(\zeta_{x_3})_{i,j,0}} & 0 & \dots & 0 & 0 \\ 0 & 0 & \dots & 0 & 0 \\ \vdots & \vdots & \ddots & \vdots & \vdots \\ 0 & 0 & \dots & 0 & 0 \\ 0 & 0 & \dots & 0 & 0 \end{bmatrix}.$$

Therefore, according to the boundary conditions at $i = 0, N_1$, $j = 0, N_2$, and $k = 0, N_3$, we have in general 24 ghost values listed in Table 4.2(a).

Table 4.2. Ghost values.

(a) 24 Ghost values			
$P_{-1,0,0}$	$P_{-1,N_2,0}$	$P_{N_1+1,0,0}$	$P_{N_1+1,N_2,0}$
$P_{-1,0,N_3}$	P_{-1,N_2,N_3}	$P_{N_1+1,0,N_3}$	P_{N_1+1,N_2,N_3}
$P_{0,-1,0}$	$P_{N_1,-1,0}$	$P_{0,N_2+1,0}$	$P_{N_1,N_2+1,0}$
$P_{0,-1,N_3}$	$P_{N_1,-1,N_3}$	P_{0,N_2+1,N_3}	P_{N_1,N_2+1,N_3}
$P_{0,0,-1}$	$P_{N_1,0,-1}$	$P_{0,0,N_3+1}$	$P_{N_1,0,N_3+1}$
$P_{0,N_2,-1}$	$P_{N_1,N_2,-1}$	P_{0,N_2,N_3+1}	P_{N_1,N_2,N_3+1}

(b) 12 Ghost values			
$P_{-1,0,0}$	$P_{-1,N_2,0}$	$P_{N_1+1,0,0}$	$P_{N_1+1,N_2,0}$
$P_{0,-1,0}$	$P_{N_1,-1,0}$	$P_{0,N_2+1,0}$	$P_{N_1,N_2+1,0}$
$P_{0,0,-1}$	$P_{N_1,0,-1}$	$P_{0,N_2,-1}$	$P_{N_1,N_2,-1}$

For Eqs. (4.69) and (4.70), we can derive eight equations at $j = 0, N_2$ and $k = 0, N_3$. Similarly, for Eqs. (4.71) and (4.72), we can derive eight equations at $i = 0, N_1$ and $k = 0, N_3$. And, finally, for Eqs. (4.73) and (4.74), we can derive eight equations at $i = 0, N_1$ and $j = 0, N_2$. For the specific boundary conditions discussed here, we have $P_{i,j,N_3} = h$. Moreover, we must also have $P_{i,j,N_3+1} = h$. Thus, we eliminate 12 unknowns, and leave 12 equations for the ghost values listed in Table 4.2(b).

To elaborate, let's express the boundary conditions (4.69)–(4.73) in details by assigning specific boundary points,

at $i = 0$:

$$\frac{(\xi_{x_2})_{0,0,0}}{2\Delta\xi}P_{-1,0,0} + \frac{(\eta_{x_2})_{0,0,0}}{2\Delta\eta}P_{0,-1,0} + \frac{(\zeta_{x_2})_{0,0,0}}{2\Delta\zeta}P_{0,0,-1}$$

$$= \frac{(\xi_{x_2})_{0,0,0}}{2\Delta\xi}P_{1,0,0} + \frac{(\eta_{x_2})_{0,0,0}}{2\Delta\eta}P_{0,1,0} + \frac{(\zeta_{x_2})_{0,0,0}}{2\Delta\zeta}P_{0,0,1},$$

$$\frac{(\xi_{x_2})_{0,N_2,0}}{2\Delta\xi}P_{-1,N_2,0} - \frac{(\eta_{x_2})_{0,N_2,0}}{2\Delta\eta}P_{0,N_2+1,0} + \frac{(\zeta_{x_2})_{0,N_2,0}}{2\Delta\zeta}P_{0,N_2,-1}$$

$$= \frac{(\xi_{x_2})_{0,N_2,0}}{2\Delta\xi}P_{1,N_2,0} - \frac{(\eta_{x_2})_{0,N_2,0}}{2\Delta\eta}P_{0,N_2-1,0} + \frac{(\zeta_{x_2})_{0,N_2,0}}{2\Delta\zeta}P_{0,N_2,1},$$

at $i = N_1$:

$$\frac{(\xi_{x_2})_{N_1,0,0}}{2\Delta\xi}P_{N_1+1,0,0} - \frac{(\eta_{x_2})_{N_1,0,0}}{2\Delta\eta}P_{N_1,-1,0} - \frac{(\zeta_{x_2})_{N_1,0,0}}{2\Delta\zeta}P_{N_1,0,-1} = \frac{(\xi_{x_2})_{N_1,0,0}}{2\Delta\xi}P_{N_1-1,0,0} - \frac{(\eta_{x_2})_{N_1,0,0}}{2\Delta\eta}P_{N_1,1,0} - \frac{(\zeta_{x_2})_{N_1,0,0}}{2\Delta\zeta}P_{N_1,0,1};$$

$$\frac{(\xi_{x_2})_{N_1,N_2,0}}{2\Delta\xi}P_{N_1+1,N_2,0} + \frac{(\eta_{x_2})_{N_1,N_2,0}}{2\Delta\eta}P_{N_1,N_2+1,0} - \frac{(\zeta_{x_2})_{N_1,N_2,0}}{2\Delta\zeta}P_{N_1,N_2,-1} = \frac{(\xi_{x_2})_{N_1,N_2,0}}{2\Delta\xi}P_{N_1-1,N_2,0} + \frac{(\eta_{x_2})_{N_1,N_2,0}}{2\Delta\eta}P_{N_1,N_2-1,0} - \frac{(\zeta_{x_2})_{N_1,N_2,0}}{2\Delta\zeta}P_{N_1,N_2,1};$$

at $j = 0$:

$$\frac{(R_1)_{0,0,0}}{2\Delta\xi}P_{-1,0,0} + \frac{(R_2)_{0,0,0}}{2\Delta\eta}P_{0,-1,0} + \frac{(R_3)_{0,0,0}}{2\Delta\zeta}P_{0,0,-1} = \frac{(R_1)_{0,0,0}}{2\Delta\xi}P_{1,0,0} + \frac{(R_2)_{0,0,0}}{2\Delta\eta}P_{0,1,0} + \frac{(R_3)_{0,0,0}}{2\Delta\zeta}P_{0,0,1} - g_o,$$

$$-\frac{(R_1)_{N_1,0,0}}{2\Delta\xi}P_{N_1+1,0,0} + \frac{(R_2)_{N_1,0,0}}{2\Delta\eta}P_{N_1,-1,0} + \frac{(R_3)_{N_1,0,0}}{2\Delta\zeta}P_{N_1,0,-1} = -\frac{(R_1)_{N_1,0,0}}{2\Delta\xi}P_{N_1-1,0,0} + \frac{(R_2)_{N_1,0,0}}{2\Delta\eta}P_{N_1,1,0} + \frac{(R_3)_{N_1,0,0}}{2\Delta\zeta}P_{N_1,0,1} - g_o;$$

at $j = N_2$:

$$\frac{(R_1)_{0,N_2,0}}{2\Delta\xi}P_{-1,N_2,0} - \frac{(R_2)_{0,N_2,0}}{2\Delta\eta}P_{0,N_2+1,0} + \frac{(R_3)_{0,N_2,0}}{2\Delta\zeta}P_{0,N_2,-1} = \frac{(R_1)_{0,N_2,0}}{2\Delta\xi}P_{1,N_2,0} - \frac{(R_2)_{0,N_2,0}}{2\Delta\eta}P_{0,N_2-1,0} + \frac{(R_3)_{0,N_2,0}}{2\Delta\zeta}P_{0,N_2,1} - g_{N_2},$$

$$
\frac{(R_1)_{N_1,N_2,0}}{2\Delta\xi}P_{N_1+1,N_2,0}+\frac{(R_2)_{N_1,N_2,0}}{2\Delta\eta}P_{N_1,N_2+1,0}
$$

$$
-\frac{(R_3)_{N_1,N_2,0}}{2\Delta\zeta}P_{N_1,N_2,-1}
$$

$$
=\frac{(R_1)_{N_1,N_2,0}}{2\Delta\xi}P_{N_1-1,N_2,0}+\frac{(R_2)_{N_1,N_2,0}}{2\Delta\eta}P_{N_1,N_2-1,0}
$$

$$
-\frac{(R_3)_{N_1,N_2,0}}{2\Delta\zeta}P_{N_1,N_2,1}+g_{N_2};
$$

at $k=0$:

$$
\frac{(\xi_{x_3})_{0,0,0}}{2\Delta\xi}P_{-1,0,0}+\frac{(\eta_{x_3})_{0,0,0}}{2\Delta\eta}P_{0,-1,0}+\frac{(\zeta_{x_3})_{0,0,0}}{2\Delta\zeta}P_{0,0,-1}
$$

$$
=\frac{(\xi_{x_3})_{0,0,0}}{2\Delta\xi}P_{1,0,0}+\frac{(\eta_{x_3})_{0,0,0}}{2\Delta\eta}P_{0,1,0}+\frac{(\zeta_{x_3})_{0,0,0}}{2\Delta\zeta}P_{0,0,1};
$$

$$
-\frac{(\xi_{x_3})_{N_1,0,0}}{2\Delta\xi}P_{N_1+1,0,0}+\frac{(\eta_{x_3})_{N_1,0,0}}{2\Delta\eta}P_{N_1,-1,0}+\frac{(\zeta_{x_3})_{N_1,0,0}}{2\Delta\zeta}P_{N_1,0,-1}
$$

$$
=-\frac{(\xi_{x_3})_{N_1,0,0}}{2\Delta\xi}P_{N_1-1,0,0}+\frac{(\eta_{x_3})_{N_1,0,0}}{2\Delta\eta}P_{N_1,1,0}+\frac{(\zeta_{x_3})_{N_1,0,0}}{2\Delta\zeta}P_{N_1,0,1};
$$

$$
\frac{(\xi_{x_3})_{0,N_2,0}}{2\Delta\xi}P_{-1,N_2,0}-\frac{(\eta_{x_3})_{0,N_2,0}}{2\Delta\eta}P_{0,N_2+1,0}+\frac{(\zeta_{x_3})_{0,N_2,0}}{2\Delta\zeta}P_{0,N_2,-1}
$$

$$
=\frac{(\xi_{x_3})_{0,N_2,0}}{2\Delta\zeta}P_{1,N_2,0}-\frac{(\eta_{x_3})_{0,N_2,0}}{2\Delta\eta}P_{0,N_2-1,0}+\frac{(\zeta_{x_3})_{0,N_2,0}}{2\Delta\zeta}P_{0,N_2,1};
$$

$$
-\frac{(\xi_{x_3})_{N_1,N_2,0}}{2\Delta\xi}P_{N_1+1,N_2,0}-\frac{(\eta_{x_3})_{N_1,N_2,0}}{2\Delta\eta}P_{N_1,N_2+1,0}
$$

$$
+\frac{(\zeta_{x_3})_{N_1,N_2,0}}{2\Delta\zeta}P_{N_1,N_2,-1}=-\frac{(\xi_{x_3})_{N_1,N_2,0}}{2\Delta\xi}P_{N_1-1,N_2,0}
$$

$$
-\frac{(\eta_{x_3})_{N_1,N_2,0}}{2\Delta\eta}P_{N_1,N_2-1,0}+\frac{(\zeta_{x_3})_{N_1,N_2,0}}{2\Delta\zeta}P_{N_1,N_2,1}.
$$

Thus, we have the following linear equations,

$$\mathbf{U}\mathbf{P}^g = \mathbf{V}\mathbf{P}^i + \mathbf{W}, \tag{4.76}$$

with

$$\begin{aligned}
(\mathbf{P}^g)^T &= \langle P_{-1,0,0}, P_{-1,N_2,0}, P_{N_1+1,0,0}, P_{N_1+1,N_2,0}, P_{0,-1,0}, P_{N_1,-1,0},\\
&\quad P_{0,N_2+1,0}, P_{N_1,N_2+1,0}, P_{0,0,-1}, P_{0,N_2,-1}, P_{N_1,0,-1}, P_{N_1,N_2,-1}\rangle,\\
(\mathbf{P}^i)^T &= \langle P_{1,0,0}, P_{0,1,0}, P_{0,0,1}, P_{1,N_2,0}, P_{0,N_2-1,0}, P_{0,N_2,1},\\
&\quad P_{N_1-1,0,0}, P_{N_1,1,0}, P_{N_1,0,1}, P_{N_1-1,N_2,0}, P_{N_1,N_2-1,0}, P_{N_1,N_2,1}\rangle,\\
(\mathbf{W})^T &= \langle 0,0,0,0,g_o,g_o,g_{N_2},g_{N_2},0,0,0,0,\rangle,
\end{aligned}$$

where the entities of the matrix **S** are directly derived.

Note that, Eq. (4.76) is solved exactly (direct solver) for every time step for explicit scheme, or every iteration for implicit scheme. Assume that the solution of Eq. (4.76) is expressed as $\mathbf{P}^g = \mathbf{T}\mathbf{P}^i + \mathbf{S}$. The coefficients are used to modify the assembled matrix.

The detailed matrix formulation enables the use of the direct linear solver as well as the implementation of algebraic multigrid iterative solver. However, to implement a geometrical multigrid solver based on Jacobi, Gauss–Seidel, or Successive Overrelaxation (SOR) iterative schemes, named after Carl Gustav Jacob Jacobi (1804–1851), Johann Carl Friedrich Gauss (1777–1855), and Philipp Ludwig von Seidel (1821–1896), we do not have to formulate such matrices. Moreover, such a geometrical multigrid iterative solver enables us to vary effectively the discretization schemes by altering the iterative procedures.

In general, we call the stations from 1 to $N_1 - 1$, $N_2 - 1$, and $N_3 - 1$ in three directions, as the domain proper illustrated in Fig. 4.18. Moreover, also depicted in Fig. 4.18, the stations at 0, N_1, N_2, and N_3 as the ghost region for Neumann or periodic boundary conditions and the boundary points for Dirichlet boundary conditions, respectively. In short, ghost cells refer to $i = -1, N_1 + 1$, $j = -1, N_2 + 1$, and $k = -1, N_3 + 1$. The subtle change of the grid point definitions for different boundary conditions are illustrated in Fig. 4.19. For example, if both ends of one dimension length

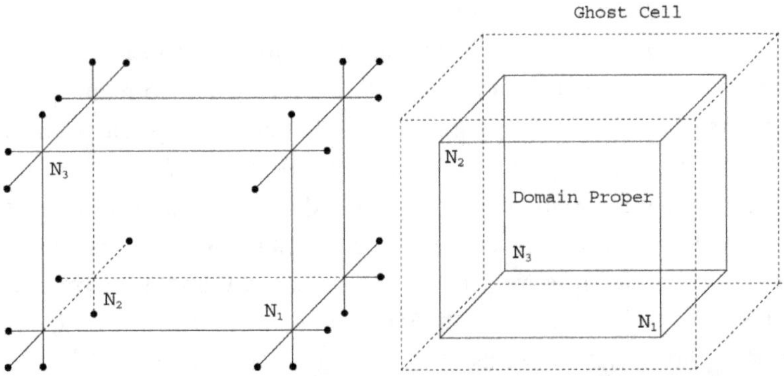

(a) Ghost Cell (b) Domain Proper

Fig. 4.18. Domain proper and ghost cell regions.

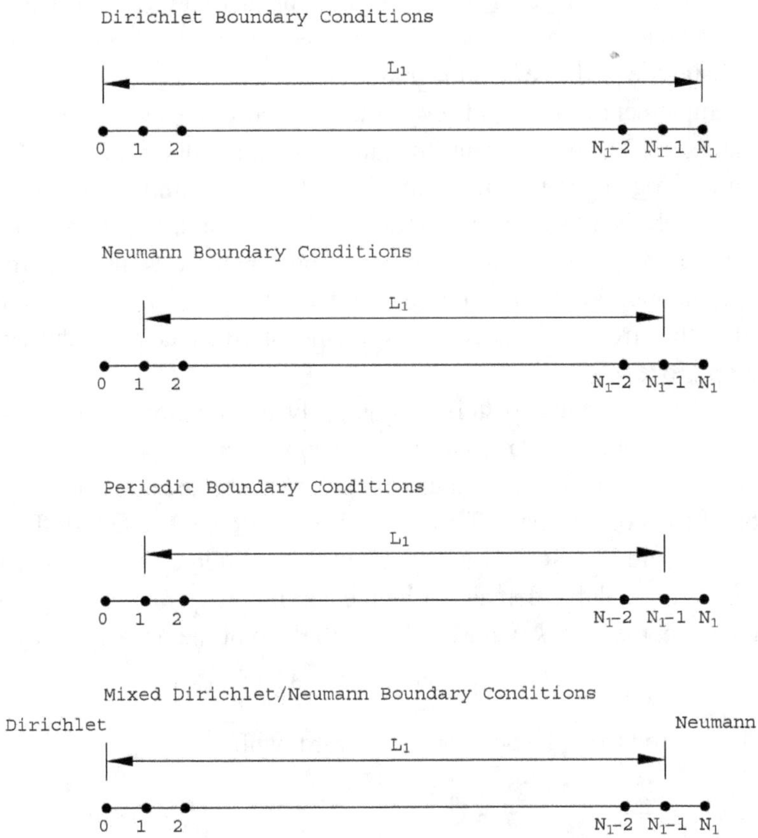

Fig. 4.19. Dirichlet, Neumann, and periodic boundary treatments.

L_1 have Dirichlet boundary conditions, with $N_1 = 2^{n_1}$ divisions, the grid size h_1 is defined as L_1/N_1. In the same computer program, in order to assign the special values to the points 0 and N_1, hence, to retain the same numbering structure in the program, i.e., the known or given points are 0 and N_1 points and the station unknowns are from 1 to $N_1 - 1$. Therefore, if both ends have Neumann boundary conditions, the physical length in the x direction starts at 1 and ends at N_1. Thus, we have the mesh size $h_1 = L_1/(N_1 - 2)$. Moreover, the station values at 0 and N_1 correspond to 2 and $N_1 - 2$, hence, named as the ghost points, whereas 1 and $N_1 - 1$ are defined as the boundary points. Naturally, as N_1 increases, such a discrepancy becomes insignificant. Likewise, for periodic boundary conditions, ghost points 0 and N_1 are assigned with the station values at $N_1 - 1$ and 1, respectively. For the case with mixed Dirichlet and Neumann boundary conditions, for instance, 0 is the Dirichlet boundary and $N_1 - 1$ is the Neumann boundary which provides a link between the station values at N_1 and $N_1 - 2$, and the grid size is defined as $h_1 = L_1/(N_1 - 1)$.

To implement a multigrid program, we need to choose a manageable data structure. The overall data structure of our three-dimensional V-Cycle multigrid program is illustrated in Fig. 4.20. In the implementation, we store the right-hand side source or boundary terms in the working array $\mathbf{f}$. In order to use various iteration or relaxation schemes, we use the array $\mathbf{v}$ to store the iteration results $\mathbf{u}$ based on $\mathbf{Au} = \mathbf{f}$, and the buffer array $\mathbf{w}$ to calculate the error $\mathbf{r} = \mathbf{f} - \mathbf{Au}$ and to store the intermediate result during the iteration process.

It is the convention to store contiguously the solutions and the right-hand side vectors on the nested grids in single arrays. The data structures of $\mathbf{f}$, $\mathbf{w}$, and $\mathbf{v}$ are identical with the cascade of hierarchical meshes stacked on top of each other. The buffer array $\mathbf{w}$ is in particular needed for the iterative procedure like SOR. We implement three different mesh coarsening strategies. For a cubic domain with an equal number of divisions in three directions, at the level k, with $1 \leq k \leq n$, the size of the working array is

$$(2^{n-k+1} + 1)(2^{n-k+1} + 1)(2^{n-k+1} + 1),$$

which is stored from the position ntt onward, with

$$ntt = \frac{2^{3n-3k+3} - 2^3}{7} + 2^{2n-2k+2} - 2^2 + 3(2^{n-k+1} - 2) + n - k.$$

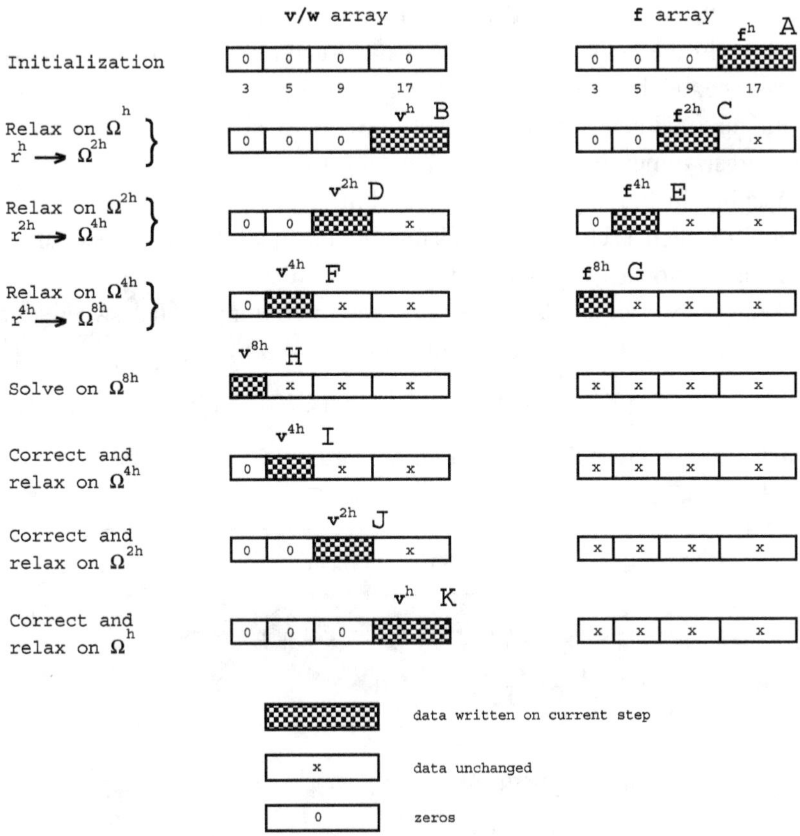

Fig. 4.20. Data structures of V-cycle multigrid procedures.

If the computational domain is a rectangular box, we have two different approaches. Suppose the size of the domain is

$$(2^{n_1}+1) \times (2^{n_2}+1) \times (2^{n_3}+1),$$

with $n_1 \leq n_2 \leq n_3$, in the first approach, we coarsen the mesh uniformly in three dimensions from the level 1 to n_1.

In general, if the division numbers in three directions are not extremely different, such a procedure will be very efficient. For example, if the second dimension has eight times the division number and the third dimension has 128 times the division number in comparison with the first dimension, we only need to solve a linear system of equations with 3825 unknowns.

The other approach to dealing with a rectangular box domain is more involved in the construction of the data structures of the working arrays. In essence, such a coarsening procedure includes a coarsening uniformly in all three dimensions from level 1 to n_1, a coarsening uniformly in two larger dimensions from level n_1+1 to n_2, if $n_2 > n_1$, and a final coarsening in the largest dimension from level n_2+1 to n_3, if $n_3 > n_2$. Therefore, such a scheme includes one-dimensional, two-dimensional, and three-dimensional V-Cycle multigrid procedures in the same program.

With respect to the data structures, the three-dimensional sweep at the level k, with $1 \le k \le n_1$, the size of the working array is

$$(2^{n_1-k+1}+1)(2^{n_2-k+1}+1)(2^{n_3-k+1}+1),$$

which is stored from the position ntt onward, with

$$\begin{aligned} ntt = nttb &+ \frac{2^{n_1+n_2+n_3-3k+3} - 2^{n_2+n_3-2n_1+3}}{7} + \frac{2^{n_2+n_1-2k+2} - 2^{n_2-n_1+2}}{3} \\ &+ \frac{2^{n_3+n_1-2k+2} - 2^{n_3-n_1+2}}{3} + \frac{2^{n_2+n_3-2k+2} - 2^{n_3+n_2-2n_1+2}}{3} \\ &+ 2^{n_1-k+1} - 2 + 2^{n_2-k+1} - 2^{n_2-n_1+1} + 2^{n_3-k+1} \\ &- 2^{n_3-n_1+1} + n_1 - k, \end{aligned}$$

with

$$\begin{aligned} nttb &= ntta + 2^{n_3+n_2-2n_1+2} - 2^{n_3-n_2+2} + 3(2^{n_2-n_1+1} - 2^1) \\ &\quad + 3(2^{n_3-n_1+1} - 2^{n_3-n_2+1}) + 3(n_2 - n_1), \\ ntta &= 9(2^{n_3-n_2+1} - 2 + n_3 - n_2). \end{aligned}$$

Likewise, the two-dimensional sweep at the level n_1+k, if $n_2 > n_1$, with $1 \le k \le n_2 - n_1$, the size of the working array is

$$(2^1+1)(2^{n_2-n_1-k+1}+1)(2^{n_3-n_1-k+1}+1),$$

which is stored from the position ntt onward, with

$$\begin{aligned} ntt = ntta &+ 2^{n_3+n_2-2n_1-2k+2} - 2^{n_3-n_2+2} + 3(2^{n_2-n_1-k+1} - 2) \\ &+ 3(2^{n_3-n_1-k+1} - 2^{n_3-n_2+1}) + 3(n_2 - n_1 - k). \end{aligned}$$

Finally, the one-dimensional sweep at the level n_2+k, if $n_3 > n_2$, with $1 \le k \le n_3 - n_2$, the size of the working array is

$$(2^1+1)(2^1+1)(2^{n_3-n_2-k+1}+1),$$

which is stored from the position *ntt* onward, with

$$ntt = 9(2^{n_3-n_2-k+1} - 2 + n_3 - n_2 - k).$$

Consider a typical V-cycle multigrid scheme, assign ν_1 and ν_2 as the iteration numbers of the down and up cycle sweeps, the computational flow can be illustrated as follows:

$$\mathbf{v}^h \leftarrow MV^h(\mathbf{v}^h, \mathbf{f}^h).$$

(1) Relax ν_1 times on $\mathbf{K}^h\mathbf{b}^h = \mathbf{f}^h$ with given initial guess $\mathbf{v}^h = \mathbf{0}$.
(2) If Ω^h coarsest grid, then go to (4), else

$$\begin{aligned}
&\mathbf{f}^{2h} \leftarrow \mathbf{I}_h^{2h}(\mathbf{f}^h - \mathbf{K}^h\mathbf{v}^h)\\
&\mathbf{v}^{2h} \leftarrow \mathbf{0}\\
&\mathbf{v}^{2h} \leftarrow MV^{2h}(\mathbf{v}^{2h}, \mathbf{f}^{2h})
\end{aligned}$$

(3) Convert $\mathbf{v}^h \leftarrow \mathbf{v}^h + \mathbf{I}_{2h}^h\mathbf{v}^{2h}$.
(4) Relax ν_2 times on $\mathbf{K}^h\mathbf{b}^h = \mathbf{f}^h$ with initial guess $\mathbf{v}^h$.

Assuming $n_1 = n_2 = n_3 = n$, we assign the grid number N as 2^n and consequently, as we move from level 1 to level n, the grid size evolves from h to $2^{n-1}h$. The multigrid scheme can be illustrated as follows:

(1) Relax on $\mathbf{K}^{2^{l-1}h}\mathbf{u}^{2^{l-1}h} = \mathbf{f}^{2^{l-1}h}$ ν_1 times with initial guess $\mathbf{v}^{2^{l-1}h} = \mathbf{0}$.
(2) Employ the full weighting restriction operator $\mathbf{I}_{2^{l-1}h}^{2^lh}$ and compute $\mathbf{f}^{2^lh} = \mathbf{I}_{2^{l-1}h}^{2^lh}\mathbf{r}^{2^{l-1}h}$.
(3) Repeat (1) and (2) from level 1 to $n-1$.
(4) Solve exactly $\mathbf{K}^{2^{n-1}h}\mathbf{u}^{2^{n-1}h} = \mathbf{f}^{2^{n-1}h}$.
(5) Employ the interpolation operator $\mathbf{I}_{2^lh}^{2^{l-1}h}$ and correct $\mathbf{v}^{2^{l-1}h}$ with $\mathbf{v}^{2^{l-1}h} + \mathbf{I}_{2^lh}^{2^{l-1}h}\mathbf{v}^{2^lh}$.
(6) Relax on $\mathbf{A}^{2^{l-1}h}\mathbf{u}^{2^{l-1}h} = \mathbf{f}^{2^{l-1}h}$ ν_2 times with initial guess $\mathbf{v}^{2^{l-1}h}$.
(7) Repeat (5) and (6) from level $n-1$ to 1.

In the three-dimensional case, the full weighting operator $\mathbf{I}_h^{2h}$ is defined as

$$\begin{aligned}24P^{2h}_{i,j,k} = {} & P^h_{2i-1,2j-1,2k-1} + P^h_{2i+1,2j-1,2k-1} + P^h_{2i-1,2j+1,2k-1}\\ & + P^h_{2i+1,2j+1,2k-1} + P^h_{2i-1,2j-1,2k+1} + P^h_{2i+1,2j-1,2k+1}\\ & + P^h_{2i-1,2j+1,2k+1} + P^h_{2i+1,2j+1,2k+1} + 2P^h_{2i-1,2j-1,2k}\\ & + 2P^h_{2i+1,2j-1,2k} + 2P^h_{2i-1,2j+1,2k} + 2P^h_{2i+1,2j+1,2k}\\ & + 2P^h_{2i-1,2j,2k-1} + 2P^h_{2i+1,2j,2k-1} + 2P^h_{2i,2j-1,2k-1} + 2P^h_{2i,2j+1,2k-1}\\ & + 2P^h_{2i-1,2j,2k+1} + 2P^h_{2i+1,2j,2k+1} + 2P^h_{2i,2j-1,2k+1} + 2P^h_{2i,2j+1,2k+1}\\ & + 4P^h_{2i+1,2j,2k} + 4P^h_{2i,2j-1,2k} + 4P^h_{2i-1,2j,2k} + 4P^h_{2i,2j+1,2k}\\ & + 4P^h_{2i,2j,2k-1} + 4P^h_{2i,2j,2k+1} + 8P^h_{2i,2j,2k},\end{aligned}$$

whereas the interpolation operator $\mathbf{I}_{2h}^h$ is provided as follows:

$$\begin{aligned}P^h_{2i,2j,2k} &= P^{2h}_{i,j,k}; \quad 2P^h_{2i+1,2j,2k} = P^{2h}_{i,j,k} + P^{2h}_{i+1,j,k},\\ 2P^h_{2i,2j+1,2k} &= P^{2h}_{i,j,k} + P^{2h}_{i,j+1,k}, \quad 2P^h_{2i,2j,2k+1} = P^{2h}_{i,j,k} + P^{2h}_{i,j,k+1},\\ 4P^h_{2i+1,2j+1,2k} &= P^{2h}_{i,j,k} + P^{2h}_{i+1,j,k} + P^{2h}_{i,j+1,k} + P^{2h}_{i+1,j+1,k},\\ 4P^h_{2i+1,2j,2k+1} &= P^{2h}_{i,j,k} + P^{2h}_{i+1,j,k} + P^{2h}_{i,j,k+1} + P^{2h}_{i+1,j,k+1},\\ 4P^h_{2i,2j+1,2k+1} &= P^{2h}_{i,j,k} + P^{2h}_{i,j+1,k} + P^{2h}_{i,j,k+1} + P^{2h}_{i,j+1,k+1},\\ 8P^h_{2i+1,2j+1,2k+1} &= P^{2h}_{i+1,j,k} + P^{2h}_{i,j+1,k} + P^{2h}_{i,j,k+1} + P^{2h}_{i,j,k} + P^{2h}_{i+1,j+1,k}\\ &\quad + P^{2h}_{i+1,j,k+1} + P^{2h}_{i,j+1,k+1} + P^{2h}_{i+1,j+1,k+1}.\end{aligned}$$

In the two-dimensional case, the full weighting operator $\mathbf{I}_h^{2h}$ is defined as

$$\begin{aligned}16P^{2h}_{j,k} = {} & P^h_{2j-1,2k-1} + P^h_{2j-1,2k+1} + P^h_{2j+1,2k-1} + P^h_{2j+1,2k+1}\\ & + 2(P^h_{2j,2k-1} + P^h_{2j,2k+1} + P^h_{2j-1,2k} + P^h_{2j+1,2k}) + 4P^h_{2j,2k},\end{aligned}$$

whereas the interpolation operator $\mathbf{I}_{2h}^h$ is provided as follows:

$$\begin{aligned}P^h_{2j,2k} &= P^{2h}_{j,k}, \quad 2P^h_{2j+1,2k} = P^{2h}_{j,k} + P^{2h}_{j+1,k},\\ 2P^h_{2j,2k+1} &= P^{2h}_{j,k} + P^{2h}_{j,k+1}, \quad 4P^h_{2j+1,2k+1} = P^{2h}_{j,k} + P^{2h}_{j+1,k} + P^{2h}_{j,k+1} + P^{2h}_{j+1,k+1}.\end{aligned}$$

In the one-dimensional case, the full weighting operator $\mathbf{I}_h^{2h}$ is defined as

$$4P^{2h}_k = P^h_{2k-1} + 2P^h_{2k} + P^h_{2k+1},$$

whereas the interpolation operator $\mathbf{I}_{2h}^{h}$ is provided as follows:

$$P_{2k}^{h} = P_{k}^{2h},$$
$$2P_{2k+1}^{h} = P_{k}^{2h} + P_{k+1}^{2h}.$$

4.3 Verifications and Validations

4.3.1 *Poisson equation models*

To illustrate the basic procedures of multigrid solvers, we present first a simple one-dimensional model problem,

$$-\frac{\partial^2 u(x)}{\partial x^2} + \sigma u(x) = f(x), \tag{4.77}$$

with $0 \leq x \leq l$ and $\sigma > 0$.

If we have the Dirichlet boundary condition $u(0) = u(l) = 0$ and $f(x) = C\sin(k\pi x/l)$, the analytical solution for Eq. (4.77) can be expressed as

$$u(x) = \frac{C}{k^2\pi^2/l^2 + \sigma}\sin(k\pi x/l),$$

where k stands for the wave number.

Suppose we discretize the domain into N segments, namely, $N+1$ points, the boundary points $x = 0$ and $x = l$ correspond to $j = 0$ and $j = N+1$, respectively, the grid size h can be expressed as $h = l/N$. Thus, the discretized $f(x)$ can be represented as $f_j^h = C\sin(jk\pi/N)$ and the exact solution is simply written as

$$u_j = Cl^2\sin(jk\pi/N)/(k^2\pi^2 + l^2\sigma).$$

Replacing the derivative of Eq. (4.77) with a second-order finite difference leads to the following system of linear equations

$$-\frac{u_{j-1} + u_{j+1} - 2u_j}{h^2} + \sigma u_j = f_j, \quad \text{with } 1 \leq j \leq N-1. \tag{4.78}$$

Naturally, from the lexicographic ordering, such a system of equations can be written in a tridiagonal matrix form and the direct solver can be very effectively used. Denote the linear system in a matrix form $\mathbf{Ku} = \mathbf{f}$, the matrix $\mathbf{K}$ can be decomposed as

$$\mathbf{K} = \mathbf{D} - \mathbf{L} - \mathbf{U},$$

where $\mathbf{D}$, $-\mathbf{L}$, and $-\mathbf{U}$ are the diagonal of $\mathbf{K}$, strictly lower, and strictly upper triangular parts of $\mathbf{K}$, respectively.

Furthermore, Eq. (4.78) can be rewritten as

$$\mathbf{u} = \mathbf{P}_J\mathbf{u} + \mathbf{D}^{-1}\mathbf{f}, \tag{4.79}$$

with $\mathbf{P}_J = \mathbf{D}^{-1}(\mathbf{L}+\mathbf{U})$.

Equation (4.79) is the starting point of various relaxation schemes. In this section, for demonstration purposes, we also illustrate a weighted Jacobi iteration method, one of the relaxation schemes, for the solution of the system equation (4.78),

$$\mathbf{v}^{(i+1)} = [(1-\omega)\mathbf{I} + \omega\mathbf{P}_J]\mathbf{v}^{(i)} + \omega\mathbf{D}^{-1}\mathbf{f}, \tag{4.80}$$

where ω is a weighting factor.

Note that with $\omega = 1$, we have the original Jacobi iteration. Of course, as the iteration number i increases, the approximation $\mathbf{v}^i$ approaches the solution $\mathbf{u}$ and the residual vector $\mathbf{r}^i = \mathbf{f} - \mathbf{K}\mathbf{v}^i$ approaches zero. The relaxation scheme like Eq. (4.80) can be easily changed in the multigrid solution scheme.

Employ the discrete L_2 norm, we define the error as

$$||\mathbf{u}^h - \mathbf{v}^h|| = \sqrt{\sum_{j=1}^{N-1} (u_j^h - v_j^h)^2/N}.$$

Note that even in the case of direct solvers, we can seldom solve the discrete problem exactly. In general, the quantity or error norm measuring the difference between the direct solver solution and the exact solution is called the global error. The global error measures the truncation error of the discretization process as well as the round-off error associated with the computation precision. Moreover, the error norm measuring the difference between the relaxation or multigrid solution and the exact solution is called the algebraic error. Naturally, the algebraic error measures how well our iterative approximations agree with the exact solution.

We intend to use Poisson equations for one-dimensional (1D), two-dimensional (2D), and three-dimensional (3D) to verify the multigrid solvers. For the one-dimensional problem presented in this section, we compare the performance of the tridiagonal, Jacobi method, and multigrid method for modes 1, 2, and 4. Without detriment to the illustrative purpose, we choose, for convenience, $l = 1$, $C = 1$, $\sigma = 2$, $\nu_1 = \nu_2 = 2$, and $\omega = 1$.

In addition, the grid size h is assigned as $1/2^5$ and the number of iterations of the Jacobi method is set to be 10.

It is clearly shown in Fig. 4.21 that for both low and high wave numbers the direct solver provides very accurate solutions, namely, the global errors are small. To be more specific, in this set of examples, the global errors are 3.979984×10^{-5}, 5.222148×10^{-5}, and 5.654029×10^{-5} with respect to modes 1, 2, and 4, respectively. Note that the global error slightly increases for higher modes. Furthermore, for 10 iterations, the algebraic errors of the Jacobi method are 5.621204×10^{-2}, 1.389512×10^{-2}, and 1.952700×10^{-3} with respect to modes 1, 2, and 4. Consistent with the theoretical prediction, the convergence rate of the iteration solution procedures for the lower modes is significantly lower than that of the higher modes. Although this trend still exists, since the multigrid relaxation sweep is the Jacobi method, the algebraic errors of the multigrid, 3.466863×10^{-3}, 9.529109×10^{-4}, and 2.1873990×10^{-4} with respect to modes 1, 2, and 4, are much lower than those of the Jacobi method.

We now move on to the following three-dimensional model problem,

$$\frac{\partial^2 u}{\partial x^2} + \frac{\partial^2 u}{\partial y^2} + \frac{\partial^2 u}{\partial z^2} = f(x,y,z), \tag{4.81}$$

with $0 \le x \le l_1$, $0 \le y \le l_2$, and $0 \le z \le l_3$.

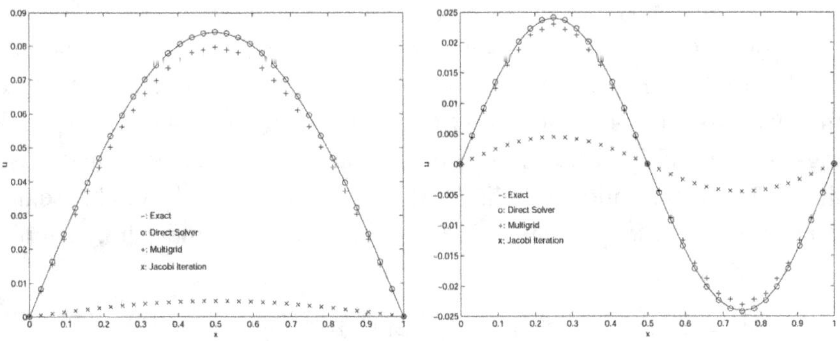

(a) Mode 1 (b) Mode 2

Fig. 4.21. Comparison of results of mode 1 and 2.

Suppose we have again the Dirichlet boundary condition $u = 0$ at 0, l_1, l_2, and l_3 and

$$f(x,y,z) = C\sin(k_1\pi x/l_1)\sin(k_2\pi y/l_2)\sin(k_3\pi z/l_3),$$

thus the analytical solution for Eq. (4.81) can be expressed as

$$u = -\frac{C}{k_1^2\pi^2/l_1^2 + k_2^2\pi^2/l_2^2 + k_3^2\pi^2/l_3^2}\sin(k_1\pi x/l_1)\sin(k_2\pi y/l_2)\sin(k_3\pi z/l_3),$$

where k_1, k_2, and k_3 stand for the wave numbers in x, y, and z directions, respectively.

Suppose we discretize the domain into $N_1 \times N_2 \times N_3$ segments, namely, $(N_1+1)\times(N_2+1)\times(N_3+1)$ points, the boundary points 0, l_1, l_2, and l_3 correspond to 0, N_1, N_2, and N_3, respectively, the discretized $f(x,y,z)$ can be represented as

$$f_{i,j,k}^h = C\sin(ik_1\pi/N_1)\sin(jk_2\pi/N_2)\sin(kk_3\pi/N_3),$$

and the exact solution is simply written as

$$\begin{aligned} u_{i,j,k} &= -\frac{C}{(k_1^2\pi^2/l_1^2 + k_2^2\pi^2/l_2^2 + k_3^2\pi^2/l_3^2)} \\ &\quad \times \sin(ik_1\pi/N)\sin(jk_2\pi/N)\sin(kk_3\pi/N). \end{aligned}$$

Replacing the derivative of Eq. (4.81) with the 19-point finite difference scheme as discussed in Section 4.2.2, the system of equations can be written in a tridiagonal block matrix form. However, in this case, the traditional tridiagonal solver will not be applicable, since the tridiagonal proper will break down in the course of the solution process. In this section, we employ the successive over-relaxation (SOR) iteration method and the corresponding V-cycle multigrid iteration method. Due to the sequence of updating, the SOR scheme could be slightly different. For example, if we choose the data structure to loop from j, via k, to i, the SOR scheme can be simply illustrated as

$$\begin{aligned} u_{i,j,k}^{(m+1)} &= \frac{\omega}{C_1}(f_{i,j,k} - C_{10}u_{i-1,j+1,k}^{(m+1)} - C_{14}u_{i-1,j,k+1}^{(m+1)} - C_3u_{i-1,j,k}^{(m+1)} \\ &\quad - C_{11}u_{i-1,j-1,k}^{(m+1)} - C_{15}u_{i-1,j,k-1}^{(m+1)} - C_7u_{i,j,k-1}^{(m+1)} - C_{13}u_{i+1,j,k-1}^{(m+1)} \end{aligned}$$

$$
\begin{aligned}
&-C_{17}u_{i,j+1,k-1}^{(m+1)} - C_{19}u_{i,j-1,k-1}^{(m+1)} - C_{18}u_{i,j-1,k+1}^{(m+1)} \\
&-C_5u_{i,j-1,k}^{(m+1)} - C_2u_{i+1,j,k}^{(m)} - C_4u_{i,j+1,k}^{(m)} - C_6u_{i,j,k+1}^{(m)} \\
&-C_8u_{i+1,j+1,k}^{(m)} - C_{12}u_{i+1,j,k+1}^{(m)} - C_{16}u_{i,j+1,k+1}^{(m)} \\
&-C_9u_{i+1,j-1,k}^{(m)}) + (1-\omega)u_{i,j,k}^{(m)},
\end{aligned}
$$

where ω is a weighting factor.

Note that with $\omega = 1$, we have the original Jacobi iteration. For the three-dimensional problem presented in this section, we compare the performance of the SOR method and multigrid method for modes 1 and 2. Without detriment to the illustrative purpose, we choose, for convenience, $l = 1, C = 1, \nu_1 = \nu_2 = 2$, and $\omega = 1.1$. In addition, the grid size h_1, h_2, and h_3 are assigned as $1/2^6$ and the number of iterations of the SOR method is set to be 100.

For low wave number, mode 1, it takes 570 seconds to obtain the algebraic error 9.385221×10^{-3} using the SOR iterative method, whereas it takes 37 seconds to reach 1.721074×10^{-3} accuracy. For mode 2, with the same number of iterations, the SOR method yields an accuracy of 1.140422×10^{-3}, which is better than 2.342916×10^{-3} produced with the multigrid solver. However, the multigrid solver takes 36 second with the same number of sweeps (2), whereas the SOR method takes 566 seconds. To be more specific, in this set of examples, the global errors are 3.979984×10^{-5}, 5.222148×10^{-5}, and 5.654029×10^{-5} with respect to modes 1, 2, and 10, respectively. Furthermore, for 100 iterations, the algebraic errors of the Jacobi method are 5.621204×10^{-2}, 1.389512×10^{-2}, and 1.952700×10^{-3} with respect to modes 1, 2, and 10, respectively. Consistent with the theoretical prediction, the convergence rates for the lower modes is significantly lower than that of the higher modes.

Finally, as shown in Figs. 4.22–4.24, the multigrid solution closely resembles the exact solution and the computation is more than 30 times more efficient than the Jacobi iterative solver given a comparable accuracy. Of course, as predicted in theory, the Jacobi iteration provides better solution for higher modes than lower modes given the same number of iterations. In this study, we confirm that the geometrical multigrid solver is much more efficient than the conventional iterative methods. This speed

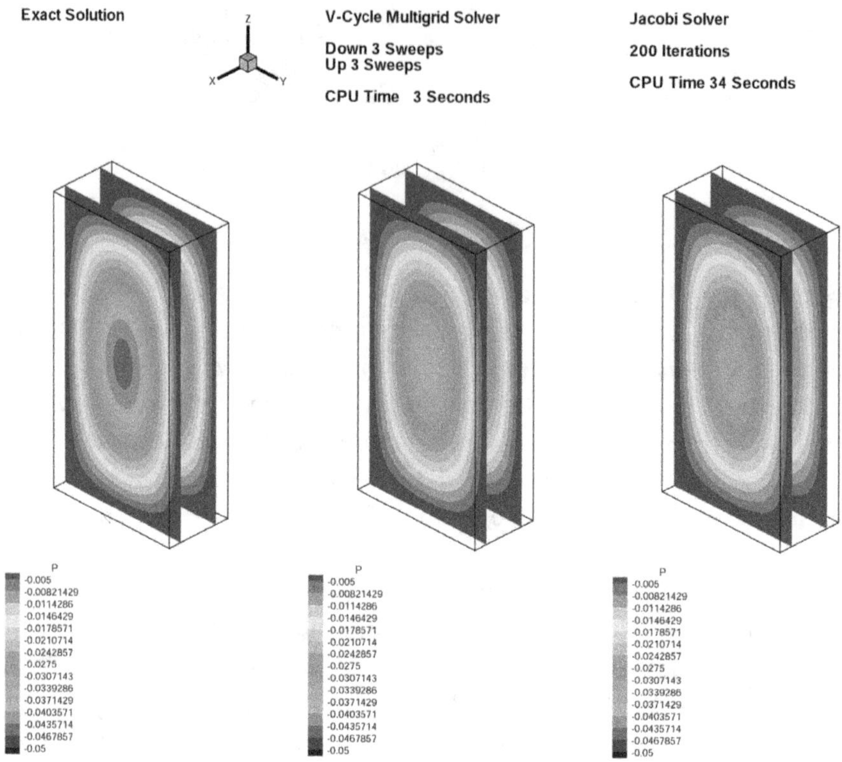

Fig. 4.22. First-mode rectangular box solution with V-cycle multigrid solver.

up in the solution of linear systems of equations is especially true for subsonic pressure solutions governed by elliptical partial differential equations. However, unlike the black box solver, the geometrical multigrid solver requires the connection with the grid coefficient generator. Nevertheless, such a cascade of grid coarsening steps only doubles the computation effort for the original problem setup, which is negligible in comparison with the saving in solution time. Furthermore, for the rectangular box computational domain, a uniform mesh coarsening coupled with the final exact solution using either tridiagonal or nonsymmetric column solver is found to be slightly more efficient than a sequential three-dimensional coarsening coupled with two-dimensional then one-dimensional coarsening.

In the first test problem of the combined scheme with ADI compact scheme based computational fluid dynamics and immersed methods.

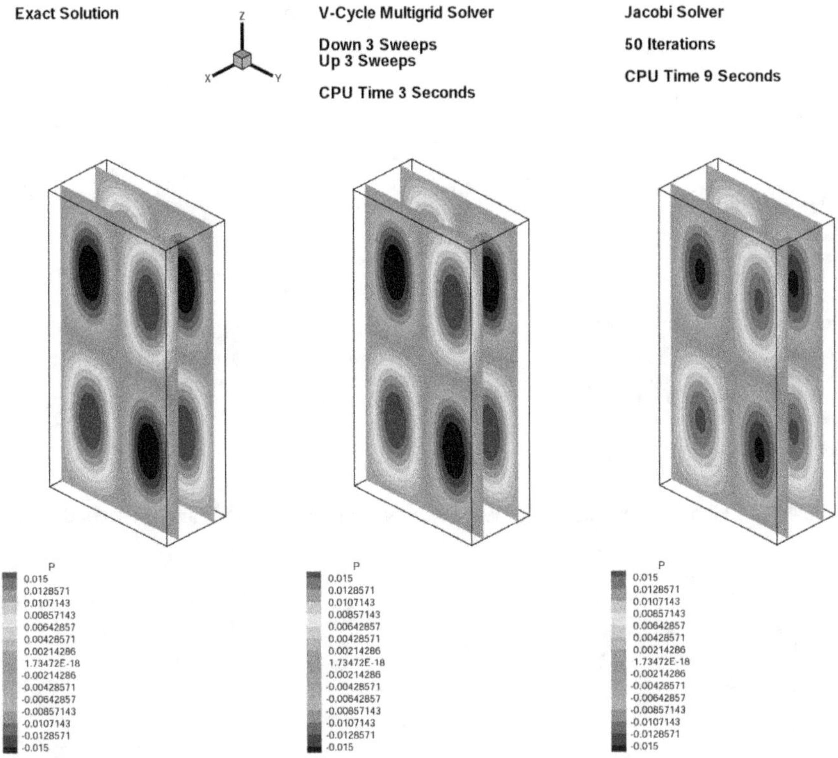

Fig. 4.23. Second-mode rectangular box solution with V-cycle multigrid solver.

A mass-spring system was studied. The mass point was tethered at the point $(2, 2.5)$. Pressure distributions of this system are depicted. It seems the pressure waves and the complex eddies can both be captured. In traditional staggered FSI analysis, there is a time step lag between fluid and solid domains. In the implementation, a subiteration is introduced to eliminate the time step. In fact, this subiteration actually represents the second best approach other than a monolithic implicit FSI analysis. The number of subiterations is studied. Ultimately, for an immersed cylinder mounted on a spring mass system along with a dashpot, the displacement, velocity, and acceleration histories along with the respective lift and drag coefficients are documented in Figs. 4.25–4.28. Furthermore, as depicted in Fig. 4.29, the so-called Karman vortex has been developed and eventually form the steady vortex shedding state.

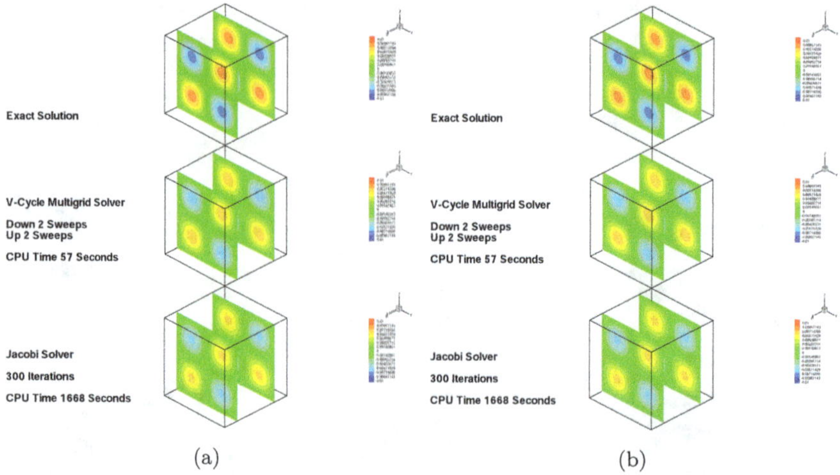

Fig. 4.24. Cubic box solution with V-cycle multigrid: (a) first mode and (b) second mode.

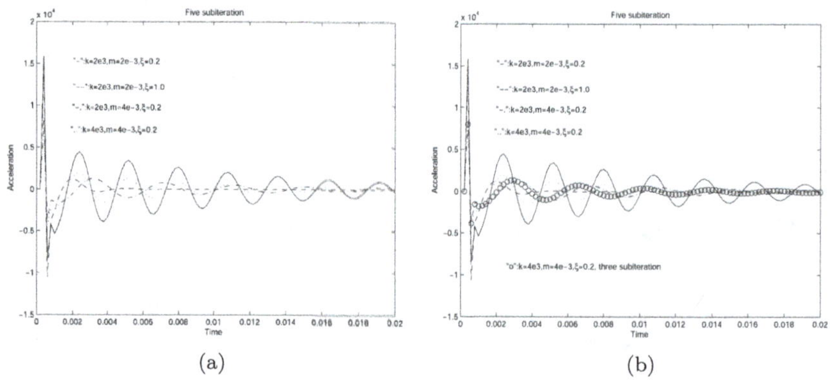

Fig. 4.25. IB methods with compact scheme flow-induced vibration example: (a) five subiteration and (b) three subiteration.

In this section, we also employ SD7003 airfoil as an example for the flexible airfoil. For simplicity, we assume the airfoil is made of hyperelastic material which allows both geometric and material nonlinearities. The entire airfoil is supported at two or three points. If two-point support is used, we choose the front part of the airfoil indicating that we define the shear center through the use of stiffeners and compartments near these two points. The preliminary two-dimensional or

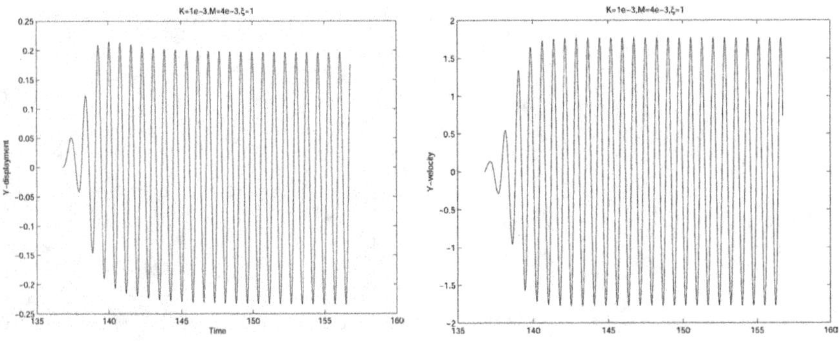

(a) (b)

Fig. 4.26. Motion of a cylinder immersed in viscous and compressible fluid: (a) vertical displacement and (b) vertical velocity.

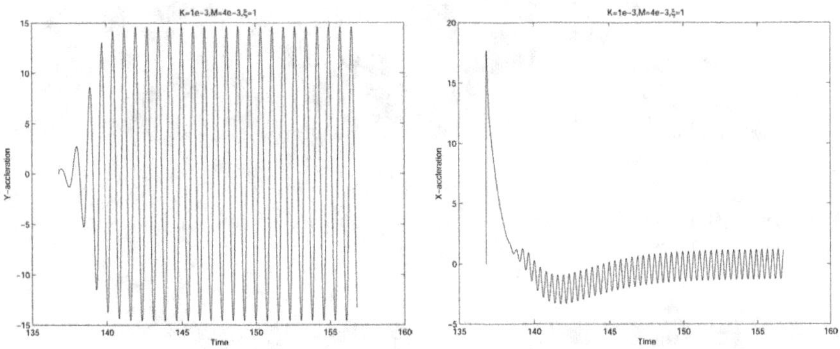

(a) (b)

Fig. 4.27. Acceleration of a cylinder immersed in viscous and compressible fluid: (a) vertical acceleration and (b) horizontal acceleration.

three-dimensional (2D+) model has the following scaled physical parameters: $\rho_f = 1$; $\rho_s = 2$; $C_1 = 4e5$; $C_2 = 2e3$; $\kappa_f = 2.1e5$; $\kappa_s = 1e6$; and $U(t) = 10(1+0.02\sin(2\pi/0.02t))$. As shown in Figs. 4.30 and 4.31, the vorticities develop initially on both sides of airfoil which occupies a finite physical space and later instabilities are triggered which lead to the vortex shedding as well as the corresponding tip motion.

Moreover, for a thin volumeless structure, namely, a sail, as shown in Fig. 4.32, the instability will lead to the camber development and eventually form the steady vortex shedding. In these cases, time step $\Delta t = 1^{-5}$ sec,

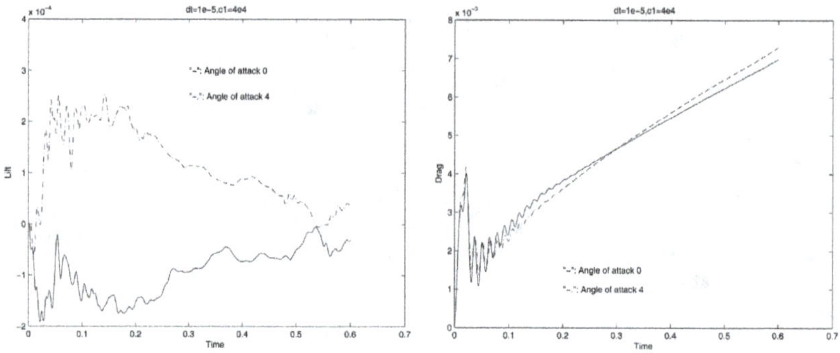

(a) (b)

Fig. 4.28. Lift and drag as functions of time for different angle of attack: (a) lift and (b) drag.

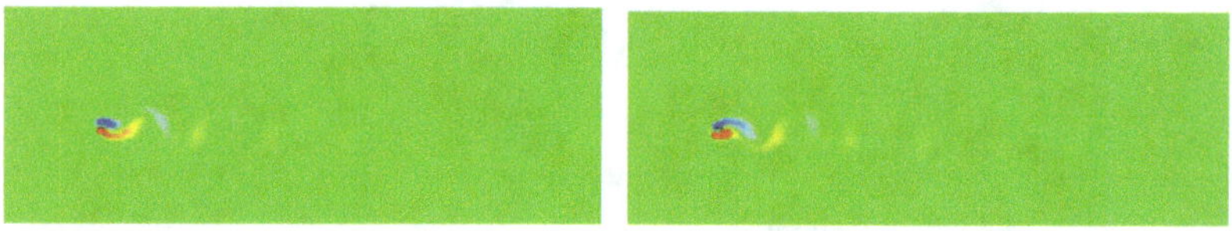

(a) (b)

Fig. 4.29. IB methods and compact schemes for a spring mass system. (a) Karman vortex and (b) steady state.

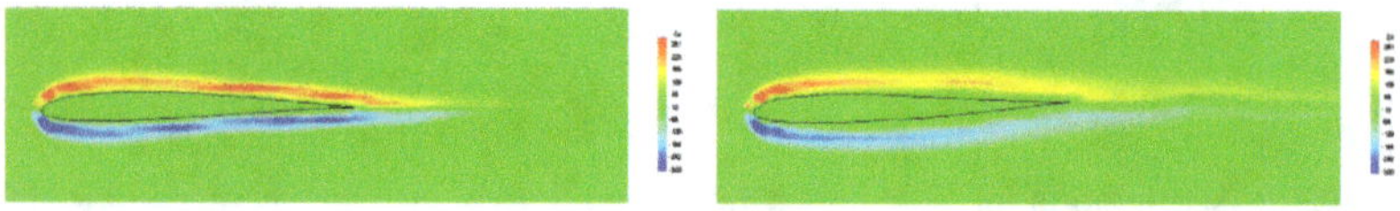

(a) (b)

Fig. 4.30. IB methods and compact schemes for a flexible airfoil No. 1. (a) Vortex shedding and (b) tip motion.

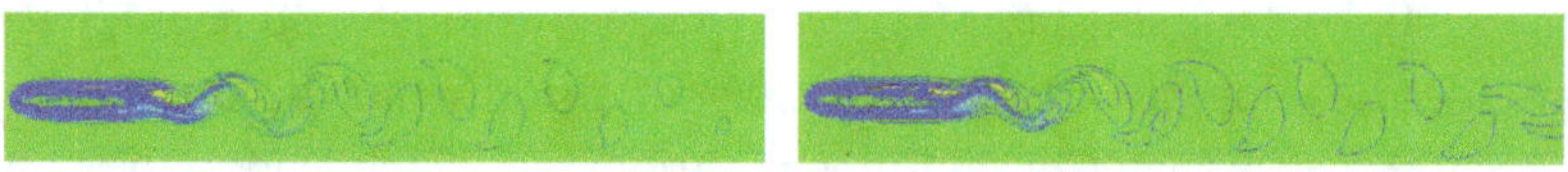

(a) (b)

Fig. 4.31. IB methods and compact schemes for a flexible airfoil No. 2. (a) Karman vortex and (b) steady state.

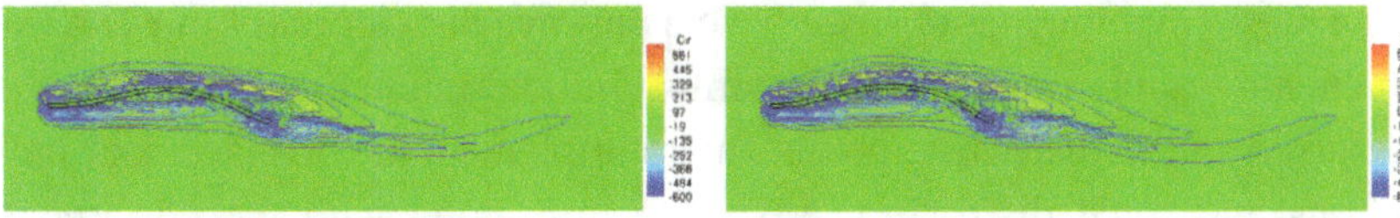

(a) (b)

Fig. 4.32. IB methods and compact schemes for a sail. (a) Entrance and (b) vortex shedding.

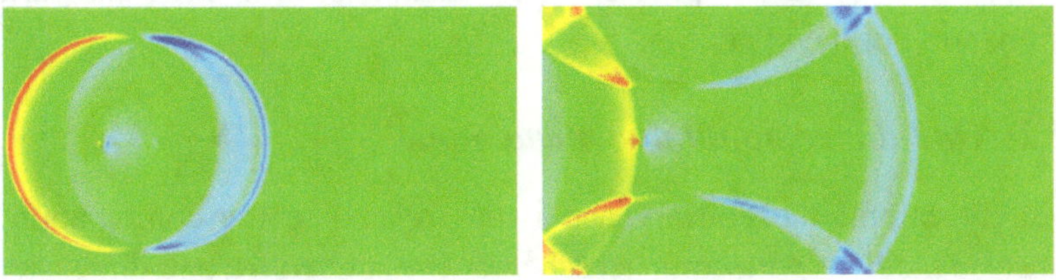

(a) (b)

Fig. 4.33. Compressible IB methods and compact schemes with shock waves: (a) time one and (b) time two.

the dynamic viscosity $\mu = 0.001$ Pa $\cdot$ s, incoming far field flow velocity 10 m/s, the flexible nonlinear rubber like material $C_1 = 2.93 \times 10^4$ Pa and $C_2 = 1.77 \times 10^4$ Pa, tether point for the rigid boundary stiffness $k_s = 1.41 \times 10^5$ N/m and tether point for flexible wing support $k_t = 2.0 \times 10^2$ N/m with the center position $(0.75, 0.75)$ and the mesh density 128×1024.

In Fig. 4.33, also depicted in Wang and Gordnier (2009), we present two time snapshots taken from a case study reported in Wang and Gordnier (2009) in which a submerged flexible structure is interacting with the surrounding compressible fluid, more specifically, air. In this case, full-fledged energy conservation form of governing equations in aerodynamics based on conformal mapping and compact schemes have been coupled with non-oscillatory time incremental schemes (Shu, 2020; Wang and Gordnier, 2009). The specific details are embedded in the proprietary code in Wright-Patterson Air Force Base. In order to illustrate the efficacy of immersed boundary methods for compact schemes, we intensionally set the boundaries to reflect the shock waves. It is clear that shocks are captured nicely as well as their interactions with the immersed structure.

In these implementations, we have a uniform background (ghost) mesh which allows us to employ the simple modulo function to locate the immersed solid or structure points and to implement the displacement projections and force distributions. This feature along with the concept of immersed methods enable us to avoid the topologically challenging mesh adjustments which are often prohibitive. The uniform background (ghost) mesh adopted for the compact scheme is also the important attributes for high accuracy and high computational efficiency. These features also naturally fit into parallel processing commands such as MPI and further vectorization of the operation.

4.3.2 *One-dimensional analytical model*

To study careful the convergence issues related to the combination of immersed methods with compact schemes, we start with a one-dimensional model problem with a solid immersed in a one-dimensional solid as illustrated in Fig. 4.34.

Assume both domains with a constant cross section A are subject to the gravitational constant g in x direction the governing equation for regions $(0, L_1) \cup (L_2, L)$ and (L_1, L_2) are written as

$$\frac{\partial}{\partial x}\left(E_f \frac{\partial u}{\partial x}\right) = \rho_f \frac{\partial^2 u}{\partial t^2} - \rho_f g, \quad \frac{\partial}{\partial x}\left(E_s \frac{\partial u}{\partial x}\right) = \rho_s \frac{\partial^2 u}{\partial t^2} - \rho_s g. \tag{4.82}$$

Equation (4.82) represent second-order partial differential equations for three connected regions. Therefore, the following six boundary conditions are needed

$$\begin{aligned} u(0) &= 0, & u(L) &= 0, \\ u(L_1^-) &= u(L_1^+), & u(L_2^+) &= u(L_2^-), \\ E_f \frac{\partial u}{\partial x}(L_1^-) &= E_s \frac{\partial u}{\partial x}(L_1^+), & E_f \frac{\partial u}{\partial x}(L_2^+) &= E_s \frac{\partial u}{\partial x}(L_2^-). \end{aligned} \tag{4.83}$$

If we consider L_1 and L_2 variables dependent of time and the state variable u, this problem will not be trivial. To start, let's look at the static equation in which Eq. (4.82) represent second-order differential equations.

Fig. 4.34. One-dimensional model problem.

Overall, six constants c_i will be introduced for the solution $u(x)$ for all three regions, namely, $(0,L_1)$, (L_2,L), and (L_1,L_2), and the solution $u(x)$ can be simply expressed as

$$
\begin{aligned}
u &= -\frac{\rho_f}{E_f} g \frac{x^2}{2} + \frac{c_1}{E_f} x + c_4, \quad x \in (0, L_1), \\
u &= -\frac{\rho_f}{E_f} g \frac{x^2}{2} + \frac{c_2}{E_f} x + c_5, \quad x \in (L_2, L), \\
u &= -\frac{\rho_s}{E_s} g \frac{x^2}{2} + \frac{c_3}{E_s} x + c_6, \quad x \in (L_1, L_2).
\end{aligned}
$$

The comparisons illustrated in Figs. 4.35–4.37 confirm the validity of immersed methods. In fact, this particular model problem represents a subtle shift to a continuum immersed in another continuum. In matrix form, the six constants can be directly solved with the linear system of equations $\mathbf{Ac} = \mathbf{b}$, represented with the boundary conditions in Eq. (4.83) with

$$
\mathbf{A} = \begin{bmatrix}
1 & 0 & -1 & 0 & 0 & 0 \\
0 & 1 & -1 & 0 & 0 & 0 \\
-\frac{L_1}{E_f} & 0 & \frac{L_1}{E_s} & -1 & 0 & 1 \\
0 & 0 & 0 & 1 & 0 & 0 \\
0 & \frac{L}{E_f} & 0 & 0 & 1 & 0 \\
0 & -\frac{L_2}{E_f} & \frac{L_2}{E_s} & 0 & -1 & 1
\end{bmatrix} \quad \text{and}
$$

$$
\mathbf{b} = \begin{bmatrix}
(\rho_f - \rho_s) g L_1 \\
(\rho_f - \rho_s) g L_2 \\
\left(\frac{\rho_s}{E_s} - \frac{\rho_f}{E_f}\right) g \frac{L_1^2}{2} \\
0 \\
\frac{\rho_f}{E_f} g \frac{L^2}{2} \\
\left(\frac{\rho_s}{E_s} - \frac{\rho_f}{E_f}\right) g \frac{L_2^2}{2}
\end{bmatrix}. \tag{4.84}
$$

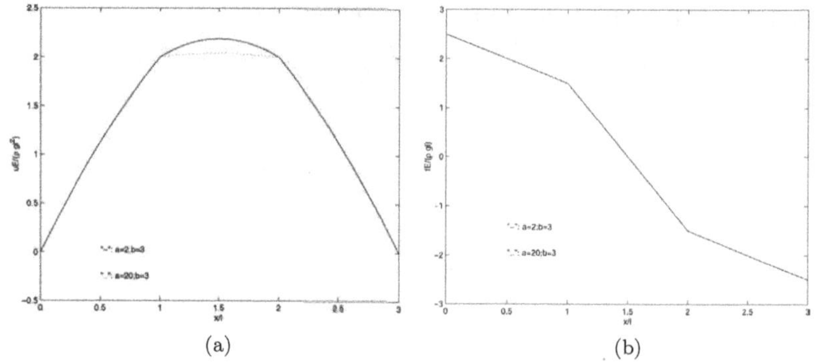

Fig. 4.35. Expected analytical solutions: (a) displacement and (b) force.

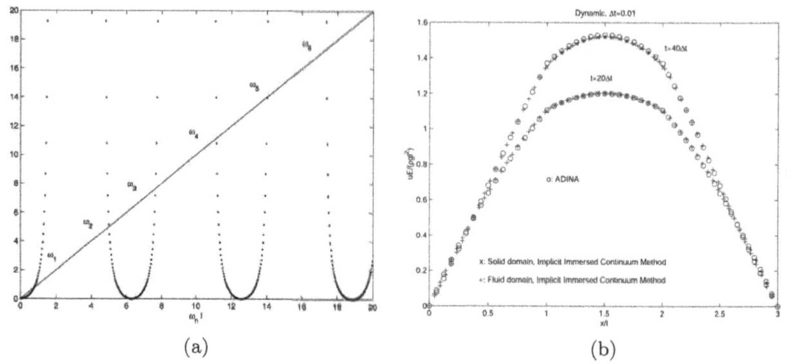

Fig. 4.36. Implicit immersed continuum solution at two time instances: (a) eigenvalues and (b) two media.

To further simplify the study, we assign $L_1 = L/3 = l$, $L_2 = 2L/3 = 2l$, $E_s = aE_f = aE$, and $\rho_s = b\rho_f = b\rho$. In addition, re-scale c_i, $i = 1,2,3$ by l/E, and define $\beta = \rho g l$, we have the following simplified form of Eq. (4.84)

$$\mathbf{A} = \begin{bmatrix} 1 & 0 & -1 & 0 & 0 & 0 \\ 0 & 1 & -1 & 0 & 0 & 0 \\ -a & 0 & 1 & -a & 0 & a \\ 0 & 0 & 0 & 1 & 0 & 0 \\ 0 & 3 & 0 & 0 & 1 & 0 \\ 0 & -2a & 2 & 0 & -a & a \end{bmatrix} \quad \text{and} \quad \mathbf{b} = -\beta \begin{bmatrix} (b-1) \\ 2(b-1) \\ \frac{1}{2}(a-b) \\ 0 \\ -\frac{9}{2} \\ 2(a-b) \end{bmatrix}. \tag{4.85}$$

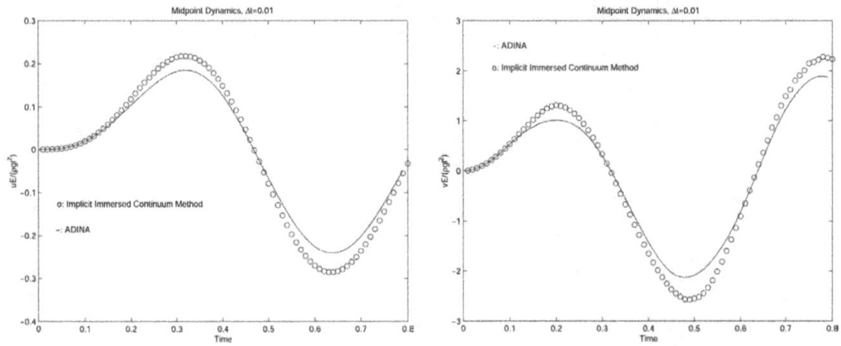

(a) (b)

Fig. 4.37. Implicit immersed continuum solution of the midpoint. (a) Displacement and (b) velocity.

Therefore, six constants c_i will be derived based on the following six boundary conditions. To further simplify the study, we assign $L_1 = L/3 = l$, $L_2 = 2L/3 = 2l$, $E_s = aE_f = aE$, and $\rho_s = b\rho_f = b\rho$. In addition, scale c_i, $i = 1,2,3$ by l/E, and define $\beta = \rho g l$, the solution vector $\mathbf{c}$ for this simplified problem can be given as

$$\mathbf{c}^T = \frac{\beta}{2}\left\langle b+2, -b+4, 3b, 0, 3b-3, 1+b-\frac{2b}{a}\right\rangle.$$

A few test examples are introduced to validate the concept of combining immersed methods with compact schemes. The immersed mass point, flexible cylinder, or flexible airfoil are tethered via spring and dashpot systems. The immersed solid is represented with a Mooney–Rivilin material model with density $\rho_s = 2$, bulk modulus 1×10^7, and two additional material constants $C_1 = 40$ and $C_2 = 20$. More preliminary results of the implicit immersed method for compressible media immersed in another compressible medium have also been reported in Wang (2007).

Chapter 5

Immersed Beams and Shells

In Chapter 2, the key ideas of the original immersed boundary method have been presented with some of the more recent understanding based on computational fluid and solid mechanics concepts. Since its inception, the immersed boundary method has initially been introduced to handle very flexible volumeless structures interacting with surrounding viscous fluid. In early versions of immersed boundary methods, immersed elastic fibers are employed to construct various solid objects. Recently, as depicted in Fig. 5.1, torsion, bending, and transverse shear effects have also been added to immersed codimension two structures, namely, beams, rods, and columns instead of fibers, trusses, and cables (Lim *et al.*, 2008). Snapshots at different times are depicted in Fig. 5.2.

One of the most important revelations of the original immersed boundary method has been the concept of volumeless fiber, truss, or cable, an essentially codimension two structure represented with only one parameter, often the arc length s, which is necessary and sufficient for the discussion of local geometrical positions. The complex interactions of this seemly simple volumeless structural model with the surrounding or rather background viscous fluid as immersed boundaries have been presented in Chapters 2 and 3. Based on the virtual work of the interior and exterior fluid domain separated by the immersed boundary in two-dimensional (2D) or the immersed surface in three-dimensional (3D) cases along with the equivalent FSI body force satisfying the respective kinematic and dynamic conditions, the entire immersed boundary method can be perfectly matched with the traditional finite element methods and computational fluid dynamics. Furthermore,

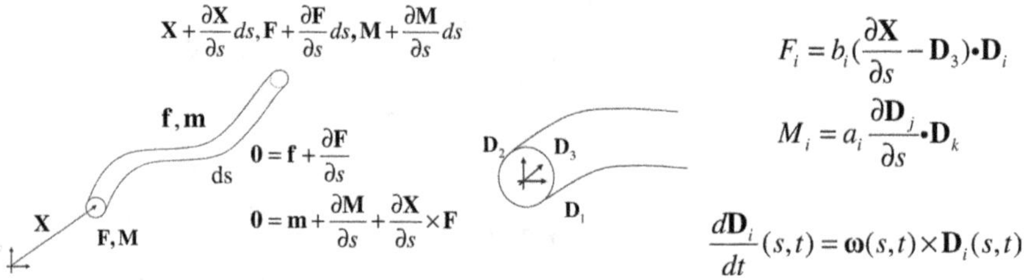

(a) External (b) Internal

Fig. 5.1. Immersed codimension two structures, namely truss, beam, and rod.

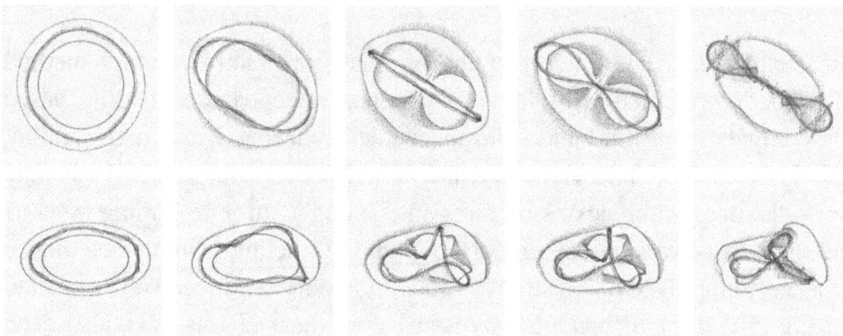

Fig. 5.2. Ring and vector snapshots of the triads with fluid markers.

the treatment of the additional mass and the finite volume of the immersed interfaces has prompted other general expansions of the original immersed boundary method in the context of immersed finite element methods and immersed continuum methods that have been elaborated for a continuum (fluid or solid) immersed in another continuum (fluid or solid) in Chapters 4 and 5.

Additional difficulties brought along with these expansions such as the finite volumes of the immersed continua, compatibility of incompressible or compressible issues, the fictitious domains occupied by these immersed continua, and the requirement of implicit and matrix-free Newton–Raphson iterative solution strategies have also been discussed in Chapters 3–5. In this chapter, we will present the key expansion of the immersed boundary method into another direction, namely, the application of volumeless codimension two structures such as beams, rods, or columns;

and codimension one structure such as plates and shells. The simple volumeless fibers, trusses, or cables for the original immersed boundary can be seamlessly expanded to the respective volumeless beams, rods, or columns. With these additions, the equivalent FSI body forces and moments related to the transverse shear forces and bending moments in two transverse directions along with a twisting moment or torque in the axial direction can be introduced and the structural response of the immersed interfaces can be more accurately and realistically represented.

5.1 Codimension Two Structures

Many different physical and biological systems, e.g., super-coiled DNA molecules, have been successfully modeled as elastic curves, ribbons, or rods. Since it is virtually impossible to model the entire molecules in an aqueous environment with *ab initio* or molecular dynamic modeling (Wu *et al.*, 2010, 2011; Wang, 2025), it is a common practice to describe such systems as framed curves. The interaction of such molecules with surrounding viscous fluid becomes a minimization problem in which a three-dimensional framed curve has an associated energy that is to be minimized subject to the constraint of there being no self-intersection.

For closed curves the knot type may be specified *a priori*. Depending on the energy form and respective boundary conditions, local minima of both open and closed framed curves often appear to involve regions of self-contact, that is, regions in which points that are separated along the one-dimensional curve are close in space. While this phenomenon of self-contact is familiar through everyday experience with string, rope, and wire, the idea is surprisingly difficult to define in a way that is physically reasonable, mathematically precise, and analytically tractable. In this chapter, we employ the notion of global radius of curvature of a spatial curve in a new formulation of the self-contact constraint, and exploit our formulation to derive existence results for minimizers, in the presence of self-contact, of a range of elastic energies that define various framed curve models. As a special case we establish the existence of ideal shapes of knots.

In current research approaches, such codimension one structures such as membranes or plates and shells are either modeled with fully three-dimensional solid models or networks of codimension two structures, namely, trusses or beams. In Section 2.3.2, we have presented the finite element representation of a truss model. Thus, in this section, we focus our attention on beams codimension two structures with bending and torsional rigidities. This structure model will enable more accurate and efficient modeling of immersed structures with one dominant dimension.

5.1.1 *Kinematic descriptions*

In this section, as depicted in Lim *et al.* (2008), all vectors and matrices can be eventually expressed in the fixed Cartesian reference coordinate system represented by its set of orthonormal vectors (triad) attached with the immersed codimension two structures, $\mathbf{e}_i,\ i = 1,2,3$. Consider a segment of a three-dimensional curve beam, $s(t) \in [s_1(t), s_2(t)]$, the time t and the arc length s are used to represent the temporal and spatial variations. Naturally, for the original configuration, we have $s(0)$ or $s_o \in [s_1(0), s_2(0)]$ or $[s_1^o, s_2^o]$. To represent the position of the curved and twisted beam at time t, we introduce $\bar{\mathbf{x}}(t,s) \in R^3$ to denote the midline or neutral axis which passes through the geometry centers of the cross sections. For simplicity, we consider the cross sections with two symmetric axes and consisting of homogeneous materials, e.g., rectangular or circular cross sections. In this case, the geometry center also coincides with the mass center, and more importantly the shear center.

Furthermore, to properly represent the deformation within a typical cross section, employ the common notations as presented in Chapter 1, we define a local orthonormal triad $\hat{\mathbf{e}}_i(t,s),\ i = 1,2,3$, a set of moving coordinate system base vectors. Since the moving coordinate system follows the material motion of the cross section, it is often called the co-rotational coordinate system. The corresponding transformation matrix $\mathbf{R}$ is defined to transfer vectors from the fixed Cartesian coordinate system $\mathbf{e}_i$ to the moving coordinate system $\hat{\mathbf{e}}_i$, namely, $\hat{\mathbf{e}}_i = \mathbf{R}\mathbf{e}_i$ and $\mathbf{e}_i = \mathbf{R}^T\hat{\mathbf{e}}_i$. Therefore, based on the orthonormal property of the transformation or rotation matrix $\mathbf{R}$, i.e., $\mathbf{R}\mathbf{R}^T = \mathbf{I}$, where $\mathbf{I}$ is an identity matrix. It is clear that the temporal and spatial derivatives of the identity matrix are zero. Utilizing the following property of the skew or anti symmetric matrix $\mathbf{S}$,

$$\mathbf{S} + \mathbf{S}^T = \mathbf{0},$$

we can easily prove that $\det \mathbf{S} = 0$ and for any vector $\mathbf{b}$, we have

$$\mathbf{S}(\mathbf{a})\mathbf{b} = \mathbf{a} \times \mathbf{b},$$

with $\mathbf{a} = \langle \xi, \eta, \zeta \rangle$, and

$$\mathbf{S} = \begin{pmatrix} 0 & -\zeta & \eta \\ \zeta & 0 & -\xi \\ -\eta & \xi & 0 \end{pmatrix}. \tag{5.1}$$

Consequently, we can derive the following skew symmetric tensors or matrices,

$$\boldsymbol{\Omega} = \frac{\partial \mathbf{R}}{\partial t}\mathbf{R}^T \quad \text{and} \quad \boldsymbol{\Gamma} = \frac{\partial \mathbf{R}}{\partial s}\mathbf{R}^T. \tag{5.2}$$

Hence, the temporal and spatial derivatives of the moving coordinate system can be summarized as

$$\frac{\partial \hat{\mathbf{e}}_i}{\partial t} = \boldsymbol{\Omega}\hat{\mathbf{e}}_i \quad \text{and} \quad \frac{\partial \hat{\mathbf{e}}_i}{\partial s} = \boldsymbol{\Gamma}\hat{\mathbf{e}}_i. \tag{5.3}$$

Moreover, following the definition of the skew symmetric matrix in Eq. (5.1), we can also introduce an angular velocity vector $\boldsymbol{\omega}$ and the curvature vector $\boldsymbol{\gamma}$ to represent the skew symmetric tensors $\boldsymbol{\Omega}$ and $\boldsymbol{\Gamma}$, respectively, such that $\forall \mathbf{a} \in R^3$, we have

$$\boldsymbol{\Omega}\mathbf{a} = \boldsymbol{\omega} \times \mathbf{a} \quad \text{and} \quad \boldsymbol{\Gamma}\mathbf{a} = \boldsymbol{\gamma} \times \mathbf{a}.$$

Alternatively, we can also relate the second-order skew symmetric tensor with the first-order axial vector as follows:

$$\boldsymbol{\Omega} = \boldsymbol{\omega} \times \mathbf{I} \quad \text{and} \quad \boldsymbol{\Gamma} = \boldsymbol{\gamma} \times \mathbf{I};$$

or

$$\Omega_{ij} = -\varepsilon_{ijk}\omega_k \quad \text{and} \quad \Gamma_{ij} = -\varepsilon_{ijk}\gamma_k,$$

where the three-dimensional permutation symbol ε_{ijk} is defined as 1 if the indices ijk are in the the sequence of 123; -1 if the indices ijk are in the sequence of 321; and 0 otherwise.

Using the tensorial equality $\varepsilon_{ijm}\varepsilon_{klm} = \delta_{ik}\delta_{jl} - \delta_{il}\delta_{jk}$, we can easily prove the following:

$$\omega_i = -\frac{1}{2}\varepsilon_{ijk}\Omega_{jk} \quad \text{and} \quad \gamma_i = -\frac{1}{2}\varepsilon_{ijk}\Gamma_{jk}.$$

Note that for a given configuration of the curved beam represented with the midline or neutral axis $\bar{\mathbf{x}}$ and the moving coordinate system $\hat{\mathbf{e}}_i$, or $\mathbf{D}_i$, $\boldsymbol{\Omega}$ and $\boldsymbol{\Gamma}$ can be uniquely defined. In practice, $\boldsymbol{\Omega}$ and $\boldsymbol{\Gamma}$ are often called the spin and curvature-twist matrices, representing the temporal and spatial changes of the beam body coordinates, respectively. In this section, we adhere to the convention that a typical vector $\mathbf{a}$ or matrix $\mathbf{A}$ can be expressed in different coordinate systems as $\mathbf{a} = a_i\mathbf{e}_i = \hat{a}_i\hat{\mathbf{e}}_i$ or $\mathbf{A} = A_{ij}\mathbf{e}_i \otimes \mathbf{e}_j = \hat{A}_{ij}\hat{\mathbf{e}}_i \otimes \hat{\mathbf{e}}_j$. Therefore, in the moving coordinate system, the angular velocity vector $\boldsymbol{\omega}$ and the curvature vector $\boldsymbol{\gamma}$ can be simply expressed as $\boldsymbol{\omega} = \hat{\omega}_i\hat{\mathbf{e}}_i$ and $\boldsymbol{\gamma} = \hat{\gamma}_i\hat{\mathbf{e}}_i$.

In particular, the axial component of the angular velocity $\boldsymbol{\omega}$ stands for the angular velocity with respect to the axial vector $\hat{\mathbf{e}}_3$ and the normal components are angular velocities with respect to two transverse directions $\hat{\mathbf{e}}_1$ and $\hat{\mathbf{e}}_2$. Moreover, $\hat{\gamma}_1$ and $\hat{\gamma}_2$ are the curvatures which correspond to the axial strain dependent of the distances from the neutral axis or rather body attached coordinates with respect to $\hat{\mathbf{e}}_1$ and $\hat{\mathbf{e}}_2$. Finally, the axial component $\hat{\gamma}_3$ is the twist angle per unit length which corresponds to the torsional strain dependent of the axial coordinate $\hat{\mathbf{e}}_3$.

Assume that the initial configuration of the curved beam is in the stress-free state, the cross-sectional plane of the curved beam must be perpendicular to the axial direction of the body coordinate triad denoted as $\hat{\mathbf{e}}_3(0,s)$ and aligned with the tangential direction $\frac{\partial \bar{\mathbf{x}}(0,s)}{\partial s}$ or $\frac{\partial \bar{\mathbf{X}}(s)}{\partial s}$. However, during the course of evolution, such an alignment will no longer be true due to the shear strains in the two transverse or normal directions at the cross-sectional plane, namely, $\hat{\mathbf{e}}_1$ and $\hat{\mathbf{e}}_2$. In Euler–Bernoulli assumption for thin beams, or Kirchhoff–Love assumption for thin shells, transverse shear strains are ignored. In Euler–Bernoulli beam, the axial direction of the beam will remain perpendicular to the cross-sectional plane at all time, namely, $\hat{\mathbf{e}}_3(t,s) = \frac{\partial \bar{\mathbf{x}}(t,s)}{\partial s}$. Note that the norm of the vector $\frac{\partial \bar{\mathbf{x}}}{\partial s}$ is always a unit with the parameter s as the real arc length. In Timoshenko assumption for thick beams, or Mindlin–Reissner assumption for thick shells, transverse shear strains introduced by transverse shear stresses can no longer be ignored and hence the axial direction $\frac{\partial \bar{\mathbf{x}}}{\partial s}$ will not coincide with the normal direction of the cross-sectional plane $\hat{\mathbf{e}}_3$. It is clear that for Euler–Bernoulli beams and Kirchhoff–Love shells we have three C^1 displacement unknowns while for Timoshenko beams and Mindlin–Reissner shells we have three C^0 displacement unknowns and three (for three-dimensional beams) or two (for three-dimensional shells) rotation unknowns. The significant impact on the finite element formulations of these two beam assumptions will be further elaborated when we discuss the beam spatial discretizations.

In order to properly represent the strain and stress definitions as well as the strain energy or strain energy variations of the entire beam structure, we represent any point within a typical cross-sectional plane as

$$\mathbf{x} = \bar{\mathbf{x}} + \sum_{i=1}^{2} d_i \hat{\mathbf{e}}_i, \tag{5.4}$$

where d_1 and d_2 are the local coordinates along the base vectors $\hat{\mathbf{e}}_1$ and $\hat{\mathbf{e}}_2$, or $\mathbf{D}_1$ and $\mathbf{D}_2$, and the translational displacement of the midline can be expressed as

$$\mathbf{u} = \mathbf{x}(t) - \mathbf{x}(0) = \mathbf{x} - \mathbf{X}. \tag{5.5}$$

It is clear that once the midline is defined along with the moving coordinate system, with Eq. (5.5), the entire beam is defined with respect to its three-dimensional geometrical positions. In practice, the lines with constant s are called fibers, and the lines with constant local coordinates d_i are called laminae. Furthermore, the translational velocity of the midline can be written as

$$\mathbf{v} = \bar{\mathbf{v}} + \sum_{i=1}^{2} d_i \boldsymbol{\omega} \times \hat{\mathbf{e}}_i; \tag{5.6}$$

whereas the acceleration is also given as

$$\dot{\mathbf{v}} = \dot{\bar{\mathbf{v}}} + \sum_{i=1}^{2} d_i [\dot{\boldsymbol{\omega}} \times \hat{\mathbf{e}}_i + \boldsymbol{\omega} \times (\boldsymbol{\omega} \times \hat{\mathbf{e}}_i)]. \tag{5.7}$$

5.1.2 *Immersed beams*

The advantage of immersed methods is to allow independent Lagrangian mesh of the immersed continuum to move on top of a fixed background

Eulerian mesh. For codimension two structures, the overall volume is assumed to be zero, and this type of fluid–structure interaction systems is very suitable for immersed methods. Because of this structural simplification, the complete fluid domain can stay the same regardless of the position of the immersed structure, and a fixed background mesh becomes possible without the fictitious domain adjustments as presented in Chapters 2–4. This structure simplification very naturally circumvents the difficulties that emerge from the required resolution for a full-fledged solid model in comparison with that of the surrounding viscous fluid.

The key of immersed methods is the energy or power conservation; namely, the energy input to the fluid domain from the immersed solid/structure is the same as that from the equivalent body force. Moreover, the same delta function must be used in both the distribution of the resultant nodal force and the interpolation of the solid velocity based on the surrounding fluid velocities. In fact, this treatment in immersed methods can be viewed as the synchronization of the fluid motion with the solid motion within the immersed solid domain Ω_s, namely, $\mathbf{v}^s = \mathbf{v}^f$ (Van Loon *et al.*, 2007; Yu, 2005). The difference between shell and membrane elements is the same as that between fiber and beam elements. To be more specific, in beam elements, there exist two bending moments and one torsion moment along with two transverse shear forces and one axial force. In shell elements, two transverse shear forces and one in-plane or membrane force must be accounted for. However, to bypass the complication due to wrinkles (Veldin *et al.*, 2019), in general, shell models include two bending moments and disregard the torsion with respect to the out-of-plane shell surface normal direction. In order to handle large displacements and large deformations, a triad is attached as the local coordinate system defining in-plane and transverse directions. The kinematic evolution of three direction vectors and the position vector yields the instantaneous curvature and hence moments and forces of a typical point within flexible shells. Finally, to couple the immersed shells/beams with the surrounding viscous fluid, the virtual power of both forces and moments exerted on the fluid will be incorporated.

Both dynamic and kinematic descriptions are presented in Fig. 5.1. The two transverse directions are denoted with the unit vectors $\mathbf{D}_1$ and $\mathbf{D}_2$ aong with the axial direction marked with the unit vector $\mathbf{D}_3$. Moreover, the internal descriptions have been used to relate the kinematic positions and configurations with the two bending moments $\mathbf{M}_1$ and $\mathbf{M}_2$, one torque $\mathbf{M}_3$, two transverse shear forces $\mathbf{F}_1$ and $\mathbf{F}_2$, and one axial force $\mathbf{F}_3$, respectively.

Coupled with fluid motions, immersed shells/beams undergo large deformations in three spatial dimensions and consequently exert both forces and moments on the surrounding fluid that resist the structural stretching, shearing, bending, and twisting (Peskin, 2002; Goriely and Tabor, 1997a, 1997b). The kinematics of the immersed beam is represented by the centerline $\mathbf{X} = \mathbf{X}(s,t)$ parameterized by arc length s and time t, and a local triad $(\mathbf{D}_1(s,t), \mathbf{D}_2(s,t), \mathbf{D}_3(s,t))$ associated with this curve (Goriely and Tabor, 1997a). Denote the structural internal forces and moments as $\mathbf{F}(s,t) = F_i\mathbf{D}_i$ and $\mathbf{M}(s,t) = M_i\mathbf{D}_i$, and the interaction forces and moments as $\mathbf{f}$ and $\mathbf{m}$, the equilibrium equations for the neutrally buoyant or quasi-static immersed beams, can be established as

$$\mathbf{0} = \mathbf{f} + \frac{\partial \mathbf{F}}{\partial s} \quad \text{and} \quad \mathbf{0} = \mathbf{m} + \frac{\partial \mathbf{M}}{\partial s} + \frac{\partial \mathbf{X}}{\partial s} \times \mathbf{F}. \tag{5.8}$$

For a given configuration of the fiber depicted with the centerline position $\mathbf{X}$ and the attached cross-sectional body coordinates $\mathbf{D}_i$, namely, $(\mathbf{X}, \mathbf{D}_i)$, Eq. (5.8) provides enough information to compute the distributed force $\mathbf{f}$ and the distributed moment $\mathbf{m}$ to be spread into the fluid domain. A preliminary result as shown in Fig. 5.2 from an implementation of immersed beams has been reported in Lim *et al.* (2008). In this section, the aforementioned nonlinear beam is immersed in viscous fluid the effects of which are included in the distributed force $\mathbf{f}$ and moment $\mathbf{m}$. As a result of the fluid motion, the filament undergoes large displacements and rotations in three spatial dimensions, and in the meantime also applies forces and moments to the surrounding fluid that resist the stretching, shearing, bending, and twisting of the filament. From this section, in order to delineate the valuables for fluids and solids, we introduce a superscript or subscript s to depict the solid domain.

The process of "spreading" influence involves the $3D$-Dirac delta function and its gradient. The influence of the filament on the fluid is expressed in terms of a singular body force $\mathbf{f}^{fsi}$, which is given by

$$\mathbf{f}^{fsi}(\mathbf{x},t) = \int -\mathbf{f}(s,t)\delta(\mathbf{x} - \mathbf{X}(s,t))\,ds + \int \frac{1}{2}\mathbf{m}(s,t) \times \nabla\delta(\mathbf{x} - \mathbf{X}(s,t))ds.$$

A procedure dual to the above updates the position of the curve and the triad by interpolating the fluid velocity $\mathbf{v}$ and the angular velocity $\boldsymbol{\omega} = (\nabla \times \mathbf{v})/2$:

$$\frac{d\mathbf{X}}{dt}(s,t) = \int \mathbf{v}(\mathbf{x},t)\delta(\mathbf{x} - \mathbf{X}(s,t))\, d\mathbf{x}, \tag{5.9}$$

$$\omega(s,t) = \frac{1}{2}\int (\nabla \times \mathbf{v})(\mathbf{x},t)\delta(\mathbf{x} - \mathbf{X}(s,t))\, d\mathbf{x}, \tag{5.10}$$

$$\frac{d\mathbf{D}_i}{dt}(s,t) = \omega(s,t) \times \mathbf{D}_i(s,t). \tag{5.11}$$

In the first attempt as presented in Lim *et al.* (2008), the two transverse forces F_1 and F_2 and the axial forces F_3 are depicted as a force vector $\mathbf{F}$

$$\mathbf{F} = F_1\mathbf{D}_1 + F_2\mathbf{D}_2 + F_3\mathbf{D}_3;$$

along with the two bending moment M_1 and M_2 and the torque M_3 expressed as a moment vector $\mathbf{M}$

$$\mathbf{M} = M_1\mathbf{D}_1 + M_2\mathbf{D}_2 + M_3\mathbf{D}_3.$$

Moreover, employ the Euler–Bernoulli beam assumption, namely, the transverse shear strain will not significantly alter the orientation of the neutral axial and the normal directions, we have the following constitutive laws for the forces and moments, respectively,

$$F_1 = b_1\mathbf{D}_1 \cdot \frac{\partial \mathbf{X}}{\partial s}, \quad F_2 = b_2\mathbf{D}_2 \cdot \frac{\partial \mathbf{X}}{\partial s}, \quad \text{and} \quad F_3 = b_3\left(\mathbf{D}_3 \cdot \frac{\partial \mathbf{X}}{\partial s} - 1\right);$$
$$M_1 = a_1\frac{\partial \mathbf{D}_2}{\partial s} \cdot \mathbf{D}_3, \quad M_2 = a_2\frac{\partial \mathbf{D}_3}{\partial s} \cdot \mathbf{D}_1, \quad \text{and} \quad M_3 = a_3\frac{\partial \mathbf{D}_1}{\partial s} \cdot \mathbf{D}_2. \tag{5.12}$$

Note that in the triad the transverse direction $\mathbf{D}_1$ and $\mathbf{D}_2$ are always perpendicular to the axial direction $\mathbf{D}_3$, therefore the force definitions in Eq. (5.12) is equivalent to the expressions in Fig. 5.1. Moreover, with the Euler–Bernoulli beam assumption, the axial direction $\mathbf{D}_3$ is the same as the tangent vector defined as $\frac{\partial \mathbf{X}}{\partial s}$. However, for deep beams or anisotropic materials such as composites, so-called Timoshenko beam assumption must be adopted, in which the axial direction of the cross sectional triad is no longer the same as the tangent vector $\frac{\partial \mathbf{X}}{\partial s}$, and a difference due to the finite shear strains γ_1 and γ_2 in the two transverse directions, defined as $\mathbf{D}_1$ and $\mathbf{D}_2$, respectively, must be considered.

In structural mechanics, for the transverse forces $\mathbf{F}_1$ and $\mathbf{F}_2$ with circular cross-sectional area A, taking into consideration of the distributions of the shear force over the cross section, we have $b_1 = b_2 = GAk$, where G is the shear modulus and k is the effective area coefficient, in this case, $k = 3/4$. For the axial force $\mathbf{F}_3$, the constant b_3 is EA, where E is the Young's modulus, named after Thomas Young (1773–1829). For linear elastic materials, the shear modulus and the Young's modulus are related with $E = 2G(1+\nu)$ where ν is the so-called Poisson's ratio, named after Simeon Denis Poisson (1781–1840). Moreover, the bending rigidities represented by a_1 and a_2 are expressed as EI for the circular cross section with the area moment of inertia $I = \pi R^4/4$, where R stands for the cross-sectional radius. and the torsional rigidity b_3 is denoted as GJ with the polar moment of inertial $J = 2I = \pi R^4/2$.

5.2 Codimension One Structures

In fluid–structure interaction systems, thin materials with two dominant dimensions are often encountered. Just as in aeronautic and aerospace structures, computer hard disk drives, and rotating blades in turbomachinery, biological systems also involve many such thin materials such as lipids and cell membranes. To effectively model the wrinkle, flutter, and other structural deformation with respect to the surrounding fluid, sophisticated mixed finite element formulations for plates and shells must be used.

5.2.1 *Membranes and plates*

Since the inception of finite element methods, many plate bending formulations and elements have been investigated; a proper review of such topics is available in (Babuška and Suri, 1994, Hughes and Tezduyar, 1981, Stenberg and Suri, 1997, and Suri *et al.*, 1995). It has been widely recognized that mixed plate formulation based on the Mindlin/Reissner theory can effectively eliminate "shear locking" and predict accurately and reliably stress levels of thin plates (Bathe and Dvorkin, 1986; Brezzi *et al.*, 1989; Dvorkin and Bathe, 1984). Recently, the Mixed Interpolation of Tensorial Components (MITC) elements have been proven, numerically, to satisfy the Inf-Sup condition (Iosilevich *et al.*, 1997). The basic difficulty in choosing proper orders of interpolation for out-of-plane displacement, section rotations, and transverse shear strains which result in nonlocking behavior and optimal convergence of the element is summarized in Brezzi *et al.* (1989) and Bathe and Dvorkin (1986).

In a three-dimensional elasticity theory, there are six stress components that are expressed in terms of six strain components through material constitutive laws. For the general plate section illustrated in Fig. 5.3, the top and bottom faces of the plate (at $z = \pm d/2$) are considered to be free from tangential traction, but under normal pressures p^+ and p^-, i.e.,

$$\tau_{xz}|_{z=\pm d/2} = 0, \quad \tau_{yz}|_{z=\pm d/2} = 0; \quad \text{and}$$
$$\tau_{zz}|_{z=+d/2} = -p^+, \quad \tau_{zz}|_{z=-d/2} = -p^-.$$

In general, $p = p^- - p^+$, yet such an assumption can be adjusted to incorporate the surrounding air shear boundary layer and porosity. Corresponding to the five stress components, the bending and twisting moments, and the transverse shearing forces, all measured per unit of length, are defined as

$$(M_x, M_y, M_{xy}) = \int_{-d/2}^{+d/2} (\tau_{xx}, \tau_{yy}, \tau_{xy}) z dz \quad \text{and} \quad (Q_x, Q_y) = \int_{-d/2}^{+d/2} (\tau_{xz}, \tau_{yz}) dz. \tag{5.13}$$

Also shown in Fig. 5.3, $\boldsymbol{\beta} = (\beta_x, \beta_y)$ is the vector whose components are, respectively, the rotations of the normal to the undeformed middle surface in the x,z and y,z planes; it is equivalent to $\theta^{\perp} = (-\theta_y, \theta_x)$,

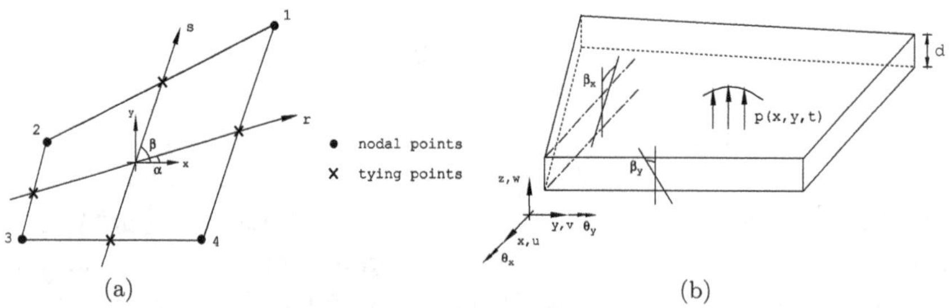

Fig. 5.3. MITC4 finite element for plates and shells. (a) MITC4 middle plane and (b) MITC4 definition.

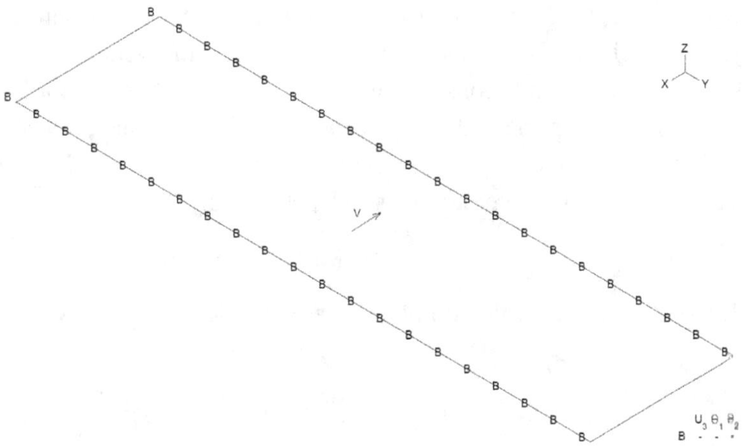

Fig. 5.4. An orthotropic plate (2 × 8 m) moving between two supports.

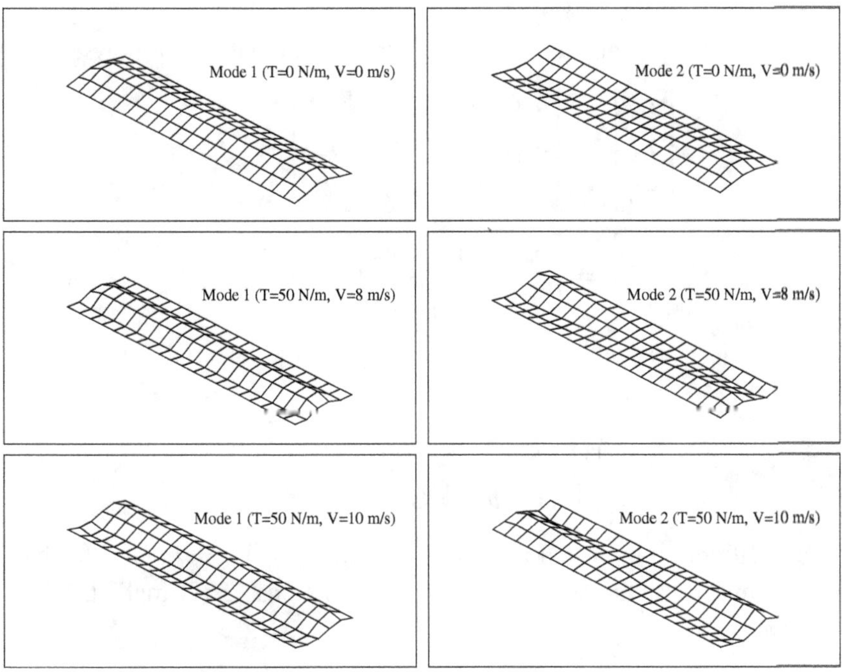

Fig. 5.5. Out-of-plane displacement mode shapes (5 × 20 plate elements).

which is normal to the vector θ of rotations about the Cartesian axes x and y. In the classical Kirchhoff plate theory, the transverse strains and stresses are ignored and, consequently, $\nabla w = \boldsymbol{\beta}$. However, according to the Mindlin/Reissner plate theory (Mindlin, 1951; Reissner, 1979), the x-direction and y-direction displacements u and v are proportional to z, and the z-direction displacement w is independent of z, i.e.,

$$u = -z\beta_x(x,y,t), \quad v = -z\beta_y(x,y,t), \quad w = w(x,y,t).$$

Moreover, the five strain components are introduced as follows:

$$\begin{bmatrix} \varepsilon_{xx} \\ \varepsilon_{yy} \\ \gamma_{xy} \end{bmatrix} = -z \begin{bmatrix} \dfrac{\partial \beta_x}{\partial x} \\ \dfrac{\partial \beta_y}{\partial y} \\ \dfrac{\partial \beta_x}{\partial y} + \dfrac{\partial \beta_y}{\partial x} \end{bmatrix} = -z\boldsymbol{\kappa}, \quad \begin{bmatrix} \gamma_{xz} \\ \gamma_{yz} \end{bmatrix} = \begin{bmatrix} \dfrac{\partial w}{\partial x} - \beta_x \\ \dfrac{\partial w}{\partial y} - \beta_y \end{bmatrix} = \boldsymbol{\gamma}. \quad (5.14)$$

The governing equation for an orthotropic material is as follows:

$$\begin{bmatrix} \tau_{xx} \\ \tau_{yy} \\ \tau_{xy} \end{bmatrix} = \begin{bmatrix} a_{11} & a_{12} & 0 \\ a_{21} & a_{22} & 0 \\ 0 & 0 & a_{33} \end{bmatrix} \begin{bmatrix} \varepsilon_{xx} \\ \varepsilon_{yy} \\ \gamma_{xy} \end{bmatrix} = -\mathbf{C}_\varepsilon z\boldsymbol{\kappa} \quad (5.15)$$

$$\begin{bmatrix} \tau_{xz} \\ \tau_{yz} \end{bmatrix} = \begin{bmatrix} a_{44} & 0 \\ 0 & a_{55} \end{bmatrix} \begin{bmatrix} \gamma_{xz} \\ \gamma_{yz} \end{bmatrix} = -\mathbf{C}_\gamma z\boldsymbol{\gamma} \quad (5.16)$$

where with a constant c defined as $1 - \nu_{xy}\nu_{yx}$ the material constants are expressed as

$$a_{11} = E_x/c, \quad a_{12} = a_{21} = \nu_{xy}E_x/c, \quad a_{22} = E_y/c, \quad a_{33} = G_{xy},$$
$$a_{44} = G_{xz}, \quad \text{and} \quad a_{55} = G_{yz}.$$

In addition to the usual necessary condition $E_x/\nu_{yx} = E_y/\nu_{xy}$, for biological materials, the so-called Campbell relationship (Campbell, 1961) also holds

$$1/G_{xy} = (1+\nu_{yx})/E_x + (1+\nu_{xy})/E_y.$$

Of course, isotropic materials yields the special case, $E_x = E_y = E$, $\nu_{xy} = \nu_{yx} = \nu$, and $G_{xy} = G_{xz} = G_{yz} = E/2(1+\nu)$.

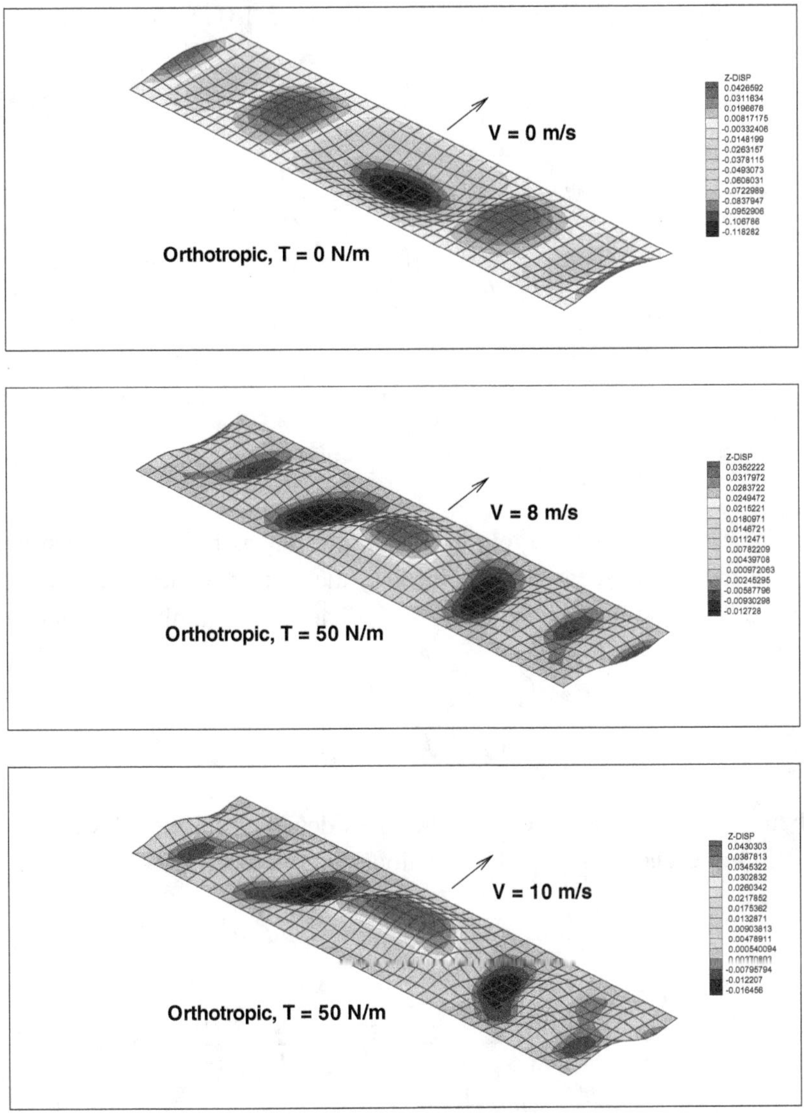

Fig. 5.6. Out-of-plane displacement w at $t = 1.2$ s (10×40 plate elements).

Substituting Eqs. (5.16) and (5.15) into Eq. (5.13) results

$$M_x = -\left(a_{11}\frac{\partial \beta_x}{\partial x} + a_{12}\frac{\partial \beta_y}{\partial y}\right)\frac{d^3}{12} \tag{5.17}$$

$$M_y = -\left(a_{21}\frac{\partial \beta_x}{\partial x} + a_{22}\frac{\partial \beta_y}{\partial y}\right)\frac{d^3}{12} \tag{5.18}$$

$$M_{xy} = -a_{33}\left(\frac{\partial \beta_x}{\partial y} + \frac{\partial \beta_y}{\partial x}\right)\frac{d^3}{12} \tag{5.19}$$

and

$$Q_x = kd\left(\frac{\partial w}{\partial x} - \beta_x\right)a_{44} \quad \text{and} \quad Q_y = kd\left(\frac{\partial w}{\partial y} - \beta_y\right)a_{55}. \tag{5.20}$$

The constant k plays a role in accounting for the nonuniform distribution of transverse shear stresses through the plate thickness, and has the value $5/6$ (Mindlin, 1951; Bathe, 1996). The force equilibrium equations are expressed as follows,

$$\tau_{ij,j} + f_i = \rho\frac{d^2 u_i}{dt^2},$$

where f_i is the body force; ρ is the mass density; and the indices i ($i = 1,2,3$) represent the x, y, and z directions, and moreover

$$\frac{\partial M_x}{\partial x} + \frac{\partial M_{xy}}{\partial y} - Q_x = -\frac{\rho d^3}{12}\frac{d^2\beta_x}{dt^2} \tag{5.21}$$

$$\frac{\partial M_{xy}}{\partial x} + \frac{\partial M_y}{\partial y} - Q_y = -\frac{\rho d^3}{12}\frac{d^2\beta_y}{dt^2} \tag{5.22}$$

$$\frac{\partial Q_x}{\partial x} + \frac{\partial Q_y}{\partial y} + p - \rho g d = \rho d\frac{d^2 w}{dt^2}. \tag{5.23}$$

Of course, as indicated in Eq. (5.23), the only body force considered here is the gravitational force in the negative z direction. Fluid induced body forces will be introduced here for immersed structures. Furthermore, in terms of out-of-plane displacement w, and two rotations β_x and β_y. Note here that when the plate is stationary, the material derivative $\frac{d}{dt}$ is the same

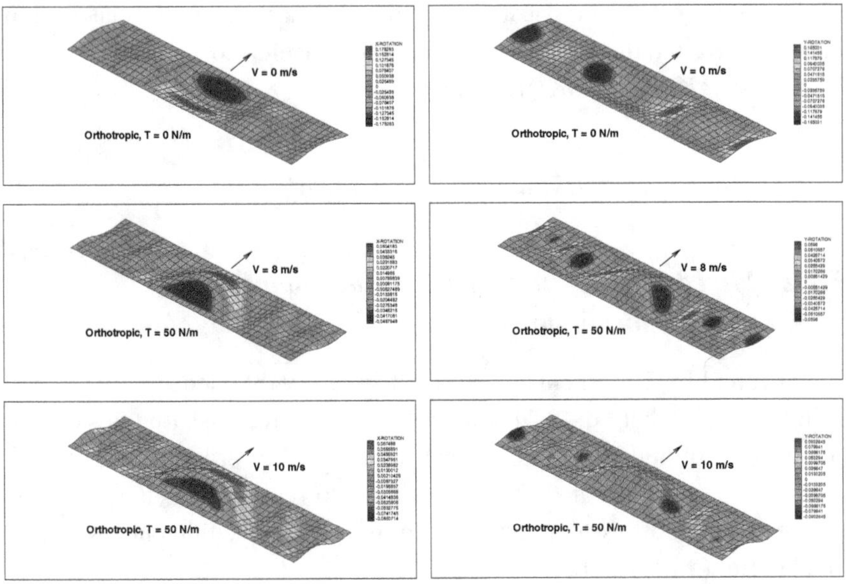

(a) (b)

Fig. 5.7. Rotation with (a) θ_x at $t = 1.2$ s (10×40 plate elements) and (b) θ_y at $t = 1.2$ s (10×40 plate elements).

as the spatial derivative $\frac{\partial}{\partial t}$; however, when the plate is moving with a velocity $\mathbf{v}$, namely, for general moving materials, we have

$$\frac{d}{dt} = \frac{\partial}{\partial t} + \mathbf{v} \cdot \frac{\partial}{\partial \mathbf{x}},$$

or expressed in compact form,

$$\frac{d}{dt} = \frac{\partial}{\partial t} + \mathbf{V} \cdot \mathbf{V}.$$

For an isotropic and stationary plate, the following governing equations presented by Mindlin (1951) are recovered,

$$\frac{D}{2}\left[(1-\nu)\nabla^2\beta_x + (1+\nu)\frac{\partial\phi}{\partial x}\right] + kGd\left(\frac{\partial w}{\partial x} - \beta_x\right) = \frac{\rho d^3}{12}\frac{\partial^2 \beta_x}{\partial t^2},$$

$$\frac{D}{2}\left[(1-\nu)\nabla^2\beta_y + (1+\nu)\frac{\partial\phi}{\partial y}\right] + kGd\left(\frac{\partial w}{\partial y} - \beta_y\right) = \frac{\rho d^3}{12}\frac{\partial^2 \beta_y}{\partial t^2},$$

$$kGd\left(\nabla^2 w - \phi\right) + p - \rho g d = \rho d\frac{\partial^2 w}{\partial t^2},$$

where ϕ is defined as $\frac{\partial \beta_x}{\partial x} + \frac{\partial \beta_y}{\partial y}$, and the bending stiffness D equals $\frac{Ed^3}{12(1-\nu^2)}$.

Furthermore, the classical theory of stationary isotropic elastic thin plates based on the Kirchhoff assumptions (refer to Timoshenko and Woinowsky–Krieger (1987)) yields

$$D\nabla^4 w + \rho d \frac{\partial^2 w}{\partial t^2} = p - \rho g d. \tag{5.24}$$

5.2.2 *Mixed Interpolation of Tensorial Components (MITC) formulation*

It was found by Timoshenko (1922), Uflyand (1948), Hencky (1947), and Mindlin (1951) that Eq. (5.24) cannot be used to predict sharp transients or frequencies of high-order vibrational modes, and in order to predict such effects, transverse shear deformations have to be included. Based on the classical Kirchhoff assumptions, for a moving isotropic thin plate with tension T, Eq. (5.24) is modified as

$$D\nabla^4 w + \rho d \left(\frac{\partial}{\partial t} + V \frac{\partial}{\partial x} \right)^2 w - T \frac{\partial^2 w}{\partial x^2} = p - \rho g d \tag{5.25}$$

and this equation has been introduced in the analysis of moving tapes, saw bands, and plastic films (Heinrich and Connolly, 1992; Lacey and Talke, 1990; Müftü and Benson, 1995), and the subsonic panel flutter problems (Dugundji *et al.*, 1963).

Nevertheless, it has been recently recognized that if the axial tension is not predominant or as the axial moving speed increases, a reliable mixed plate theory is needed to predict accurately the stress distributions within a moving thin plate (Wang and Renner, 1998). In the mixed formulation based on the Mindlin/Reissner plate theory, the effects of inertia are accounted by applying the Hamilton's principle with the following definition of the kinetic energy,

$$\int_\Omega \left[\frac{\rho d^3}{24} \left(\frac{d\beta_x}{dt} \right)^2 + \frac{\rho d^3}{24} \left(\frac{d\beta_y}{dt} \right)^2 + \frac{\rho d}{2} \left(\frac{dw}{dt} \right)^2 \right] d\Omega, \tag{5.26}$$

where Ω represents the middle surface area. In actuality, in the body forces the following inertia terms are included

$$\rho d\left(\frac{\partial}{\partial t}+V\frac{\partial}{\partial x}\right)^2 w, \quad \frac{\rho d^3}{12}\left(\frac{\partial}{\partial t}+V\frac{\partial}{\partial x}\right)^2 \boldsymbol{\beta}.$$

Defining the spaces $\Theta = (H_0^1(\Omega))^2$ and $W = H_0^1(\Omega)$, and assuming that a force function f_w is given in $L^2(\Omega)$, and a moment function $\mathbf{f}_\beta$ is given in $(L^2(\Omega))^2$, our problem can be presented as follows:

$$\inf_{\boldsymbol{\beta}\in\Theta, w\in W} \frac{d^3}{2}a(\boldsymbol{\beta},\boldsymbol{\beta})+\frac{kd}{2}\|\boldsymbol{\beta}-\nabla w\|_0^2 - d^3(f_w,w)-d^3(\mathbf{f}_\beta,\boldsymbol{\beta}),$$

where the body forces (or moments) have the form

$$f_w = \left(p-\rho g d-\rho d\left(\frac{\partial}{\partial t}+V\frac{\partial}{\partial x}\right)^2 w\right)\bigg/ d^3, \tag{5.27}$$

$$\mathbf{f}_\beta = -\frac{\rho}{12}\left(\frac{\partial}{\partial t}+V\frac{\partial}{\partial x}\right)^2 \boldsymbol{\beta}, \tag{5.28}$$

and the bending strain energy $\frac{d^3}{2}a(\boldsymbol{\beta},\boldsymbol{\beta})$, which includes the contribution of the normal stress $\sigma_T(y)$ due to the axial tension from the effective immersed body force, is expressed by

$$\frac{d^3}{2}a(\boldsymbol{\beta},\boldsymbol{\beta}) = \frac{d^3}{24}\int_\Omega \boldsymbol{\kappa}^T\mathbf{C}_\varepsilon\boldsymbol{\kappa}\, d\Omega + \int_\Omega \frac{d\sigma_T(y)}{2}\left(\frac{\partial w}{\partial x}\right)^2 d\Omega.$$

Given the finite element subspaces $\Theta_h \subset \Theta$ and $W_h \subset W$, the discretized problem may be written as

$$\inf_{\boldsymbol{\beta}_h\in\Theta_h, w_h\in W_h} \frac{d^3}{2}a(\boldsymbol{\beta}_h,\boldsymbol{\beta}_h)+\frac{kd}{2}\|R\boldsymbol{\beta}_h-\nabla w_h\|_0^2 - d^3(f_w,w_h)-d^3(\mathbf{f}_\beta,\boldsymbol{\beta}_h),$$

where the linear operator R satisfies

$$R\nabla w_h = \nabla w_h, \quad \forall w_h \in W_h,$$

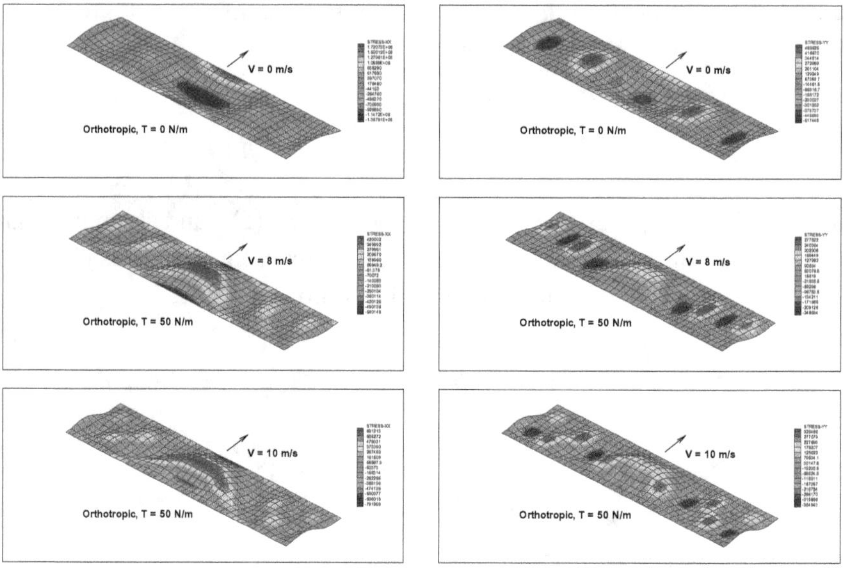

(a) (b)

Fig. 5.8. Normal stress with (a) τ_{xx} at $t = 1.2$ s (10×40 plate elements) and (b) normal stress τ_{yy} at $t = 1.2$ s (10×40 plate elements).

and

$$\|R\boldsymbol{\beta}_h\|_1 \leq c\|\boldsymbol{\beta}_h\|_1, \quad \forall \boldsymbol{\beta}_h \in \Theta_h,$$

and the transverse shear strains $\boldsymbol{\gamma}_h$ are expressed as $kd^{-2}(R\boldsymbol{\beta}_h - \nabla w_h)$.

Note that the shear strain energy is given by $\frac{kd}{2}\|R(\boldsymbol{\beta}_h - \nabla w_h)\|_0^2$ instead of $\frac{kd}{2}\|\boldsymbol{\beta}_h - \nabla w_h\|_0^2$. In essence, the successful selection of R ensures the reliability of the mixed formulation (Bathe and Dvorkin, 1986; Dvorkin and Bathe, 1984). In fact, the need for such a mixed formulation is confirmed in Wang (2008), Bathe (1996), and Bao *et al.* (2001) when the shear terms are predominant just as the bulk modulus for almost incompressible materials. When the plate is moving, as occurred during a transport process or during the rotation of blades, computer hard disks, and space structures, additional difficulties arise when the Coriolis force breaks the self-adjoint property and, moreover, along with the centrifugal force, introduces destabilizing effects.

For the limiting case in which $d \to 0$, the limit w corresponds to the Kirchhoff model, and the variational problem assumes the form

$$a(\boldsymbol{\beta}_h, \bar{\boldsymbol{\beta}}_h) + (\boldsymbol{\gamma}_h, R\bar{\boldsymbol{\beta}}_h - \nabla \bar{w}_h) = (f_w, \bar{w}_h) + (\mathbf{f}_\beta, \bar{\boldsymbol{\beta}}_h),\ \forall \bar{\boldsymbol{\beta}}_h \in \Theta_h,\ \forall \bar{w}_h \in W_h,$$
$$R\boldsymbol{\beta}_h = \nabla w_h.$$

For a typical MITC4 element as shown in Fig. 5.3, the linear operator R is selected in such a way that at the tying points (midpoints of each edge) yield $R\boldsymbol{\beta}_h = \boldsymbol{\beta}_h$ (Bathe, 1996). Also as shown in Fig. 5.3, the angles between the r and x axes and s and x axes are defined as α and β, and the shear strains are expressed as

$$\gamma_{xz} = \gamma_{rz} \sin\beta - \gamma_{sz} \sin\alpha \quad \text{and} \quad \gamma_{yz} = -\gamma_{rz} \cos\beta + \gamma_{sz} \cos\alpha,$$

with

$$\gamma_{rz} = \lambda_1 \left\{ (1+s) \left[\frac{w_1 - w_2}{2} - \frac{x_1 - x_2}{4}(\beta_x^1 + \beta_x^2) - \frac{y_1 - y_2}{4}(\beta_y^1 + \beta_y^2) \right] \right.$$
$$+ (1-s) \left[\frac{w_4 - w_3}{2} - \frac{x_4 - x_3}{4}(\beta_x^4 + \beta_x^3) \right.$$
$$\left. \left. - \frac{y_4 - y_3}{4}(\beta_y^4 + \beta_y^3) \right] \right\} \Big/ 8\det\mathbf{J},$$

$$\gamma_{sz} = \lambda_2 \left\{ (1+r) \left[\frac{w_1 - w_4}{2} - \frac{x_1 - x_4}{4}(\beta_x^1 + \beta_x^4) - \frac{y_1 - y_4}{4}(\beta_y^1 + \beta_y^4) \right] \right.$$
$$+ (1-r) \left[\frac{w_2 - w_3}{2} - \frac{x_2 - x_3}{4}(\beta_x^2 + \beta_x^3) \right.$$
$$\left. \left. - \frac{y_2 - y_3}{4}(\beta_y^2 + \beta_y^3) \right] \right\} \Big/ 8\det\mathbf{J},$$

$$\mathbf{J} = \begin{bmatrix} \dfrac{\partial x}{\partial r} & \dfrac{\partial y}{\partial r} \\[2ex] \dfrac{\partial x}{\partial s} & \dfrac{\partial y}{\partial s} \end{bmatrix},$$

$$\lambda_1 = \sqrt{(C_x + rB_x)^2 + (C_y + rB_y)^2}, \quad \lambda_2 = \sqrt{(A_x + sB_x)^2 + (A_y + sB_y)^2};$$
$$A_x = x_1 - x_2 - x_3 + x_4, \quad A_y = y_1 - y_2 - y_3 + y_4;$$
$$B_x = x_1 - x_2 + x_3 - x_4, \quad B_y = y_1 - y_2 + y_3 - y_4;$$
$$C_x = x_1 + x_2 - x_3 - x_4, \quad C_y = y_1 + y_2 - y_3 - y_4.$$

The moving plates considered here have two built-in edge supports at $x = 0$ and $x = L$, and two free edges (zero shear forces, bending moments, and twisting moments) at $y = 0$ and $y = B$. For the mixed thin plate theory, general specific boundary considerations are discussed in Häggblad and Bathe (1990). With the standard procedure of interpolating, a typical finite element formulation is expressed as

$$w = \mathbf{H}_w \widehat{\mathbf{W}}, \quad \frac{\partial w}{\partial x} = \mathbf{B}_w \widehat{\mathbf{W}}, \quad R\boldsymbol{\beta} - \nabla w = \mathbf{B}_{\gamma w} \widehat{\mathbf{W}} + \mathbf{B}_{\gamma x} \hat{\boldsymbol{\beta}}_x + \mathbf{B}_{\gamma y} \hat{\boldsymbol{\beta}}_y$$

$$\beta_x = \mathbf{H}_x \hat{\boldsymbol{\beta}}_x, \quad \frac{\partial \beta_x}{\partial x} = \mathbf{B}_{x1} \hat{\boldsymbol{\beta}}_x, \quad \frac{\partial \beta_x}{\partial y} = \mathbf{B}_{x2} \hat{\boldsymbol{\beta}}_x$$

$$\beta_y = \mathbf{H}_y \hat{\boldsymbol{\beta}}_y, \quad \frac{\partial \beta_y}{\partial x} = \mathbf{B}_{y1} \hat{\boldsymbol{\beta}}_y, \quad \frac{\partial \beta_y}{\partial y} = \mathbf{B}_{y2} \hat{\boldsymbol{\beta}}_y$$

where $\mathbf{H}_w$, $\mathbf{H}_x$, and $\mathbf{H}_y$ are the interpolation matrices for w, β_x, and β_y, respectively, and the solution vector is denoted as $\widehat{\mathbf{U}} = (\widehat{\mathbf{W}}, \boldsymbol{\beta} h_x, \boldsymbol{\beta} h_y)^T$.

The matrix equations derived from the standard Galerkin formulation yield

$$\mathbf{M}\ddot{\widehat{\mathbf{U}}} + \mathbf{C}\dot{\widehat{\mathbf{U}}} + \mathbf{K}\widehat{\mathbf{U}} = \mathbf{R},$$

with

$$\mathbf{M} = \begin{bmatrix} \mathbf{M}_{ww} & \mathbf{0} & \mathbf{0} \\ \mathbf{0} & \mathbf{M}_{xx} & \mathbf{0} \\ \mathbf{0} & \mathbf{0} & \mathbf{M}_{yy} \end{bmatrix}, \quad \mathbf{C} = \begin{bmatrix} \mathbf{C}_{ww} & \mathbf{0} & \mathbf{0} \\ \mathbf{0} & \mathbf{C}_{xx} & \mathbf{0} \\ \mathbf{0} & \mathbf{0} & \mathbf{C}_{yy} \end{bmatrix}$$

$$\mathbf{K} = \begin{bmatrix} \mathbf{K}_{ww} & \mathbf{K}_{wx} & \mathbf{K}_{wy} \\ \mathbf{K}_{xw} & \mathbf{K}_{xx} & \mathbf{K}_{xy} \\ \mathbf{K}_{yw} & \mathbf{K}_{yx} & \mathbf{K}_{yy} \end{bmatrix}, \quad \mathbf{R} = \begin{Bmatrix} \mathbf{R}_w \\ \mathbf{R}_x \\ \mathbf{R}_y \end{Bmatrix}$$

and

$$\mathbf{K}_{xx} = \int_\Omega \frac{d^3}{12} (a_{11} \mathbf{B}_{x1}^T \mathbf{B}_{x1} + a_{33} \mathbf{B}_{x2}^T \mathbf{B}_{x2}) d\Omega + \int_\Omega kd \mathbf{B}_{\gamma x}^T \mathbf{C}_\gamma \mathbf{B}_{\gamma x} d\Omega$$
$$- \int_\Omega \frac{\rho d^3 V^2}{12} \mathbf{B}_{x1}^T \mathbf{B}_{x1} d\Omega,$$

$$\mathbf{K}_{yy} = \int_\Omega \frac{d^3}{12}(a_{22}\mathbf{B}_{y2}^T\mathbf{B}_{y2} + a_{33}\mathbf{B}_{y1}^T\mathbf{B}_{y1})d\Omega + \int_\Omega kd\mathbf{B}_{\gamma y}^T\mathbf{C}_\gamma\mathbf{B}_{\gamma y}d\Omega,$$

$$-\int_\Omega \frac{\rho d^3V^2}{12}\mathbf{B}_{y1}^T\mathbf{B}_{y1}d\Omega,$$

$$\mathbf{K}_{xy} = \int_\Omega \frac{d^3}{12}(a_{12}\mathbf{B}_{x1}^T\mathbf{B}_{y2} + a_{33}\mathbf{B}_{x2}^T\mathbf{B}_{y1})d\Omega + \int_\Omega kd\mathbf{B}_{\gamma x}^T\mathbf{C}_\gamma\mathbf{B}_{\gamma y}d\Omega$$

$$\mathbf{K}_{yx} = \int_\Omega \frac{d^3}{12}(a_{21}\mathbf{B}_{y2}^T\mathbf{B}_{x1} + a_{33}\mathbf{B}_{y1}^T\mathbf{B}_{x2})d\Omega + \int_\Omega kd\mathbf{B}_{\gamma y}^T\mathbf{C}_\gamma\mathbf{B}_{\gamma x}d\Omega$$

$$\mathbf{K}_{ww} = \int_\Omega kd\mathbf{B}_{\gamma w}^T\mathbf{C}_\gamma\mathbf{B}_{\gamma w}d\Omega + \int_\Omega (\sigma_T(y)d - \rho dV^2)\mathbf{B}_w^T\mathbf{B}_w d\Omega,$$

$$\mathbf{R}_w = \int_\Omega (p - \rho g d)\mathbf{H}_w^T d\Omega,$$

$$\mathbf{K}_{xw} = \int_\Omega kd\mathbf{B}_{\gamma x}^T\mathbf{C}_\gamma\mathbf{B}_{\gamma w}d\Omega, \quad \mathbf{K}_{yw} = \int_\Omega kd\mathbf{B}_{\gamma y}^T\mathbf{C}_\gamma\mathbf{B}_{\gamma w}d\Omega;$$

$$\mathbf{K}_{wx} = \int_\Omega kd\mathbf{B}_{\gamma w}^T\mathbf{C}_\gamma\mathbf{B}_{\gamma x}d\Omega, \quad \mathbf{K}_{wy} = \int_\Omega kd\mathbf{B}_{\gamma w}^T\mathbf{C}_\gamma\mathbf{B}_{\gamma y}d\Omega;$$

$$\mathbf{C}_{xx} = \int_\Omega \frac{\rho d^3V}{6}\mathbf{H}_x^T\mathbf{B}_{x1}d\Omega, \quad \mathbf{C}_{yy} = \int_\Omega \frac{\rho d^3V}{6}\mathbf{H}_y^T\mathbf{B}_{y1}d\Omega;$$

$$\mathbf{C}_{ww} = \int_\Omega 2\rho dV\mathbf{H}_w^T\mathbf{B}_w d\Omega, \quad \mathbf{M}_{ww} = \int_\Omega \rho d\mathbf{H}_w^T\mathbf{H}_w d\Omega;$$

$$\mathbf{M}_{xx} = \int_\Omega \frac{\rho d^3}{12}\mathbf{H}_x^T\mathbf{H}_x d\Omega, \quad \mathbf{M}_{yy} = \int_\Omega \frac{\rho d^3}{12}\mathbf{H}_y^T\mathbf{H}_y d\Omega$$

and the contributions from distributed or concentrated moments in x and y directions are included in $\mathbf{R}_x$ and $\mathbf{R}_y$.

Some preliminary results, as shown in Figs. 5.4 to 5.9, have been reported to demonstrate the capabilities and reliability of the formulation based on the Mindlin/Reissner plate theory (Wang and Bathe, 1997a, 1997b) for a typical orthotropic plate with the following physical parameters: $\rho = 700\,\text{kg/m}^3$; $d = 0.7\,\text{mm}$; $L = 2\,\text{m}$; $B = 8\,\text{m}$; $E_x = 7.44\,\text{GPa}$; $E_y = 3.47\,\text{GPa}$; $\nu_{xy} = 0.149$; $G_{xy} = 2.04\,\text{GPa}$; $G_{xz} = 0.099\,\text{GPa}$; $G_{yz} = 0.137\,\text{GPa}$.

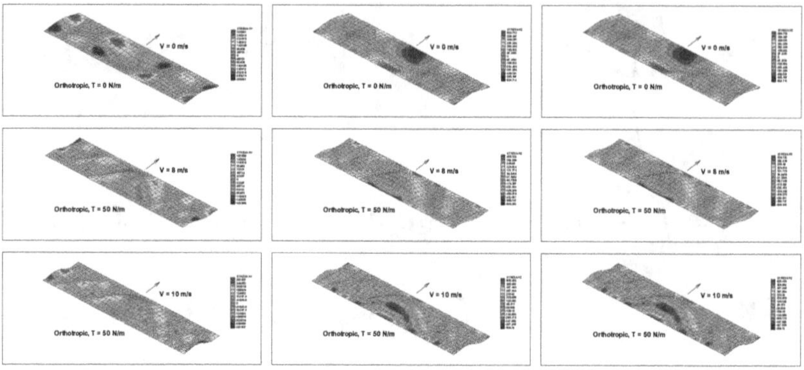

(a) (b) (c)

Fig. 5.9. (a) In-plane shear stress τ_{xy} at $t = 1.2$ s (10×40 plate elements). Transverse shear stress with (b) τ_{xz} at $t = 1.2$ s (10×40 plate elements) and (c) τ_{yz} at $t = 1.2$ s (10×40 plate elements).

Many constituents of biological systems can be represented as immersed flexible shells/beams. This research aims to establish a broad and universal methodology for the treatment of such complex fluid–structure systems. If successful, the developed modeling methods will be used to link the macroscopic mechanical behaviors and microscopic chemical/physical changes of interest in current bioengineering research. The developed methods will be eventually used to improve artificial valve and heart designs. In essence, protein, DNA, muscle fibers, and polymers can be viewed as immersed beam-like structures, and lipids and cell membranes can also be modeled as plates and shells. The preliminary results confirm that the new methods can be readily coupled with microscopic and cellular level biomechanics, and hence be employed for a better understanding of various multi-scale and multi-physics phenomena. Finally, this research will also lead to the enhancement of a multidisciplinary curriculum and the training of students, scientists and engineers in various science and engineering fields.

Chapter 6

Special Applications

Fluid–structure interactions occur in many engineering applications and systems, as well as innate bio-mechanical functions in a human body. These fluid–structure interaction continuum models can typically be categorized into two groups: one-way coupling where one field provides the solution information to another field and repeats over time cycles, and two-way coupling or fully-coupled fluid–structure interaction models where the fluid feeds the information to the solid and the solid feeds information back to the fluid, be it explicit or implicit in time. Recently, some breakthroughs have been made in the development of finite element formulations for immersed boundary or immersed continuum methods (Wang and Liu, 2004; Zhang *et al.*, 2004). In addition, a preliminary formulation for beams with both bending and torsional moments has been presented (Lim *et al.*, 2008). In this chapter, we focus on the expansion of heat and mass transfer issues using the immersed boundary (IB) concepts, a specific problem of compressible air interacting with vocal folds, and applications of IB methods in human circulatory systems.

In general, one-way coupling between the fluid and the solid is where the motion and the deformation of the solid are only determined by the fluid pressure acting on the fluid–structure interface and the solid does not provide any feedback to the fluid domain. The one-way coupling approaches include controlled/manipulated solid motion by some pre-designed spatial-temporal function and superimposed solutions of the fluid and solid (Luo *et al.*, 2008; Mittal *et al.*, 2011). It is well known that one-way coupling often leads to divergence or converging to a wrong or unphysical solution. On the contrary, a fully-coupled fluid–structure system typically yields more stable and more realistic solution comparing to that of one-way coupling. The IB method is a numerical technique that involves two-way coupling between the fluid and the solid. A first attempt to use the IB method to simulate the air flow interacting with vocal folds using a 2D model was reported and documented in Luo *et al.* (2008).

6.1 Heat and Mass Transfer

In order to include heat and mass transfer in the original IB methods, energy conservation and species conservation equations must be included in total systems. The same kernel functions will be employed to represent the equivalent sources for heat and mass transfer. Moreover, matrix-free Newton–Krylov iterative procedures with V-cycled multigrid preconditioning will be combined for general IB methods with additional unknown temperature for heat transfer and species concentration for mass transfer.

6.1.1 *Formulations and theories*

The advantage of immersed methods is to allow independent Lagrangian mesh of the immersed continuum to move on top of a fixed background Eulerian mesh. Therefore, it is necessary to combine both immersed domain (Ω_s) and exterior domain (Ω_f) into a complete domain (Ω). Because this complete domain stays the same regardless of the position of the immersed continuum, a fixed background mesh becomes possible. In practice, a MOD function can be used to efficiently identify the locations of immersed continua. Furthermore, since the immersed solids are impermeable, to have a complete background fluid domain, concepts of the fictitious domain method (Glowinski *et al.*, 2001; Van Loon *et al.*, 2007; Yu, 2005) should be employed.

The key of immersed methods is the energy conservation; namely, the energy input to the fluid domain from the immersed solid or immersed structure is the same as that from the equivalent body force. Moreover, the same delta function must be used in both the distribution of the resultant nodal force and the interpolation of the solid velocity based on the surrounding fluid velocities. In fact, this treatment in immersed methods can be viewed as the synchronization of the fluid motion with the solid motion within the immersed solid domain Ω_s, namely, $\mathbf{v}^s = \mathbf{v}^f$.

This constraint introduces the (distributed) Lagrange multiplier as the equivalent body force $\mathbf{f}^s$. In this case, the equivalent body forces can be directly calculated along with independent fluid and solid velocity vectors (Sulsky and Brackbill, 1991). Naturally, if the surrounding fluid is viscous and incompressible, the immersed solid must be incompressible in immersed methods (Wang and Liu, 2004; Zhang *et al.*, 2004; Boffi *et al.*, 2007, 2008). In this work, we consider general cases with compressible solids immersed in compressible fluid medium.

The weak form of the fluid–solid system can be written as $\forall \mathbf{w} \in [H^1_{0,\Gamma_v}(\Omega)]^d$

$$\int_{\Omega} [w_i \rho_f (\dot{v}_i - g_i) + w_{i,j}\sigma_{ij}] d\Omega - \int_{\Gamma_f} w_i f_i^{\Gamma_f} d\Gamma - \int_{\Omega_s} w_i^s f_i^s d\Omega = 0,$$

with

$$\int_{\Omega_s} w_i^s f_i^s d\Omega = -\int_{\Omega_s} [w_i(\rho_s - \rho_f)(\dot{v}_i - g_i) + w_{i,j}(\sigma_{ij}^s - \sigma_{ij}^f)] d\Omega, \quad (6.1)$$

and the stress definition

$$\sigma_{ij} = -p\delta_{ij} + \tau_{ij}, \quad (6.2)$$

along with the continuity equation

$$v_{i,i} + \frac{\dot{p}}{\kappa} = 0, \quad (6.3)$$

where for Newtonian fluid model, the pressure p is subtracted from the stress components σ_{ij} to obtain the deviatoric stress components τ_{ij} denoted as $\tau_{ij} = \mu(v_{j,i} + v_{i,j})$, and κ stands for the equivalent bulk modulus.

Moreover, the continuity equation of the compressible viscous fluid is expressed with the pressure unknown. For typical fluids, the change of the fluid density can be easily updated with the pressure change through the state equation. Likewise, the solid stress tensor is decomposed as a hydrostatic pressure p^s, and a deviatoric stress tensor τ_{ij}^s,

$$\sigma_{ij}^s = -p^s\delta_{ij} + \tau_{ij}^s. \quad (6.4)$$

In nonlinear mechanics, the solid deformation gradient $F_{ij} = \partial x_i^s(t)/\partial x_j^s(0)$ has to be introduced, from which the Green–Lagrangian strain ε_{ij} can be derived. Consequently, the energy conjugate stress S_{ij}, namely, the second Piola–Kirchhoff stress, can be derived from the elastic energy $\bar{W}$, which is often related to the three invariants of the Cauchy–Green deformation tensor $\mathbf{C}$ defined as $\mathbf{F}^T\mathbf{F}$. To better handle the compressibility issues, an additional term $-[p^s + \kappa^s(J_3 - 1)]^2/2\kappa^s$ is added to the elastic energy $\bar{W}$, and the solid unknown pressure p^s is introduced as a separate unknown, matching the treatment of the fluid, in the continuity equation, Eq. (6.3). Note that for brevity, the subscript or superscript f for fluid domain variables are ignored.

In the same framework of the IB methods, the equivalent representation of the effects from the immersed structure is represented in Eq. (6.11). In fact, q^i in Eq. (6.11) is very much the same as the equivalent body force $\mathbf{f}^s$ in Eq. (6.1), which can be interpreted as the distributed Lagrangian multiplier to enforce the governing equations of the immersed continuum. Of course, to include heat/mass transfer, additional unknowns must be introduced. For Eq. (6.11), we represent the temperature as $T^h = N_I^T T_I$, where N_I^T stands for the interpolation functions at node I for the temperature; and T_I is the nodal value of the discretized temperature unknown.

Of course, to relate to the expression in Eq. (6.4), the solid Cauchy stress is converted from the second Piola–Kirchhoff stress,

$$\sigma_{ij}^s = \frac{1}{\det(\mathbf{F})} F_{i,m} S_{mn} F_{j,n}. \tag{6.5}$$

Since the solid displacement is dependent on the fluid velocity, the primary unknowns for the coupled fluid–solid system are the fluid velocity $\mathbf{v}$, the fluid pressure p, and the solid pressure p^s. Define the Sobolev spaces, the weak form of governing equations can be expressed as: $\forall q \in L^2(\Omega)$, $q^s \in L^2(\Omega_s)$, $\mathbf{w} \in [H^1_{0,\Gamma_v}(\Omega)]^d$, which includes $\forall \mathbf{w}^s \in [H^1(\Omega_s)]^d$, and find $\mathbf{v}$ and $p \in \Omega$, $p^s \in \Omega_s$, such that

$$\begin{aligned}
&\int_\Omega w_i \rho(\dot{v}_i - g_i) d\Omega + \int_\Omega (w_{i,j}\tau_{ij} - p w_{i,i}) d\Omega - \int_{\Gamma_f} w_i f_i^{\Gamma_f} d\Gamma \\
&+ \int_{\Omega_s} \left[w_i^s(\rho_s - \rho)(\dot{v}_i - g_i) + w_{i,j}^s(\tau_{ij}^s - \tau_{ij}^f) - (p^s - p) w_{i,i}^s \right] d\Omega \\
&+ \int_\Omega q\left(v_{j,j} + \frac{p_{,t}}{\kappa}\right) d\Omega + \int_{\Omega_s} q^s \left(J_3 - 1 + \frac{p^s}{\kappa^s}\right) d\Omega = 0.
\end{aligned} \tag{6.6}$$

For the fluid domain, the following interpolations for the entire domain Ω are introduced:

$$\mathbf{v}^h = N_I^v \mathbf{v}_I, \quad \mathbf{w}^h = N_I^v \mathbf{w}_I, \quad p^h = N_I^p p_I, \quad q^h = N_I^p q_I, \tag{6.7}$$

where N_I^v and N_I^p stand for the interpolation functions at node I of the velocity vector and the pressure; and $\mathbf{v}_I$, $\mathbf{w}_I$, p_I, and q_I are the nodal values of the discretized velocity vector, admissible velocity variation, pressure, and pressure variation, respectively.

Note that in general, the interpolation functions for the velocity vector unknowns denoted by the superscript v and the pressure unknowns denoted by p are different. Likewise for the solid domain Ω_s, the discretization is based on the following:

$$\mathbf{u}^{s,h} = N_J^u \mathbf{u}_J^s, \quad \mathbf{w}^{s,h} = N_J^u \mathbf{w}_J^s, \quad p^{s,h} = N_J^{p^s} p_J^s, \quad q^{s,h} = N_J^{p^s} q_J^s, \tag{6.8}$$

where N_J^u and $N_J^{p^s}$ stand for the interpolation functions at node J for the displacement vector and the pressure unknowns; and $\mathbf{u}_J^{s,h}$, $\mathbf{w}_J^{s,h}$, $p_J^{s,h}$, and $q_J^{s,h}$ are the nodal values of the discretized displacement vector, admissible velocity variation, pressure, and pressure variation, respectively.

Substituting both discretizations (6.7) and (6.8) into Eq. (6.6), the following discretization of the weak form: $\forall q^h \in L^2(\Omega^h)$, $q^{s,h} \in L^2(\Omega_s^h)$, $\mathbf{w}^h \in [H_{0,\Gamma_v^h}^{1,h}(\Omega^h)]^d$, which includes $\forall \mathbf{w}^{s,h} \in [H^{1,h}(\Omega_s^h)]^d$,

$$\begin{aligned}
&\int_{\Omega^h} w_{iI} N_I^v \rho \dot{v}_i^h d\Omega - \int_{\Gamma_f^h} w_{iI} N_I^v f_i^{\Gamma_f^h} d\Gamma + \int_{\Omega^h} \left(w_{iI} N_{I,j}^v \tau_{ij} - p^h w_{iI} N_{I,i}^v \right) d\Omega \\
&+ \int_{\Omega_s^h} \left[w_{iJ}^s N_J^u (\rho_s - \rho)(\dot{v}_i^h - g_i) + w_{iJ}^s N_{J,i}^u (\sigma_{ij}^s - \sigma_{ij}^f) \right] d\Omega \\
&- \int_{\Omega^h} w_{iI} N_I^v \rho g_i d\Omega + \int_{\Omega^h} q_I N_I^p \left(v_{j,j}^h + \frac{p_{,t}^h}{\kappa} \right) d\Omega \\
&+ \int_{\Omega_s^h} q_J^s N_J^{p^s} \left(J_3 - 1 + \frac{p^{s,h}}{\kappa^s} \right) d\Omega = 0.
\end{aligned}$$

Therefore, for the entire domain Ω, due to the arbitrariness of the variations w_{iI}, q_I, and q_J^s, four equations at each fluid node denoted as I and one equation at each solid node denoted as J are developed,

$$r_{iI}^v = 0, \quad r_I^p = 0, \quad r_J^{p^s} = 0, \tag{6.9}$$

where the residuals are defined as

$$r_{iI}^{v} = \int_{\Omega^h} N_I^v \rho \dot{v}_i^h d\Omega + \int_{\Omega^h} (N_{I,j}^v \tau_{ij} - p^h N_{I,i}^v) d\Omega - \int_{\Gamma_f^h} N_I^{v,\Gamma_f^h} f_i^{\Gamma_f^h} d\Gamma$$

$$+ \int_{\Omega_s^h} \tilde{N} \left[N_J^u (\rho_s - \rho)(\dot{v}_i^h - g_i) + N_{J,i}^u (\sigma_{ij}^s - \sigma_{ij}^f) \right] d\Omega - \int_{\Omega^h} N_I^v \rho g_i d\Omega,$$

$$r_I^p = \int_{\Omega^h} N_I^p \left(v_{j,j}^h + \frac{p_{,t}^h}{\kappa} \right) d\Omega,$$

$$r_J^{p^s} = \int_{\Omega_s^h} N_J^{p^s} \left(J_3 - 1 + \frac{p^{s,h}}{\kappa^s} \right) d\Omega. \tag{6.10}$$

Note that the nonlinear mapping $\tilde{N}$ from $\mathbf{w}_I$ to $\mathbf{w}_J^s$, namely, from the fluid mesh to the solid mesh, depends on the choice of kernel functions which have been discussed in Chapter 2. Moreover, this isentropic compressible fluid formulation should yield the same results to mixed finite element formulations for acoustic fluid–solid interactions (Wang and Bathe, 1997a, 1997b). With sufficiently small time steps, pressure wave propagation and related scattering and radiation phenomena will be captured.

Note that the traditional assumption for acoustic FSI systems is small strain and small displacement; namely, the FSI interface does not move at all. It seems that immersed methods bear no advantages over traditional mixed finite element formulations or potential formulations (Wang, 2008; Bao *et al.*, 2001). However, acoustic fluid–solid interaction models provide a perfect venue to study, validate, and improve immersed methods.

6.1.2 *Equivalent heat and mass sources*

This proposed research aims to expand immersed methods to continuum-continuum systems with heat and mass transfers. With the proposed monolithic approach for such coupled systems, it will be possible to apply appropriate reduced order modeling procedures accordingly. The developed methods are useful for micro aerial vehicle research, soft material design and manufacturing, ultrasonic therapy, and studies of complex biological systems. This proposed research will also lead to the enhancement of a multi-disciplinary education model in various science and engineering fields.

In this work, we also adopt the conservative form for background fluid domain, namely, domain f and immersed solid domain, domain s. Define the specific heat capacity as C, and the heat sources q^f and q^s for domain f and domain s, respectively. For coupled continua, using heat transfer as an example, we have the weak form of the coupled system $\forall \delta T \in H^1_{0,\Gamma_v}(\Omega)$

$$\int_{\Omega} \left[\delta T \rho_f C_f \dot{T} + \nabla \delta T \cdot K_f \nabla T\right] d\Omega - \int_{\Gamma_f} \delta T q^{\Gamma_f} d\Gamma$$
$$- \int_{\Omega_f} \delta T q^f d\Omega - \int_{\Omega_s} \delta T q^i d\Omega = 0, \tag{6.11}$$

with

$$\int_{\Omega_s} \delta T q^i d\Omega = - \int_{\Omega_s} \delta T \left[(\rho_s C_s - \rho_f C_f)\dot{T} - q^s\right] d\Omega - \nabla \delta T \cdot (K_s - K_f) \nabla T d\Omega. \tag{6.12}$$

Again, q^i in Eqs. (6.11) and (6.12) is very much the same as the equivalent body force $\mathbf{f}^s$, which can be interpreted as the distributed Lagrangian multiplier to enforce the governing equations of the immersed continuum. Of course, to include heat/mass transfer, additional unknowns must be introduced. For Eqs. (6.11) and (6.12), we represent the temperature as $T^h = N_I^T T_I$, where N_I^T stands for the interpolation functions at node I for the temperature; and T_I is the nodal value of the discretized temperature unknown. A preliminary result of an implementation of immersed methods with heat transfer is shown in Fig. 6.1 for demonstration purposes (Wang and Lua, 2010).

6.2 Valves and Vocal Folds

The self-sustained vocal fold vibration is such an application that intrinsically involves extensive interactions between fluid (air) and structure (vocal folds). The vocal folds are a pair of flexible structures at the top of the trachea, where air travels through the larynx and produces sounds due to its fluctuations (Alipour *et al.*, 2000). Over the past two decades, a number of numerical models are built to facilitate the understanding of the self-sustained vocal fold vibration, ranging from multi-mass models in which the vocal folds are modeled as mass-spring-damper system to continuum models in which the vocal folds are modeled using realistic

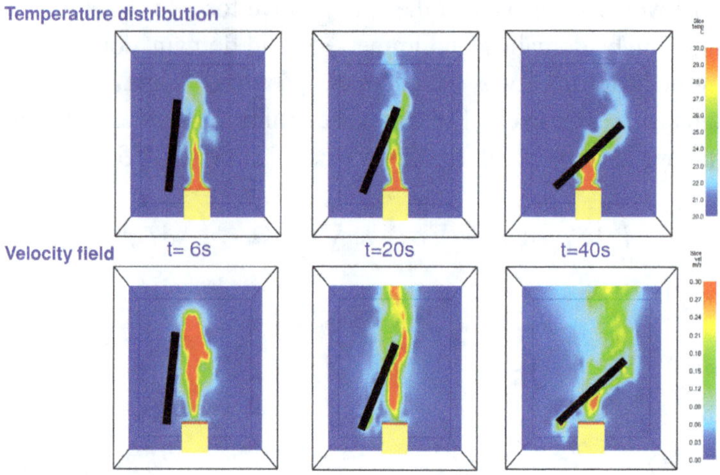

Fig. 6.1. General IB methods for couple continua with heat and mass transfer.

physical representation (Alipour and Scherer, 2000; Tao and Jiang, 2006). With the dramatic increase in computational capabilities and efficiencies in recent years, more continuum vocal fold models have been developed using 2D and 3D finite elements, some of which are coupled with Bernoulli's equation for fluid to study the mechanisms behind the self-sustained oscillation process (Tao *et al.*, 2006; Thomson *et al.*, 2005).

In the Immersed Finite Element Method (EIFEM), the fluid was incompressible and viscous while the vocal folds were modeled with a finite difference method with appropriate viscoelastic material properties. The coupling between the two domains was done via a so-called sharp interface methodology, where the solid was driven by the traction boundary condition interpolated from the previous time step and solved as a quasi-steady state. The solid was then updated to a new position and a corresponding velocity field for both interior and boundary was obtained. The velocity at the solid boundary was then interpolated back to the fluid domain and applied as the boundary condition for the fluid. This so-called boundary is not the Dirichlet or Neumann boundary of the fluid domain, but rather, the fluid–structure interface inside the fluid domain.

Using the IB method, Luo *et al.* (2009) and Zheng *et al.* (2009) further studied the effects of false vocal folds on glottal flow and vocal fold vibration. An improved solid model was also presented in Luo *et al.* (2009), where the viscoelastic solid dynamics was solved using finite element methods rather than finite difference methods as presented in Luo *et al.* (2008), which provided a better and easier way of applying solid boundary conditions, especially for complex geometries. In addition, a contact model was also included instead of the previous model in which vocal fold displacements have been manipulated to avoid overlapping with each other. Furthermore, the acoustic field has been modeled in Mittal's group with the so-called hydrodynamic/acoustic splitting method to include linearized perturbed compressible fluid equations (Mittal *et al.*, 2011). However, this acoustic field was not coupled with the dynamic vocal fold model.

Despite the challenges, the fully-coupled approach in the study of vocal fold vibrations has significantly advanced the capabilities in representing a feasible and realistic glottal system. As of date, there remains a major deficiency in all the sophisticated models to study the full interactions between aeroelasticity and aeroacoustics — they ignored the compressibility of air. The inclusion of compressibility in a fully-coupled fluid–structure glottal system is vitally important since the sound wave generation and propagation during vocal fold vibrations are induced and facilitated by the air compression.

In fact, our initial attempt to use the incompressible fluid for the coupled system resulted in a reasonable magnitude of vocal fold vibrations, but ultimately failed to reach a sustained vibration. Without a steady-state vibration, it is futile to perform any analysis of the results to understand the fundamental mechanism of sound production.

In this work, we design a new numerical tool that encompass the compressibility of the fluid, its duly interactions with aeroelastic materials, as well as other features that are necessary to successfully simulate the laryngeal system. The modified Immersed Finite Element Method (mIFEM) is built upon a fully-coupled acoustoelastic fluid–structure interaction model presented in Wang and Bathe (1997b). The major distinction between the class of Immersed Finite Element Methods and that of the IB methods is that it utilizes the concept of the principle of virtual work to remove the artificial effects from the overlapping background fluid, thus providing an independent and accurate solid material description. Both fluid and solid solvers use a finite element approach where non-uniform and unstructured meshes can be employed to handle complex geometries. The versatility of the mIFEM provides us with the perfect tool for the study of vocal fold vibrations with the consideration of vocal fold motions. Finally, the vocal fold dynamics is governed by viscoelastic constitutive law and solved on a Lagrangian mesh using the finite element method, while the fluid is governed by Navier–Stokes equations describing compressible isentropic air that is represented on an Eulerian background mesh.

We further utilize the numerical tool to perform a series of parametric studies using a range of lung pressure and vocal fold stiffness to evaluate the effects in the fundamental frequency and magnitude of the self-sustained vibration. In this chapter, we will first present the numerical method formulated for compressible fluid interacting with aeroelastic structures. The solid model also includes a contact algorithm. The two-dimensional (2D) vocal fold model is presented in Section 6.2.1. With a successful mesh convergence study, we have first verified vocal fold vibrations in terms of vibration frequency and magnitude. A parametric study is conducted to further analyze the effects of lung pressure and vocal fold stiffness. The Immersed Finite Element Method (IFEM) is a volume-based finite element FSI numerical method. Finite elements are not only used as a way of discretization, but the entire IFEM formulation is derived from the weak form, or principle of virtual work, where the total work in the entire FSI system is maintained.

Even though the concept of "immersed" is similar to that of the immersed boundary (IB) method (Peskin, 2002) where the solid or rather the volumeless structure is immersed in the fluid and the two domains are constructed independently, the major distinction between the two is that the solid is not just an interface as it is in the IB method, it actually occupies a volume in the computational domain. As such, the formulation for the IFEM must be volume-based, rather than point-based in the IB method. Similarly, the coupling scheme is derived through the balance of the total work in the entire computational space rather than a direct elastic force evaluation on each discrete point.

In IFEM, the fluid and solid domains are defined on two sets of meshes independently with their respective governing equations. The solid domain Ω^s is completely immersed in its surrounding fluid Ω^f, depicted in Fig. 6.6. Since the solid domain is immersed in the fluid domain, the solid boundary

Γ^s is also the common interface intersecting with the fluid domain, called the fluid–structure interface $\Gamma^{fsi} \equiv \Gamma^s$.

The fluid domain and the immersed solid domain jointly form the entire computational domain, $\Omega = \Omega^f \cup \Omega^s$. All variables associated with the solid are denoted with either subscript s or superscript s, while all of those associated with the fluid are denoted with either subscript f or superscript f. The key assumption of the IFEM algorithm is that the fluid exists everywhere in the entire computation domain, similar to the IB method. However, unlike the connected discrete points in representing the solid boundary in IB, the solid in IFEM occupies volume in the computational space. The part of the fluid that overlaps with the solid domain is called the artificial or fictitious or phantom fluid, $\bar{\Omega}$, since it does not physically exist. Note that by definition and implementation, the *artificial* fluid domain $\bar{\Omega}$ and the solid domain Ω^s do occupy the same space. The solid domain gets updated at every time step from its initial Ω^s_0. As a result, the artificial domain also moves in the entire computational space. In fact, it is very important to understand that by incorporating the dynamics of the solid into the coupled system, the inertial force from the acceleration must be included. In the original IB method, the immersed elastic boundary is simply experiencing a quasi-static deformation, namely the inertia of the volumeless boundary is not considered, and the boundary motion is entirely governed by the dynamics of the surrounding fluid.

In the modified IFEM (mIFEM), the algorithm is entirely reformulated to accommodate large Reynolds number flows in which the solid and the fluid have disparate density difference such as that between soft tissue and air, in addition, the solid inertia is significant along with fluid–structure interactions. The mIFEM addresses these problems by reformulating the coupling scheme and solving for the solid dynamics rather than imposing it. It has been rigorously derived and verified with 2D and 3D studies (Wang and Bathe, 1997a, 1997b; Wang and Zhang, 2015). The mathematical studies of the newly proposed Immersed Finite Element Methods have been presented in (Kim and Liu, 2006; Liu *et al.*, 2007a). Using this concept of the mIFEM, we have further incorporated the compressibility so that air can be appropriately presented as sound is produced by the compression of air, which is a necessary step in the study of aeroelastic interactions.

6.2.1 *Anatomy of vocal folds and tracks*

As shown in Fig. 4.16, even with the coarse mesh used, the implicit immersed continuum method with high-order mixed elements provides reasonable results comparable to reference solutions. It is very important to note that the discretized delta function provides the possibility of linking the immersed solid node with the surrounding fluid nodes, as done in meshless finite element methods (Boffi *et al.*, 2007). As long as the virtual power input to the surrounding fluid is preserved, the concept of immersed methods will stay the same. In fact, this procedure is the same as the standard finite element procedure to distribute concentrated forces and to interpolate displacement/velocity within the element. Whether or not this procedure is as efficient as the meshless type of communication depends very much on the search algorithm for the purpose of locating the immersed solid node. Here are a list of terminologies used in this section:

Glottis width: The glottis width is defined as the size of the narrowest glottal opening which varies with time as the vocal fold vibrates. Specifically for the half-space model, half of the glottis width, h, is defined as the minimum distance from the vocal fold surface to the center line (or the symmetric boundary), as labeled in Fig. 6.7. The full glottis width is assumed to be $2h$ whereas $h_{\max}$ is denoted as the maximum glottis opening during a vibrational cycle in a half-space model.

Volume flow rate: The volume flow rate is also measured at different cross sections. Since it is to be evaluated in a 2D model, the "volume" flow rate is effectively the rate of flow at a particular x as: $Q_C = \int_{x=C} v_x dy$. In this study, the flow rate is evaluated at three different locations (−2.2, 0, 0.8 cm), which respectively fall into the subglottal, glottal, and supraglottal regions as labeled in Fig. 6.7. The origin of the coordinates is set at the trailing edge of the vocal fold.

Reynolds number: The Reynolds number in phonation is typically defined using the peak velocity at the glottis as, $\mathrm{Re} = \rho U_{\max}(2h_{\max})/\mu$. Here, $U_{\max}$ is the maximum glottal jet velocity and $2h_{\max}$ is the maximum glottis width during the opening phase of each vibrational cycle. ρ and μ are the air density and dynamic viscosity, respectively. Since the air density only varies slightly in compressible flows, the initial constant ρ_0 is used when evaluating the Reynolds number.

Fundamental and reduced frequencies: The fundamental frequency, f, is evaluated by taking the Fast Fourier Transform (FFT) of all the directly measurable quantities, namely, glottis width, volumetric flow rate, fluid velocity and pressure. The fluid velocity and pressure at the glottis region are used for frequency analysis so that they are consistent with the glottis width and the flow rate, which are more conservative quantities. The reduced frequency is computed based on the fundamental frequency as: $f^* = fL/U_{\max}$, where L is the length of the base of the vocal fold. In the current model, we choose $L = 1.505$ cm as indicated in Fig. 6.7.

The definitions of both the Reynolds number Re and the reduced frequency f^* are suggested by Krane *et al.* (2004). For typical life-scale problems, the Re number can reach as high as 8000 with reduced or normalized frequency f^* within the range of $0.1 \sim 0.01$.

The human vocal folds are a pair of pliable structures located within the larynx at the top of the trachea. In order to produce sound in voice, the lungs blow air against the pair of closed vocal folds causing the immediate increase of the air pressure beneath the vocal folds. This pressure pushes the two vocal folds apart from each other and a jet of air, sometimes called glottal jet, escapes from the gap between the vocal folds, as illustrated in Fig. 6.2(a). As the air rapidly flows through the larynx, it induces a decrease in pressure and the vocal folds are brought together collaborated with the elastic recoil of the deformed tissue, shown in Fig. 6.2(b). Once the vocal folds close, the pressure beneath the vocal folds rises again, where the compressed air forces its way back through the open passageway to the downstream, repeating the process and driving sustained vibration of the vocal folds.

Table 6.1. Computation time with implicit procedures.

Step Size (sec)	Number of Steps	Running Time (sec)
0.5	100	26163
1	50	13452
2	25	7105

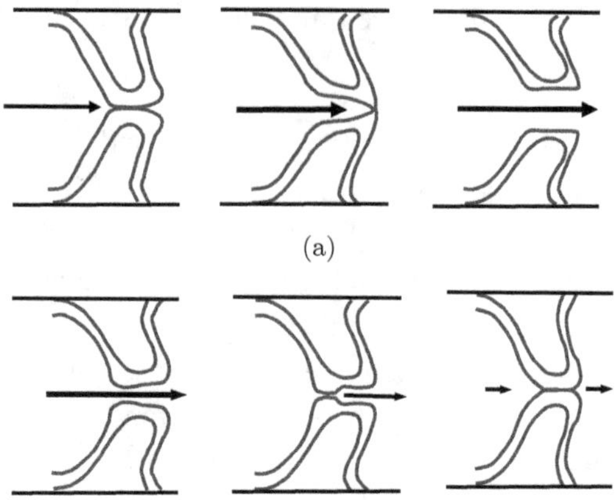

Fig. 6.2. A schematic drawing of the glottis opening and closing process. (a) From closed position to fully opened position and (b) From opened position to fully closed position.

It is important to note that here the air is compressible that undergoes an isentropic process as elaborated in Sections 1.2.3 and 4.2.2. Frequency of vocal fold vibrations are also known as the fundamental frequencies of human voice, which is in the range of 100 to 150 Hz for males and 150 to 200 Hz for females. The vocal fold vibration can be strongly disturbed by various diseases, such as laryngeal cancer, vocal fold paralysis and edema, etc., which are the causes of voice disorders. A clear understanding of the mechanism of vocal fold vibration during phonation is important for providing effective diagnosis and treatment of voice disorders.

Researchers have been studying the vocal fold vibration for decades experimentally, theoretically, and numerically. Berg *et al.* (1957) started the experimental study of the glottal flow using Bernoulli's equation to relate the transglottal pressure to the glottal volume flow. Scherer and Titze (1983) measured the pressure distribution along the walls of a static vocal fold model. Cranen (1985) performed measurements of transglottal pressure in live subjects, with associated difficulties in experimental control and characterizing glottal shape. Deverge *et al.* (2003) used scaled-up models and measured pressure in trachea and glottis, and downstream of the vocal folds while one of the model vocal folds was moved. More recently, Krane *et al.* (2004) used Digital Particle Image Velocimetry (DPIV) to obtain time-resolved measurements of the whole velocity field through a cycle of vocal fold wall motion. The purpose of these experimental findings is to have a full understanding of the physical mechanism of the phonation process and navigate ways to characterize the entire energy flow (from airstream to glottal jet dissipation and to sound production) in a glottal system. Using computational means as we demonstrate here, a more thorough analysis can be obtained. Particularly in this study, fundamental and measurable quantities such as pressure, volume flow rate, glottis opening width, are examined to show the validity of the proposed numerical approach and to be used as a baseline for more complex analysis in the future.

A two-dimensional (2D) vocal folds in a vocal tract is modeled in this study, as shown in Fig. 6.3. The numerical model setup follows a vocal fold acoustic experimental setup designed by Campo and Krane Campo (2012), which is a simplified real-size model. The vocal tract is represented by a long straight rectangular air duct that is 31.514 cm in length and 2.794 cm in width, while the two symmetric vocal folds are placed in the middle of the vocal tract giving sufficient duct length in both upstream and downstream regions for the compressive air to travel through the glottis without significant boundary effects, either physically or numerically.

The vocal fold model has a base length of 1.505 cm that is attached to the vocal tract wall and a height of 1.397 cm. The convergent angle, which is defined as the angle formed at the lips of the vocal fold is 10° at the

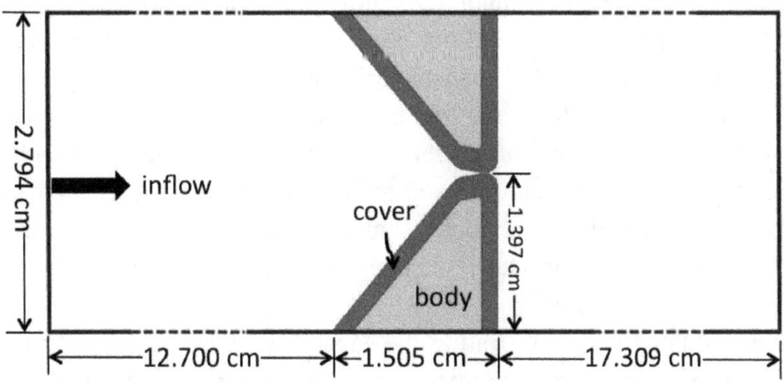

Fig. 6.3. Model setup for a 2D two-layer vocal fold in a long duct.

resting condition, based on the M5 model in Scherer and Shinwari (2001). The vocal fold in this model is composed of two parts, the cover and the body layers, both of which are considered as viscoelastic materials with the density $\rho^s = 1.0\,\text{g/cm}^3$, the Poisson's ratio $\nu = 0.3$, and the damping factor $\eta = 10$ Poise. The elastic moduli adopted for the two parts, however, are different. The cover layer is typically softer than the body (Chhetri *et al.*, 2011). Studies have measured the Young's moduli for the tissues to be in the range between 2.7 and 110 kPa (Alipour and Titze, 1991; Perlman *et al.*, 1984). Our initial study has the body elastic modulus E_b to be four times higher than that of the cover E_c, namely, $E_c = 10$ kPa and $E_b = 40$ kPa.

Since sound generation greatly involves the compression of air, the working fluid in the vocal tract is represented as compressible air governed by the ideal gas law at $273K$ under an isentropic process. The air density $\rho^f = 1.3 \times 10^{-3}\ \text{g/cm}^3$ and viscosity $\mu = 1.8 \times 10^{-4}\ \text{g/cm} \cdot \text{s}$ are used. Constant total pressure boundary condition is applied at the duct inlet on the left, and the stress-free outflow boundary condition is applied at the exit of the duct on the right, as shown in Fig. 6.3. No-slip and no-penetration boundary conditions are applied on the duct interior walls and the vocal fold surface. For the contact model the penetration tolerance $g_{\max}$ is set to be 1 μm.

It is worth mentioning that the material parameters and computational boundary conditions must be properly set. In the case of simulating vocal fold vibrations, if the material properties or input parameters are set beyond a reasonable range or improper boundary condition types are applied, the vocal folds may either cannot fully close or cannot produce sustained steady vibration, which would not be a surprising outcome if the parameters and conditions do not realistically represent the physical system. In healthy larynx physiology, the top and bottom vocal folds are symmetric about the center line and have identical material properties, although the resulting jet flow are often observed to be asymmetric in experiments and numerical simulations. One such example is the full-space model shown in Fig. 6.4, where the glottal jet is diverting towards one side rather than following the center line.

For the purpose of illustrating the capability of our numerical model, obtaining meaningful measurements, and capturing the key features of the vocal fold vibration, all the numerical results shown here are obtained by a

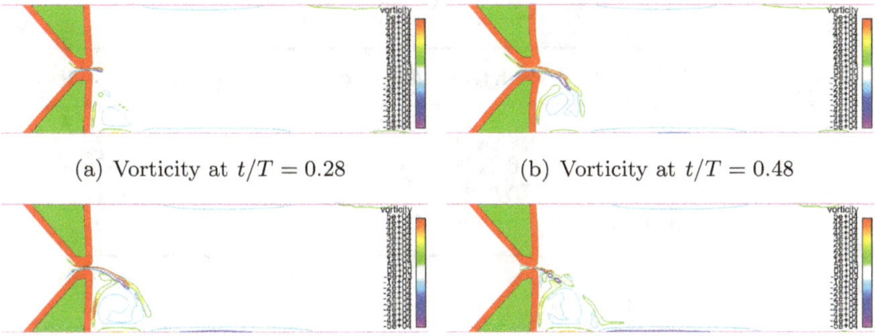

(a) Vorticity at $t/T = 0.28$

(b) Vorticity at $t/T = 0.48$

(c) Vorticity at $t/T = 0.81$

(d) Vorticity at $t/T = 0.95$

Fig. 6.4. Glottal jet vorticity contour with one vibration cycle full-space model: (a) $t/T = 0.28$, (b) $t/T = 0.48$, (c) $t/T = 0.81$, and (d) $t/T = 0.95$.

so-called half-space model. In this study we create a half-space model by assuming the fluid field is symmetric about the center line with a symmetric boundary condition. The causes of glottal jet asymmetry and its effects on the vocal fold vibration will be examined in our subsequent studies.

6.2.2 *Acoustic signatures*

Before the numerical analysis, a mesh convergence test must be performed for this model to examine the effects of mesh resolution in the coupled solution accuracy. For non-conforming meshing techniques for FSI modeling such as the IFEM used in this study, mesh convergence test can be complicated. Since there are two sets of meshes involved, incompatible meshing may result in inaccurate coupled solutions. A very careful mesh convergence study of the IFEM was previously shown in Wang and Zhang (2010). Moreover, in particular for this vocal folds study, the meshing near the glottis opening is especially critical in capturing the opening and closing processes and a sustained vibration. This is another reason why conforming meshing techniques are not suitable for this study as the closing of the vocal folds may involve frequent and intensive mesh re-construction. Here, we design four sets of background fluid meshes, with each one varying in either stream-wise (x-) or gap-width (y-) direction. Since the glottis gap width is measured in the y-direction, it is more important to study the resolution in the y-direction. Ideally, the element size is sufficiently smaller

Table 6.2. Meshing information for mesh convergence test.

	Mesh Size - Δx	Mesh Size - Δy	No. of Elements	No. of Nodes
Mesh 1	0.002 cm	0.014 cm	26992	27480
Mesh 2	0.005 cm	0.014 cm	24308	24836
Mesh 3	0.005 cm	0.028 cm	8750	9060
Mesh 4	0.010 cm	0.028 cm	6905	7215

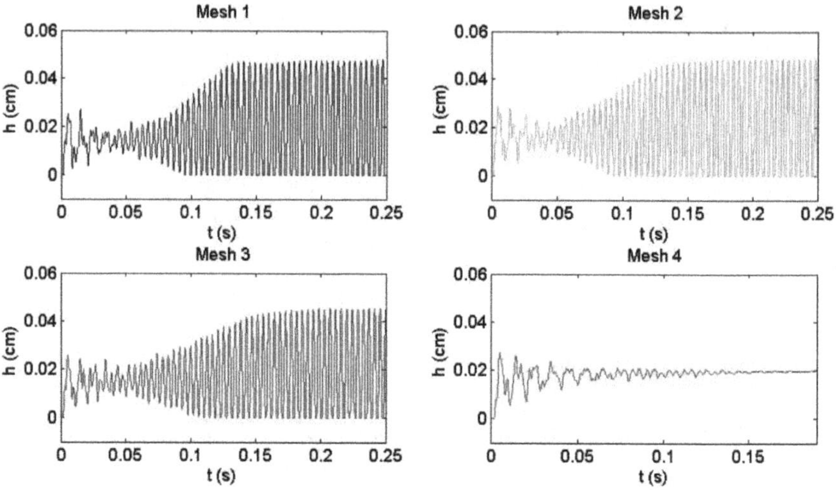

Fig. 6.5. Vocal fold models with four meshes.

than the gap size. Non-uniform meshing is used in all the meshes to accommodate the complex nature of the geometries. Therefore, the total number of nodes and elements is not proportionally varied. The mesh sizes listed here are based on the local resolution near the center line. The detailed information for the four meshes is listed in Table 6.2.

To determine the fidelity of a mesh, we examine two quantities that are directly measured from the simulation using that mesh, the distance between vocal fold and centerline as shown in Fig. 6.5 and the vocal fold vibration frequency as shown in Fig. 6.6. Despite differences in element numbers and element sizes, Meshes 1, 2, and 3 produce the same glottis gap width during the steady-state vibration and the vocal fold vibration reaches

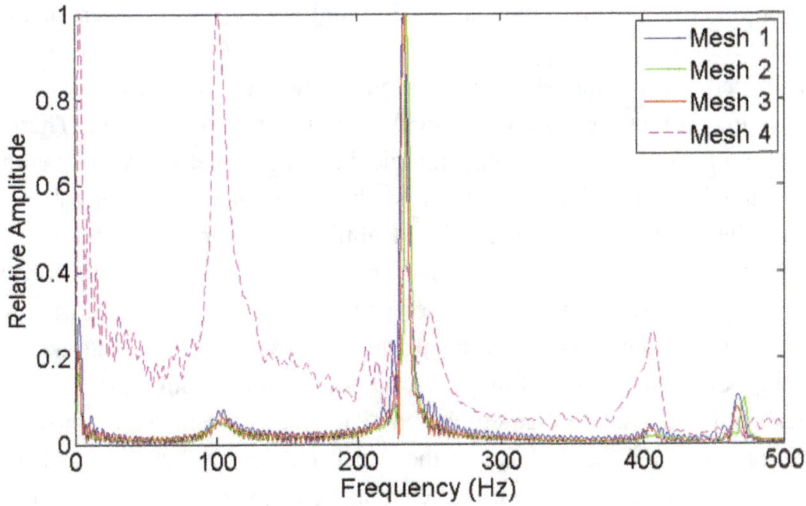

Fig. 6.6. Vocal fold models with vibration frequency produced by four meshes.

the same frequency, while Mesh 4 fails to produce a sustained vibration due to the low resolution in the vicinity of the glottis gap. This outcome indicates that there is a lower bound in the mesh size, particularly near the tip of the vocal folds where instantaneous variations of the flow needs to be captured.

Based on the outcome of the mesh convergence test, Mesh two is chosen as the optimum mesh which has $24,308$ quadrilateral elements and $24,836$ nodes in the fluid domain, and $6,476$ triangle elements and $4,306$ nodes in the solid domain. Comparing to Mesh three, finer mesh resolution is imposed near the center line. The time step size is chosen to be 2×10^{-5} s to ensure the stability of the numerical algorithm.

The computing resource to perform these computations is the IBM Blue Gene Q system located in the Center of Computation Innovations at Rensselaer Polytechnic Institute, with each computing node consisting of a 16-core 1.6 GHz A2 processor and 16 GB of DDR3 memory. The source code is efficiently parallelized with MPI, please refer to Wang and Bathe (1997a) for the computing efficiency studies. For each run, 1024 cores are utilized, which takes approximately 40 hours to complete $10,000$ time steps of simulation. With the time step we choose, it is 50 to 60 cycles of the vocal fold vibration, which is more than sufficient number of cycles to reach steady-state vibrations.

To perform the analysis on the glottal flow and the vocal fold vibration using simulations, a few physical quantities are first defined. Then, a benchmark case that uses all physiological input parameters from a healthy vocal tract is setup. This benchmark study is to demonstrate that the solution technique provides here has the capability in yielding a sustained vocal fold vibration without any additional enforcement or manipulation of the physical interactions. It also provides us a basic understanding of the fundamental mechanism between the air flow and the vocal folds interactions through detailed analysis. Further parametric studies of lung pressure and vocal fold stiffness are then employed. The goal is to understand how the inlet pressure and the stiffness of the vocal folds can affect the vocal fold vibration, which enhances our understanding of the dynamics of the coupled system. A total of nine cases are studied, with the corresponding quantities such as the peak velocity, maximum glottis width, maximum volume flow rate, fundamental and reduced frequencies, and Reynolds number at the glottis, yielded from the simulations listed in Table 6.3.

The benchmark case of this study is listed as Case 4 in Table 6.3 where the inlet pressure, namely, the lung pressure, is set to be 1 kPa, and the Young's modulus for the cover layer E_c and the body E_b are defined as 10 kPa and 40 kPa, respectively. Similar input parameters and material properties for the vocal fold are also used in other numerical studies in Alipour *et al.* (2000), Rosa *et al.* (2003), Tao *et al.* (2006), Zheng *et al.* (2009), and Luo *et al.* (2009). A few snapshots of the vocal folds from a

Table 6.3. Pressure inlet P_{in} and Young's modulus E and output quantities.

#	P_{in} (kPa)	E_c (kPa)	E_b (kPa)	f (Hz)	$U_{\max}$ (cm/s)	$h_{\max}$ (cm)	$Q_{\max}$ (cm^2/s)	f^* (10^{-3})	Re
1	0.7	10	40	234	5318.2	0.0390	84.99	66.220	2995.9
2	0.8	10	40	234	5852.5	0.0430	105.2	60.174	3635.1
3	0.9	10	40	234	6208.7	0.0461	123.5	56.722	4134.3
4	1.0	10	40	234	6658.1	0.0492	145.8	52.893	4731.7
5	1.1	10	40	234	6813.4	0.0506	156.9	51.688	4979.8
6	1.2	10	40	234	6597.5	0.0526	160.6	53.379	5012.6
7	1.3	10	40	234	7361.7	0.0534	187.2	47.838	5678.3
8	1.0	5	20	169	6543.0	0.0574	185.7	38.873	5424.9
9	1.0	20	40	258	6291.0	0.0392	113.5	61.722	3562.1

closed position to an open position as the air travels and pushes through the glottis are shown in Figs. 6.7 and 6.8. At the beginning of the simulation, the vocal fold starts at a resting position. The time histories of the volume flow rate at $x = -0.1$ cm (the tip of the vocal fold) and the glottis width are shown in Figs. 6.9 and 6.10, respectively. As shown in the two plots, the vocal fold starts its consistent vibration pattern after 0.05 s and reaches steady vibration after around 0.2 s. During the steady vibration, the maximum glottis width reaches 0.1066 cm and the minimum glottis value is zero, which indicates the vocal folds are able to shut off the air flow completely. Using FFT for both h and Q, the fundamental frequencies of the

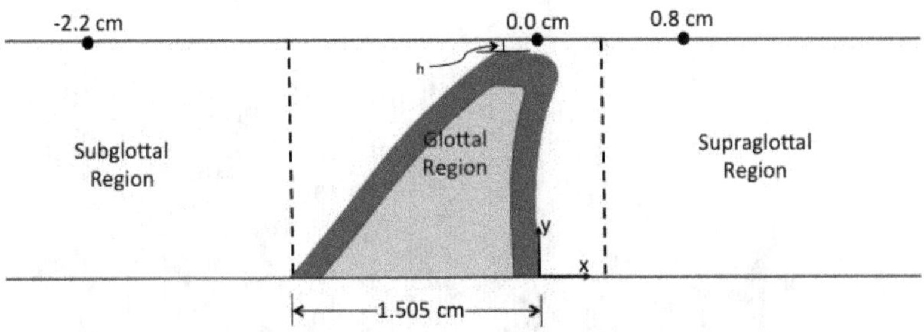

Fig. 6.7. Labels and definitions for variables of interest.

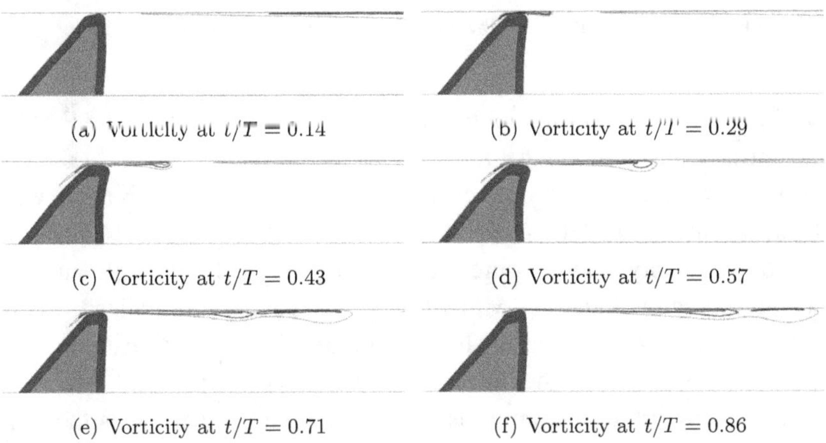

(a) Vorticity at $t/T = 0.14$ (b) Vorticity at $t/T = 0.29$

(c) Vorticity at $t/T = 0.43$ (d) Vorticity at $t/T = 0.57$

(e) Vorticity at $t/T = 0.71$ (f) Vorticity at $t/T = 0.86$

Fig. 6.8. Glottal jet vorticity contours with one vibration cycle half-space model.

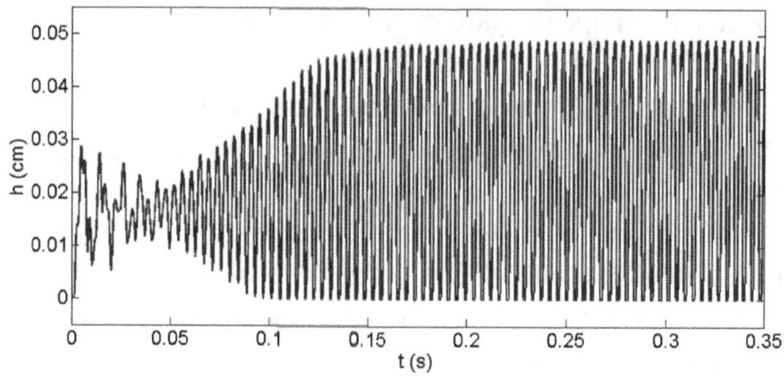

Fig. 6.9. Time history with half Glottis width h.

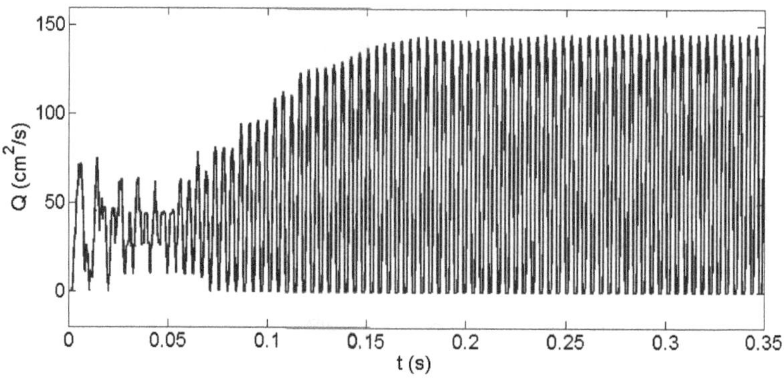

Fig. 6.10. Time history of flow rate Q at $x = -0.1$ cm and glottis width h.

vocal fold vibration, are found to be the same at $f = 234$ Hz. The Fourier transform is shown in Fig. 6.11.

A more detailed view of each cycle during the steady vibration with physical time $t = 0.2210 \sim 0.2254$ s as shown in Figs. 6.12–6.14. Volume flow rate, velocity, and pressure are closely examined at glottal, subglottal, and supraglottal regions. Moreover, the time axis is normalized, where $T^* = 0$ represents the instance when the glottis width reaches its minimum, namely, close position. The time interval when the glottis width increases from its minimum to maximum is called open quotient; whereas the time interval when the glottis width decreases from its maximum back to minimum is called close quotient. In this test example, the glottis reaches fully

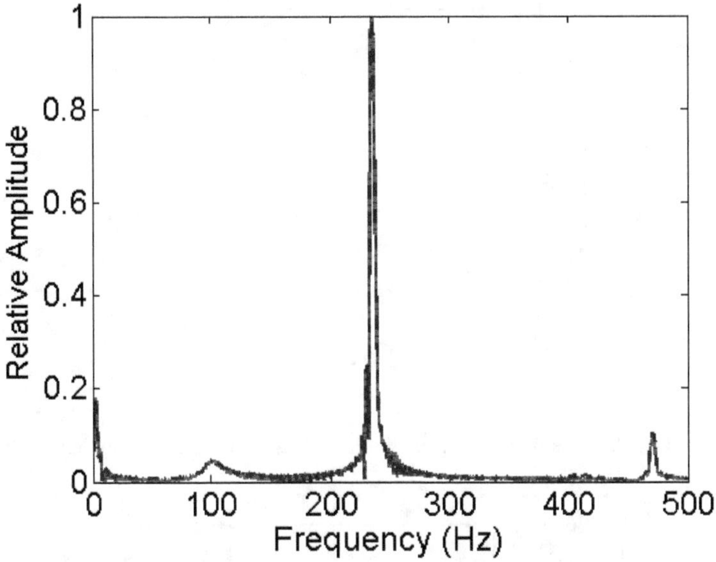

Fig. 6.11. Respective Fourier transforms.

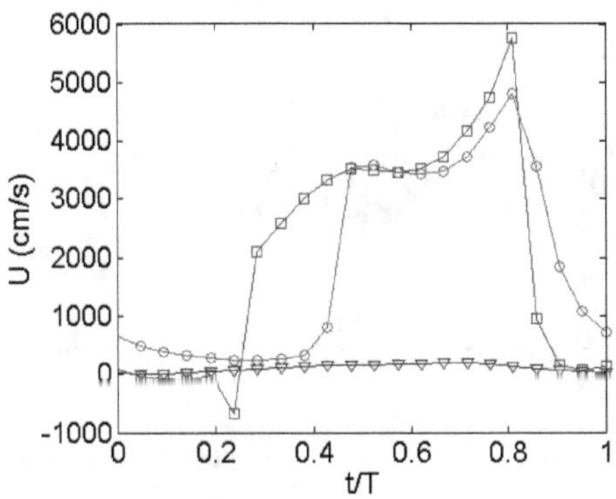

Fig. 6.12. Velocity at subglottal '▽', glottal '□', and supraglottal '○'.

open position right after $T^* = 0.5$. The volume flow rate reaches to its maximum when the glottis is fully open, as shown in Fig. 6.13.

At fully closed position, the volume flow rate is zero at the glottis ($x = 0$ cm) indicating that the vocal fold shuts off the air flow completely. Meanwhile, both subglottal ($x = -2.2$ cm) and supraglottal ($x = 0.8$ cm)

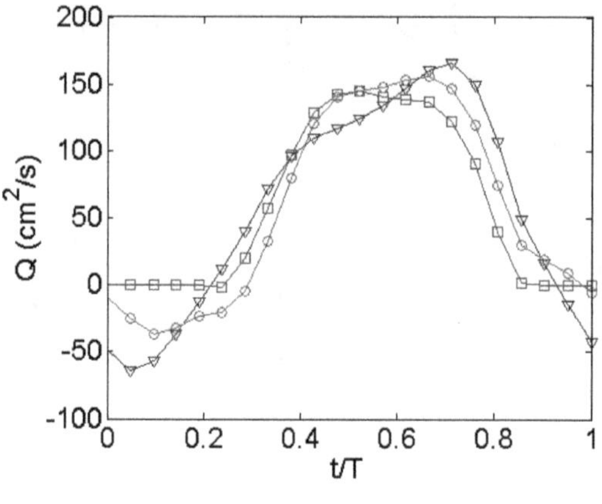

Fig. 6.13. Volume flow rate at subglottal '▽', glottal, '□', and supraglottal '○'.

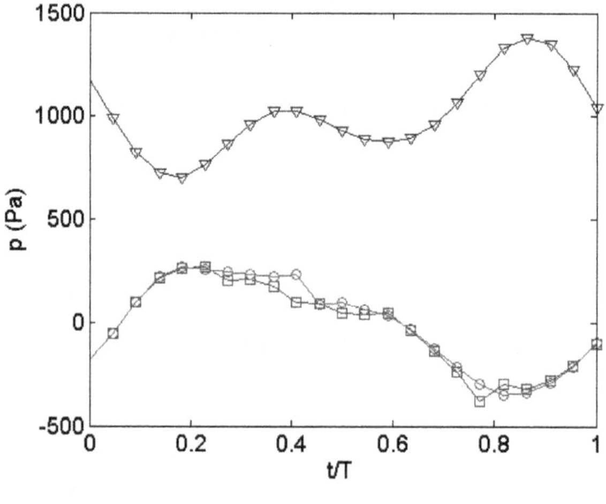

Fig. 6.14. Pressure at subglottal '▽', glottal '□', and supraglottal; '○'.

regions are observed to have negative flow rates when the glottis is closed. This phenomenon is due to the fact that even as the cover shuts off the flow, the body is still been carried backward which causes a reversed air flow, namely, a negative volume rate. Similar phenomenon is also obtained in the experimental study with a scaled-up model with prescribed vocal fold motion (Krane *et al.*, 2007). However, it has not been captured in other numerical simulations with finite element methods (Tao and Jiang, 2006;

Tao *et al.*, 2006) and the IB method (Zheng *et al.*, 2009; Luo *et al.*, 2009), where these numerical models numerically block the fluid flow by enforcing the flow rate to be zero when the glottis width is less than a critical value to avoid the contact between two vocal fold walls.

The velocity over one cycle at subglottal ($x = -2.2$ cm), glottal ($x = 0$ cm), supraglottal ($x = 0.8$ cm) regions are shown in Fig. 6.12. The glottal velocity is also considered as the glottal jet velocity and its maximum value during the vibration reaches 6658.1 cm/s. Therefore, the Reynolds number of the glottal flow is Re $= 4731.7$ and the reduce frequency is $f^* = 0.052893$, both of which fall within the range of the life-scale problems. The glottal velocity has a quick rise as the glottis starts to open, and then increases slowly before it fully opens. Just before the glottis closes, the glottal velocity has a second quick rise, and then drops to zero when the vocal folds shut off the air flow. The same waveform of the glottal velocity is also found by other numerical and experimental studies (Luo *et al.*, 2009; Krane *et al.*, 2007).

The pressure over one cycle at subglottal ($x = -2.2$ cm), glottal ($x = 0$ cm), supraglottal ($x = 0.8$ cm) regions are shown in Fig. 6.14. The subglottal pressure reaches its maximum right after the vocal fold shut offs the air flow, when the air in the subglottal region is compressed by the air flow from the upstream and the vocal fold surface. The high subglottal pressure then pushes the vocal folds apart and makes the glottis open again. After the glottis opens, the glottal pressure decreases due to the rapid increasing of the glottal velocity, from the Bernoulli's effect. These observations are also consistent with those documented in other numerical studies (Alipour *et al.*, 2000; Tao *et al.*, 2006).

The effect of the inlet pressure, namely, the lung pressure, is examined in this study. Seven different lung pressures ranging from 0.7 to 1.3 kPa (cases 1 to 7 in Table 6.3) are investigated while all the other physical parameters are set to be exactly the same. The simulations for the seven cases yield similar results, where the directly measured quantities, such as the gap width and volume flow rate, demonstrated the fact that the vocal fold vibration reaches a steady state vibration around 0.2 s, although higher inlet pressure may trigger to steady vibration slightly earlier.

The power spectra of the volume flow rates for all seven cases are found to be almost the same at 234 Hz, as shown in Fig. 6.15, which indicates

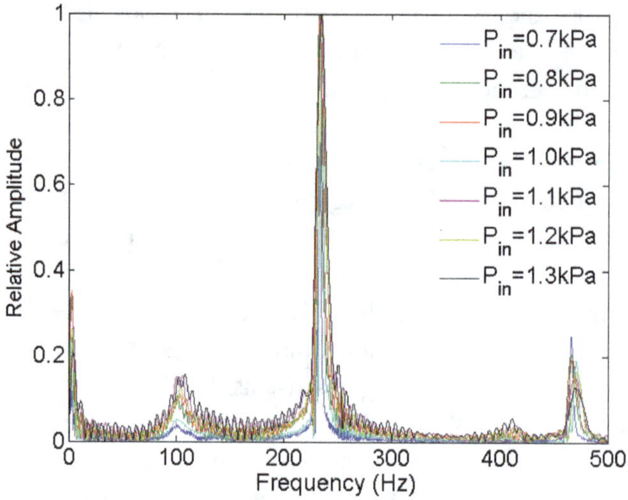

Fig. 6.15. Fourier transform of volume flow rates with different pressure inputs.

that the fundamental frequency is independent of the inlet pressure. The same conclusion was also drawn in Zhang's experimental study (Zhang *et al.*, 2006). In their single layer vocal fold acoustic model, the fundamental frequency does not change with the inlet pressure either. However, the effects of the inlet pressure on the fundamental frequency are debatable. Titze predicts that the fundamental frequency should increase with the growing inlet pressure, using small-amplitude approximation (Titze, 1989). Different numerical models tend to have different conclusions on how the inlet pressure affects the fundamental frequency. Luo *et al.* (2009) and Tao *et al.* (2007) report that there does not exist any clear relationship between the vibration frequency and lung pressure. Rosa *et al.* (2003) concludes that the vibration frequency decreases with the lung pressure grows. Therefore, further experimental and numerical studies are required to reveal the relationship between the inlet pressure and the fundamental frequency.

Moreover, the glottis width for each cycle with normalized time scale T^* using their respective fundamental frequency. As expected, the higher the pressure, the larger the maximum glottis width, which means the vibration magnitude also increases along with volume flow rates, or a stronger signal, namely, a louder voice during phonation. In fact, we also find an

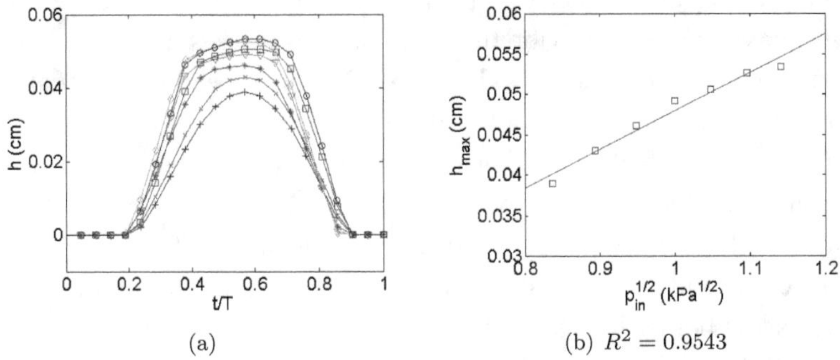

Fig. 6.16. Relationship between maximum glottis width and inlet pressure.

approximate linear relationship between the maximum glottis width and the square root of the inlet pressure. Such rough linear relationship is also predicted in analytical and numerical studies. Titze (1989) and Titze and Durham (1987) have conducted experiments with warm humidified air through the glottis and found a linear relationship between the vocal fold's oscillation amplitude and the square root of the subglottal pressure. Tao *et al.* (2006) and Jiang and Titze (1994) later used their finite element model simulation to confirm this relationship between the impact pressure and oscillation amplitude. Again, the results from our glottal pressure study are consistent with these findings.

In Fig. 6.16, we have (a) half glottis width with different pressure inputs: $P_{\text{in}} = 0.7$ ($'+'$), 0.8 ($'\times'$), 0.9 ($'*'$), 1.0 ($'\bigtriangledown'$), 1.1 ($'\square'$), 1.2 ($'\diamond'$), and 1.3 ($'\circ'$) kPa; (b) half glottis width and pressure relationship, reliability fitting $R^2 - 0.9543$. To study the effect of stiffness on the vocal fold vibration, three sets of different vocal fold stiffness are examined; they are listed in Table 6.3 as Cases 4, 8, and 9. The benchmark, Case 4, with $E_c = 10$ kPa and $E_b = 40$ kPa, is denoted as 'stiff' case. Case 8 is a 'soft' case with $E_c = 5$ kPa and $E_b = 20$ kPa. Cases 4 and 8 have the same stiffness ratio of 4. Case 9 has a lower stiffness ratio between the cover and the body with $E_c = 20$ kPa and $E_b = 40$ kPa for the body. All other numerical parameters are kept exactly the same including the same pressure input of 1 kPa, Poisson's ratio μ is 0.3 and the damping factor η is 10 Poise for both cover and body layers. Based on the results, there are no fundamental differences between Cases 4 and 9. For direct comparisons, we will closely examine Cases 4 and 8 since the stiffness ratio for both cases are the same. The stiff case starts steady vibration after 0.2 s as it takes a little longer for the soft case to get to the steady vibration. The wave speed in the vocal fold is proportional to the square root of the Young's modulus. Therefore, with the same damping factor and density, the stiffer tissue reaches the steady vibration faster than the softer one.

Higher vibration frequency is expected when large stiffness is given for both the cover and the body, as in Case 8. The power spectra of the volume flow rates for both stiff and soft cases are shown in Fig. 6.17. The fundamental frequencies are evaluated as $f_{\text{stiff}} = 234$ Hz and $f_{\text{soft}} = 169$ Hz for the stiff and soft cases, respectively. The glottis width and volume flow rate for both the soft and stiff cases during one cycle are plotted in Fig. 6.18. To compare the two cases, the time scale is again normalized by their respective fundamental frequencies.

As shown in Fig. 6.18, the maximum glottis width over one cycle for the soft case is larger than that for the stiff case. Moreover, the wave forms of the flow rate and glottis width are also very different. As shown in Fig. 6.18, the glottis width for the soft case decreases more drastically in the middle of the open quotient and then increases again. After the glottis width reaches its maximum, it drops to its minimum much faster in the soft case than in the stiff case. Such difference is caused by different vibration patterns between the soft and stiff cases. For the soft vocal folds, up-down

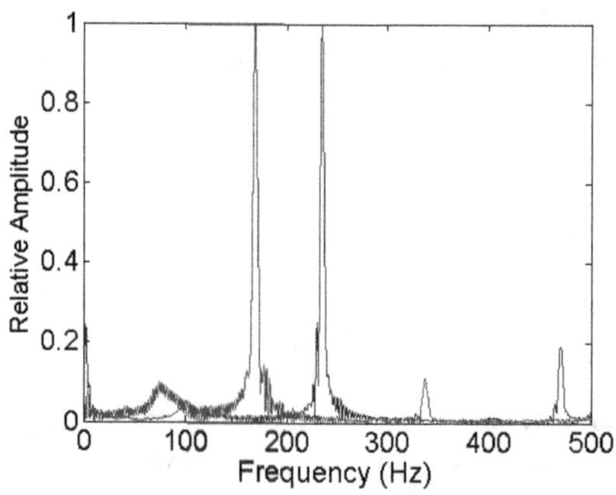

Fig. 6.17. Fourier transform of volume flow rate for stiff ('-') and soft ('-') cases.

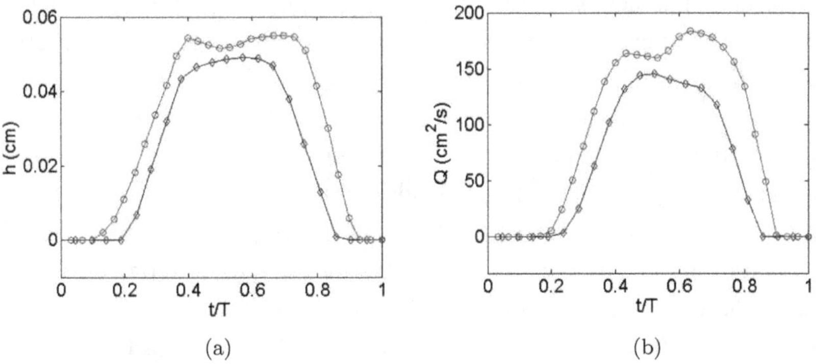

Fig. 6.18. (a) Half glottis width and (b) volume flow rate for stiff ('◇') and soft ('○') cases.

motion of the interior edge is much more pronounced, which causes the glottis width to decrease in the middle of the open quotient. In addition, the volume flow rate changes similarly to the glottis width, because during the glottis opening the flow rate directly depends on the glottis width. In Fig. 6.18, the flow rate also indicates that the time interval of air shutting off is longer for the stiff case than for the soft case, which means that it takes less time for the pressure to build up and push the vocal folds open when the vocal folds stiffness is smaller.

To reiterate, the modified Immersed Finite Element Method (mIFEM) that we have been developing over the past decade for the study of FSI systems, has two additional features are implemented: fluid compressibility and solid contact model. The governing equations for this FSI system are compressible Navier–Stokes equations for the air flow and linear viscoelastic material constitutive law with layered structure for the vocal folds. The dynamic solutions for the air flow and vocal folds are strongly coupled with each other at every time step. This mIFEM approach has two unique advantages over other FSI methods: (1) the vocal folds behavior is accurately predicted by solving the solid dynamic governing equation using the fully coupled algorithm, which is essential in these high-Reynolds-number laryngeal flows, and (2) the compressibility of the air flow and vocal folds are taken into consideration, which more accurately captures the physics of the aeroelastic coupling. In addition to the robust self-sustained vibration of the vocal folds, the flow fields (velocity and pressure) at the steady vibration agree well with measurements obtained from published experimental data (Zhang *et al.*, 2006; Krane *et al.*, 2007) and the results obtained from other numerical studies.

Our parametric studies further showed that the inlet pressure, or the lung pressure, does not affect the fundamental frequency of the vibration but instead, higher inlet pressure leads to larger vibration magnitude. An approximate linear relationship between the vibration magnitude and the square root of the inlet pressure is found, which agrees with previous experimental and analytical findings. On the other h and, the stiffness of the vocal fold seems to have a direct effect on the fundamental frequency, as expected. Low fundamental frequency and larger magnitude vibration is observed for softer vocal folds. We recognize that the simulation was not performed in 3D, therefore the vocal fold vibration pattern would be somewhat deviated from reality. However, with the concern of the tremendous computational cost and additional complexities in the analysis, it is not done in this initial study, even though the tool presented here is implemented for 3D.

We intend to further show and compare the differences in symmetric and asymmetric laryngeal air flows, and study and discuss the dynamic and energetic impact of glottal jet asymmetry on the vocal fold vibration as well as the resultant phonation. We also expect to further this study with more detailed control volume analysis to fully understand and identify the effects of fluid compressibility, friction on the vocal folds surface, as well as air viscosity. More work is to be done on carefully designing boundary treatment in the aeroacoustic model which will help in identifying acoustic sources in the phonation process.

6.3 Microcirculation and Red Blood Cells

Human blood circulatory systems have evolved to supply nutrients and oxygen to, and carry the waste away from, the cells of multicellular organisms through the transport of blood, a complex fluid composed of deformable cells, proteins, platelets, and plasma. Currently, cardiovascular diseases represent the leading cause of death in developed countries. The lack of understanding of short and long-term development and evolution of many of the arterial and vascular diseases directly limits disease diagnosis and prevention as well as the planning of therapeutic approaches. It is of significant clinical relevance to understand blood composition and its rheological behavior in the context of multi-scale and multi-physics hemodynamics. In fact, blood rheology directly relates to cardiovascular and vascular disease, sickle cell anemia, diabetic, and stroke conditions. One important application of immersed methods is the modeling of the human cardiovascular system.

The physiological functions of large blood vessels and micro- and capillary vessels are distinctly different. In general, blood is transported to critical locations within the body via large vessels. It has been recognized that flow conditions near vessel bifurcation points and low shear stress areas are among the important factors in the development of arterial and vascular diseases (Pedley, 1980). Furthermore, the modeling of macro-scale cardiovascular flows is directly beneficial to preoperative planning of surgical procedures for various cardiovascular diseases (Taylor *et al.*, 1996; Quarteroni, 2001). On the other hand, the exchange of materials through the vessel walls with surrounding cells is accomplished within micro- and capillary vessels which account for, in adult cardiovascular and pulmonary circulations, a majority of the 10^{11} or so blood vessels (Fung, 1996, 2000).

As depicted in Fahraeus–Lindqvist effects, named after Robin Sanno Fahraeus (1888–1968) and Johan Torsten Lindqvist (1906–2007), blood viscosity is lower in small vessels (diameter less than 1.5 mm) than in large vessels. In fact, the viscosity in capillaries is less than half of that in large vessels, due in part to red cells moving together in single files through small vessels. While theories of suspension rheology generally focus on homogeneous flows in infinite domains, the important phenomena of blood flows in microcirculation depend on the combined effects of vessel geometries, cell deformability, wall compliance, flow shear rates, and many micro-scale chemical, physiological, and biological factors (Huang *et al.*, 2001, Diamond, 2001).

There have been studies on shear flow effects on one or two cells (Guo *et al.*, 2001), leukocytes adhesion to vascular endothelium (Cao *et al.*, 1998), and particulate flow based on continuum enrichment methods (Wagner *et al.*, 2001, 2003). So far no mature theory has been available for the prediction of blood rheology and blood perfusion through micro-vessels and capillary networks. The different time and length scales as well as large motions and deformations of immersed solids pose tremendous challenges to the mathematical modeling of blood flow at that level (Wagner, 2001; Shyy *et al.*, 2001).

6.3.1 *Protein molecular dynamics*

Red blood cells (RBCs, or erythrocytes) are biconcave, disk-shaped deformable bodies, as shown in Fig. 6.19, and typically comprise approximately 40% of blood by volume. A single RBC is normally $8 \sim 10\ \mu$m in diameter and $7.5 \sim 10$ nm in thickness. The RBC density is approximately 15% higher than the fluid and the density of the blood in 45% of hematocrit is 1.07 g/ml. The RBC membrane has a shear modulus of 4.2×10^{-3} dyn/cm^2, a dilation modulus of 500 dyn/cm, and a bending modulus of 1.8×10^{-12} dyn/cm. As the shear modulus of the RBC membrane is much smaller than the dilation modulus, the RBC membrane is susceptible to deformation by shear with very little area change. Furthermore, the cytoplasm contained within a typical RBC is incompressible, and even with the large deformation experienced by the cell membrane, there is no excessive membrane area change.

The typical diameter of the micro-vessel is $1.5 \sim 3$ times larger than that of the cell, which is around $12 \sim 24\ \mu$m. On the contrary, in the capillary, the vessel diameter is about $2 \sim 4\ \mu$m which is significantly smaller. However, since the cell membrane is very flexible, the cells can squeeze through capillaries. Moreover, the pressure gradient which drives the flow is usually around $3.2 \sim 3.5$ kPa. For the chosen diameter and pressure, the Reynolds number in a typical capillary is around 0.01. In fact, in the process of squeezing through capillaries, large deformations of red blood cells

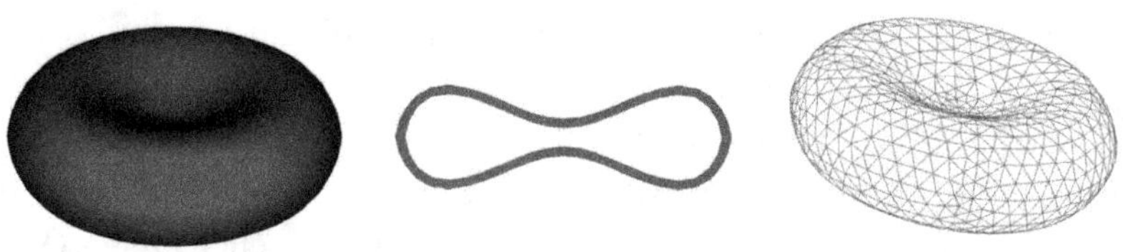

(a) (b) (c)

Fig. 6.19. A three-dimensional finite element mesh of a single RBC model. (a) 3D RBC model, (b) RBC cross-section, and (c) RBC mesh.

not only slow down the blood flow, but also enable the exchange of oxygens through capillary vessel walls. Three factors that influence the property of blood flow will be addressed in this section, namely, cell–cell interaction effect, shear rate effect, and vessel geometrical effect.

Blood plasma can be accurately modeled with the Newtonian fluid model, yet blood flows do exhibit non-Newtonian or viscoelastic behaviors, especially under low Reynolds numbers. The plasma viscosity μ is $0.01 \text{ dyn} \cdot \text{s}/\text{cm}^2$. The inner cytoplasm is assumed to be an incompressible Newtonian fluid with the same viscosity as the plasma. Moreover, human blood is a typical thixotropic material of which the viscoelastic characteristics may vary significantly with stress and strain levels as well as time. Although in most arteries, blood behaves in a Newtonian fashion with a roughly constant viscosity, in microcirculatory system, non-Newtonian behavior is prevalent, in particular at low shear rates. Direct simulations of such non-Newtonian behaviors utilizing immersed methods have been reported in Liu *et al.* (2004b) and Zhang *et al.* (2004).

As a coarse grain approach to multi-scale and multi-physics FSI systems, we introduce a so-called protein molecular dynamics Liu *et al.* (2004b, 2007b). With respect to cell–cell interaction forces, as shown in Fig. 6.20, the cut-off length is chosen to be around 3 μm, beyond which the attractive force decays very fast. This means that nodes outside the influencing domain, essentially a sphere with the cut-off length as its radius, have very little effect. Furthermore, the potential function is chosen such

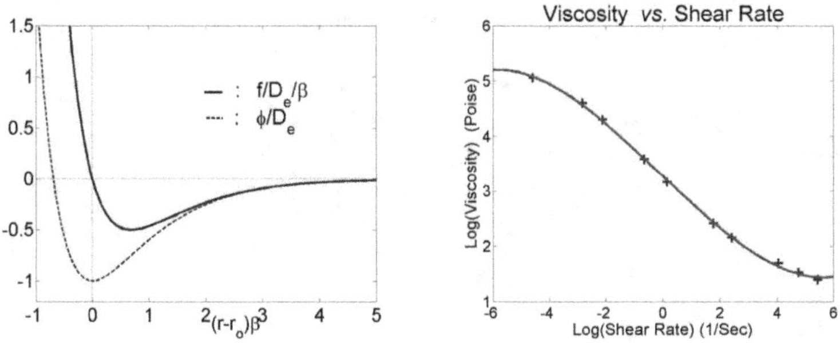

(a) (b)

Fig. 6.20. Morse potential and force and shear-rate dependent viscosity: (a) Morse potential and (b) vertical displacement.

that cells will de-aggregate at the shear rate above 0.5 1/s, according to the experimental observation (Chien, 1976). For convenience, we set the zero force length as 250 nm as suggested from the experiments. However, as we reduce the zero force length, local enrichment and multi-scale coupling must also be introduced. It is important to recognize that one microliter of normal human blood contains about 5 million red blood cells, 7 thousand white blood cells, and 300,000 platelets. At this point, even with high performance parallel computing tools, it is impossible to handle such complex fluids as a whole.

An illustration of the shear rate-dependence of blood viscosity is shown in Fig. 6.20. Note that the Morse potential and force depicted in Fig. 6.20 is non-dimensionalized. It is clear that the blood viscosity will decrease and eventually reach a plateau with the increase of shear rate. In this section, we focus on the three-dimensional Newtonian fluid interacting with a few flexible red blood cells.

Cell-cell adhesion plays an important role in various physiological phenomena including the recognition of foreign cells. Although the exact physiological mechanisms of the RBC coagulation and aggregation are still ambiguous, it has been found that both the RBC surface structure and membrane proteins are key factors in producing adhesive repulsive forces. In essence, RBC membranes held together by hydrophobic interactions are chemically and electrostatically active as depicted in Fig. 6.21. Moreover, most membrane lipids and some proteins can drift laterally within the membrane. In particular, phospholipids move quickly ($\sim 2\ \mu m/s$) along the membrane while membrane proteins drift at a slower pace.

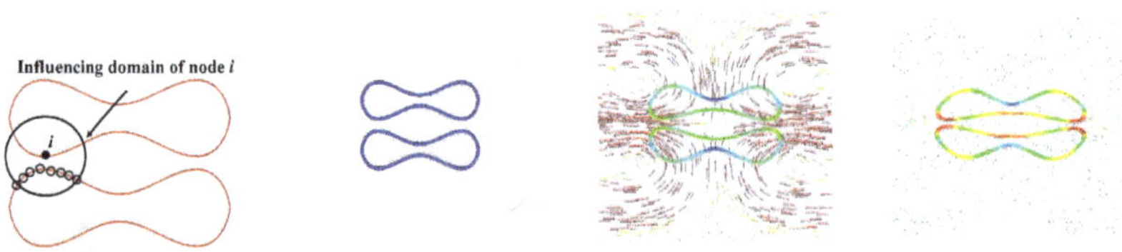

(a) (b) (c) (d)

Fig. 6.21. Two-RBC model, cell–cell contacts, and cell deformation: (a) interaction, (b) initial, (c) deformability, and (d) equilibrium.

In general, the cell–cell interaction forces are not sufficient to deform cell membranes. However, the ensuing aggregate could alter the surrounding fluid significantly. Conventional explanations in general fall into two categories: the bridging and adsorption models, with the adsorption and exclusion of the plasmatic macromolecules at RBC surfaces, respectively. The primary macromolecules that cause the RBC aggregation are the so-called Fibrinogen. The depletion layer results in a reduction of osmotic pressure in the cell gap between nearby RBCs, which consequently produces an attractive force. It can be assumed that the cell adhesion will occur only if cells are sufficiently close. Nevertheless, as illustrated in Figs. 6.20 and 6.21, when cells are too close to each other, a repulsive force will hinder contact. The repulsive forces include the steric forces due to the glycocalyx and the electrostatic repulsive force induced from the same negative fixed charges at RBC surfaces. Of course, the level of the RBC aggregation depends on the initial position of the cells and the strength of the adhesive force in comparison with other forces, such as hydrodynamic forces.

In practice, we often characterize blood flow with the strength of the shear flow, as shown in Fig. 6.22, which is measured with the shear rate $\dot{\gamma}$ within the range from 0.1 to 100 s^{-1}. It is clear that as the shear rate increases the blood viscosity decreases initially, eventually reaching a plateau marking the plasma viscosity where the blood elasticity continues to decrease. In particular, Kelvin and Maxwell models, as elaborated in Section 6.3.2, demonstrate that the blood viscoelastic behaviors can be accurately predicted via analytical or numerical experiments on the macroscopic level.

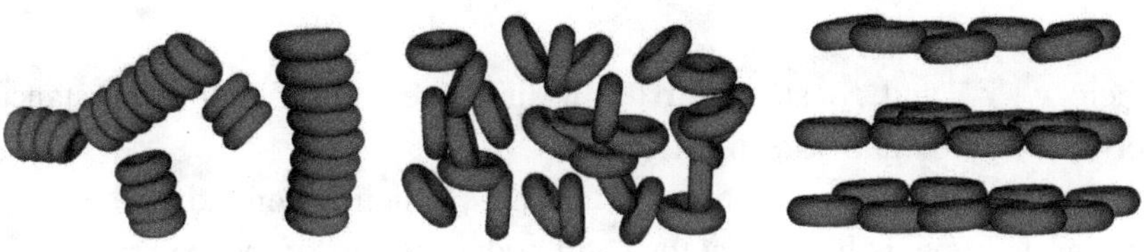

(a) (b) (c)

Fig. 6.22. Blood microscopic changes under different shear rates: (a) low shear region, (b) mid shear region, and (c) high shear region.

The basic behavior of the interaction forces between two RBCs is simply illustrated as the weak attractive force at far distances and strong repulsive force at short distances as depicted in Fig. 6.20. To illustrate the aggregation formation, we consider two red blood cells attracting each other in the quiescent fluid under the influence of the strong cell–cell interaction forces, the disassociated cells move towards each other and gradually form a doublet with moderate deformations, as shown in Fig. 6.21. Due to the complexity of the aggregation process, we accumulate the intermolecular force, electrostatic force, and protein dynamics into a potential function, similar to an intermolecular potential. Here we adopt the Morse potential, found to be capable of quantitatively predicting the aggregation behaviors consistent with experimental observations:

$$\phi(r) = D_e \left[e^{2\beta(r_o - r)} - 2e^{\beta(r_o - r)} \right], \tag{6.13}$$

$$f(r) = -\frac{\partial \phi(r)}{\partial r} = 2D_e\beta \left[e^{2\beta(r_o - r)} - e^{\beta(r_o - r)} \right], \tag{6.14}$$

where r_o and D_e stand for the zero force length and surface energy, respectively, and β is a scaling factor.

In the actual implementation, after the finite element discretization of the solid domain, a sphere with the diameter of the cut-off length is used to identify the nodes represented as j within the influencing domain around the surface node i. Hence, a typical cell–cell interaction force can be denoted as, with $\mathbf{r}_{ij} = \mathbf{r}_j - \mathbf{r}_i$,

$$\mathbf{f}_i = -\sum_j \frac{\partial \phi(r_{ij})}{\partial \mathbf{r}_{ij}}, \tag{6.15}$$

where $\mathbf{r}_i$, $\mathbf{r}_j$, and r_{ij} stand for the position vectors of and the distance between nodes i and j, respectively.

Finally, the cell–cell interaction force $\mathbf{f}^c$, which is defined as the force per unit volume, is lumped at the solid nodes, as part of the external nodal forces. This procedure is applicable to non-uniform meshes.

To summarize the coupling of the Navier–Stokes equations and protein molecular dynamics in the context of the Immersed Finite Element Method, let us consider an incompressible three-dimensional deformable structure in Ω^s completely immersed in an incompressible fluid domain Ω^f. Together, the fluid and the solid occupy a domain Ω, but they do not intersect:

$$\Omega^f \cup \Omega^s = \Omega \quad \text{and} \quad \Omega^f \cap \Omega^s = \oslash.$$

In contrast to the immersed boundary (IB) formulation, the solid domain can occupy a finite volume in the fluid domain. Since we assume both fluid and solid to be incompressible, the union of two domains can be treated as one incompressible continuum with a continuous velocity field. In the computation, the fluid spans over the entire domain Ω, thus an Eulerian fluid mesh is adopted; whereas a Lagrangian solid mesh is constructed on top of the Eulerian fluid mesh. The coexistence of fluid and solid in Ω^s requires some considerations when developing the momentum and continuity equations.

In the computational fluid domain Ω, the fluid grid is represented by the time-invariant position vector $\mathbf{x}$; while the material points of the structure in the initial solid domain Ω^s_o and the current solid domain Ω^s are represented by $\mathbf{X}^s$ and $\mathbf{x}^s(\mathbf{X}^s, t)$, respectively. The superscript or the subscript s is used in the solid variables to distinguish the fluid and solid domains.

In the fluid calculations, the velocity $\mathbf{v}$ and the pressure p are the unknown fluid field variables; whereas the solid domain involves the calculation of the nodal displacement $\mathbf{u}^s$, which is defined as the difference of the current and the initial coordinates: $\mathbf{u}^s = \mathbf{x}^s - \mathbf{X}^s$. The velocity $\mathbf{v}^s$ is the material derivative of the displacement $d\mathbf{u}^s/dt$.

As in Zhang *et al.* (2004) and Wang and Liu (2004), we define the fluid–structure interaction force within the domain Ω^s as $f_i^{fsi,s}$:

$$f_i^{fsi,s} = -(\rho^s - \rho^f)\frac{dv_i}{dt} + \sigma^s_{ij,j} - \sigma^f_{ij,j} + (\rho^s - \rho^f)g_i + f_i^c, \quad \mathbf{x} \in \Omega^s,$$

where the effective body force $\mathbf{f}^c$ includes the cell–cell adhesive or repulsive forces in a special molecular potential as well as the effects of fixed-charge density on the cell membrane, cytoplasm, and protein effects.

Naturally, the interaction force $f_i^{fsi,s}$ as in Eqs. (2.6) and (3.6) is calculated with the Lagrangian description. Moreover, a Dirac delta function δ is used to distribute the interaction force from the solid domain onto the computational fluid domain:

$$f_i^{fsi}(\mathbf{x},t) = \int_{\Omega^s} f_i^{fsi,s}(\mathbf{X}^s,t)\delta(\mathbf{x}-\mathbf{x}^s(\mathbf{X}^s,t))\,d\Omega.$$

Hence, the governing equation for the fluid can be calculated by combining the fluid terms and the interaction force as

$$\rho^f \frac{dv_i}{dt} = \sigma_{ij,j}^f + f_i^{fsi}, \quad \mathbf{x} \in \Omega. \tag{6.16}$$

Since we consider the entire domain Ω to be incompressible, we only need to apply the incompressibility constraint once in the entire domain Ω:

$$v_{i,i} = 0. \tag{6.17}$$

To delineate the Lagrangian description for the solid and the Eulerian description for the fluid, we introduce different velocity field variables v_i^s and v_i to represent the motions of the solid in the domain Ω^s and the fluid within the entire domain Ω. The coupling of both velocity fields is accomplished with the Dirac delta function:

$$v_i^s(\mathbf{X}^s,t) = \int_{\Omega} v_i(\mathbf{x},t)\ \delta(\mathbf{x}-\mathbf{x}^s(\mathbf{X}^s,t))\,d\Omega.$$

Let us assume that there is no traction applied on the fluid boundary, i.e., $\int_{\Gamma_{h_i}} \delta v_i h_i d\Gamma = 0$, applying integration by parts and the divergence theorem, following Zhang *et al.* (2002), we can easily derive the weak form:

$$\begin{aligned}&\int_{\Omega}(\delta v_i + \tau^{\mathrm{m}} v_k \delta v_{i,k} + \tau^{\mathrm{c}}\delta p_{,i})\left[\rho^f(v_{i,t}+v_j v_{i,j}) - f_i^{fsi}\right]d\Omega + \int_{\Omega}\delta v_{i,j}\sigma_{ij}^f d\Omega\\ &+\int_{\Omega}(\tau^{\mathrm{m}} v_k \delta v_{i,k} + \tau^{\mathrm{c}}\delta p_{,i})\sigma_{ij,j}^f d\Omega + \int_{\Omega}(\delta p + \tau^{\mathrm{c}}\delta v_{j,j})v_{i,i}d\Omega = 0,\end{aligned}$$

where τ^{m} and τ^{c} are scalar stabilization parameters that are in general functions of computational grid, the time step size, and the flow variables.

The nonlinear systems are solved with the Newton–Raphson method. Moreover, to improve the computation efficiency, we also employ the GMRES iterative algorithm and compute the residuals based on matrix-free techniques (Saad and Schultz, 1986; Wang, 2021, 2022). Note that for brevity we ignore the fluid stress within the solid domain.

The transformation of the weak form from the updated Lagrangian to the total Lagrangian description is to change the integration domain from Ω^s to Ω^s_o. Since we consider incompressible fluid and solid, and the Jacobian determinant is one in the solid domain, the transformation of the weak form to total Lagrangian description yields

$$\int_{\Omega^s_o} \delta u_i(\rho^s - \rho^f)(\ddot{u}^s_i - g_i)\, d\Omega^s_o + \int_{\Omega^s_o} \delta u_{i,j} P_{ij}\, d\Omega^s_o + \int_{\Omega^s_o} \delta u_i f^s_i\, d\Omega^s_o = 0,$$

where the first Piola–Kirchhoff stress P_{ij} is defined as $P_{ij} = JF^{-1}_{ik}\sigma^s_{kj}$ and the deformation gradient F_{ij} as $F_{ij} = \partial x_i/\partial X_j$.

Note again that the boundary integral terms on the fluid–structure interface for both fluid and solid domains will cancel each other and for brevity are not included in the corresponding weak forms. For structures with large displacements and deformations, the second Piola–Kirchhoff stress S_{ij} and the Green–Lagrangian strain E_{ij} are used in the total Lagrangian formulation:

$$S_{ij} = \frac{\partial W}{\partial E_{ij}} \quad \text{and} \quad E_{ij} = \frac{1}{2}(C_{ij} - \delta_{ij}), \tag{6.18}$$

where the first Piola–Kirchhoff stress P_{ij} can be obtained from the second Piola–Kirchhoff stress as $P_{ij} = S_{ik}F_{jk}$.

Moreover, a Neo–Hookean strain energy function is used to depict the material behavior of the RBC membrane:

$$W = C_1(J_1 - 3) + \frac{C_2}{2}(J_2 - 1)^2, \tag{6.19}$$

with the material properties C_1 and C_2, where J_1 and J_2 are the functions of the invariants of the deformation gradient $\mathbf{C}$.

Finally, in the interpolation process from the fluid onto the solid grid, the discretized form of Eq. (2.14) can be written as

$$v^s_{iI} = \sum_J v_{iJ}(t)\phi_J(\mathbf{x}_J - \mathbf{x}^s_I), \quad \mathbf{x}_J \in \Omega_{\phi I}. \tag{6.20}$$

Here, the solid velocity $\mathbf{v}^s_I$ at node I can be calculated by gathering the velocities at fluid nodes within the influence domain $\Omega_{\phi I}$. A dual procedure takes place in the distribution process from the solid onto the fluid grid. The discretized form of Eq. (2.13) is expressed as

$$f_{iJ}^{fsi} = \sum_I f_{iI}^{fsi,s}(\mathbf{X}^s, t)\phi_I(\mathbf{x}_J - \mathbf{x}_I^s), \quad \mathbf{x}_I^s \in \Omega_{\phi J}.$$

By interpolating the fluid velocities onto the solid particles in Eq. (6.20), the fluid within the solid domain is bounded to solid material points. This ensures not only the no-slip boundary condition on the surface of the solid, but also stops automatically the fluid from penetrating the solid, provided the solid mesh is at least two times denser than the surrounding fluid mesh. This heuristic criterion is based on the numerical evidence and needs further investigation.

As depicted in Fig. 6.23, the four-cell model is placed at the center of the fluid domain with the arrows represent the fluid-cell interaction force and a vertical center distance of 3.96 μm. Due to the biconcave shape, the adhesive or repulsive forces mainly exist around RBC perimeters. With the boundary shear velocity of 10 μm for both upper and bottom surfaces, we get a shear rate of 0.5 1/s.

On the microscopic level, deformable cells play an important role in the blood viscoelastic behaviors (Fung, 1996, 2000). In the quiescent state, normal red blood cells tend to aggregate. Under low shear rate, aggregates are mainly influenced by the cell–cell interaction forces; in the mid-shear rate region, RBC aggregates start to disintegrate and the influence of the deformability gradually increases. Under high shear rate, red blood cells tend to stretch, align with the flow, and form layers. These three different shear stress stages have been shown in the cartoon illustrations in Fig. 6.22 as well as the direct simulation results with IFEM and protein molecular

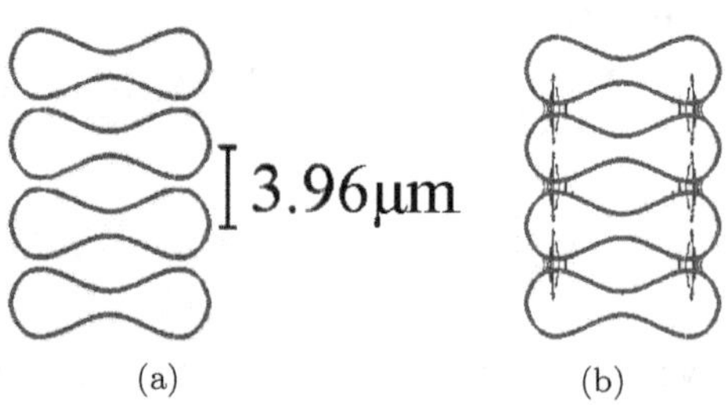

Fig. 6.23. Four-RBC model initial configurations and cell–cell interaction forces.

dynamics in Figs. 6.24(a)–6.24(d). Moreover, the simulation results also demonstrate that cell–cell interaction forces play an important roll in holding RBC aggregates together under macroscopic shear.

As expected, without significant cell–cell interaction forces, the RBCs will rotate and align with the flow separately while maintaining their vertical separations as shown in Fig. 6.24(a). On the other hand, the RBCs aggregate with cell–cell interaction forces subject to low or medium shear rates rotate initially as a bulk with varying vertical distances as shown in

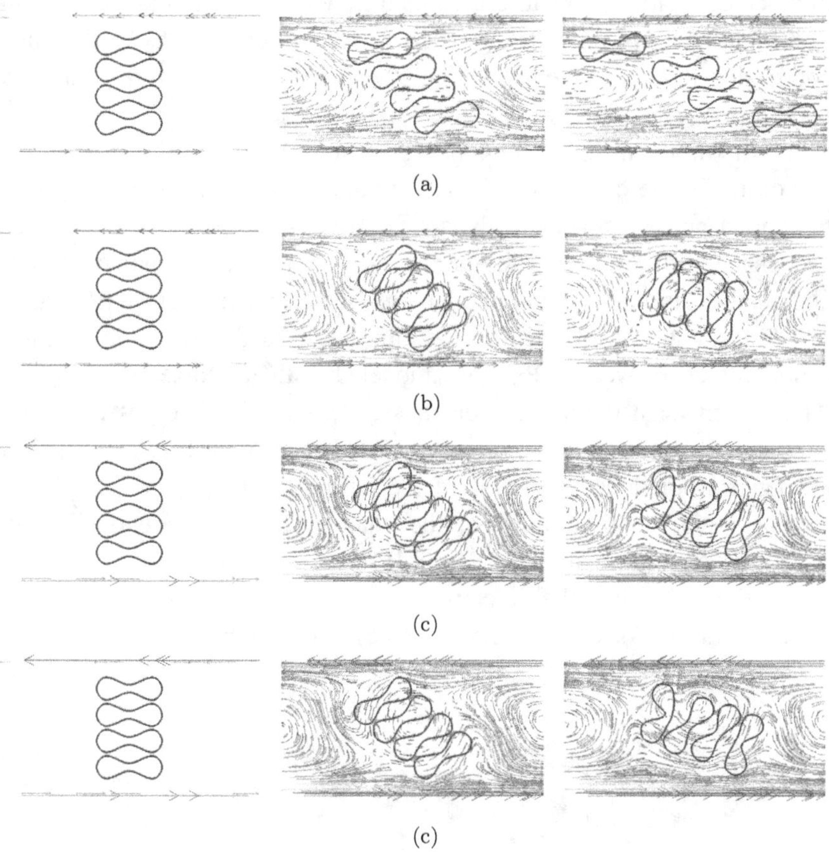

Fig. 6.24. (a) Four-RBC model at shear rate 0.25 1/s without cell–cell interaction forces. (b) Four-RBC model at shear rate 0.25 1/s with cell–cell interaction forces. (c) Four-RBC model at shear rate 0.5 1/s with cell–cell interaction forces. (d) Four-RBC model at shear rate 3.0 1/s with cell–cell interaction forces.

Figs. 6.24(b) and 6.24(c). These comparisons demonstrate that cell–cell interaction forces restrain the disintegration of RBC aggregates and introduce elasticity in the macroscopic mechanical behaviors of human blood as elaborated in Liu and Liu (2006b).

Sickle cell anemia is caused by genetic abnormalities in hemoglobin, which is a complex molecule constituting the most important component of red blood cells. When sickle hemoglobin loses oxygen, the deoxygenated molecule form rigid rods which distort the cell membrane into a sickle or crescent shape. The sickle-shaped cells are both rigid and sticky and tend to block capillaries and cause the blood flow blockage to the surrounding tissues and organs. The strong viscous shear introduced by such a flow contraction leads to some interesting phenomena of RBC aggregation with respect to cell–cell interaction forces and cell deformability.

Furthermore, the modeling of this complex fluid–solid system also demonstrates the capability of the coupling of the Navier–Stokes equations with protein molecular dynamics. We assign the sickle cell a crescent shape with higher rigidity and cell–cell adhesive or repulsive forces. Unlike the discrete-particle model presented in Dzwinel *et al.* (2003) and Tanaka and Araki (2000), in immersed methods, complex interaction forces between RBCs are modeled with a Morse potential and three-dimensional deformable biconcave cells are modeled as individual continuum objects. Thus, we can take advantage of various mature finite element formulations for both fluid and solid domains.

The initial configurations of the channel and cells are depicted in Fig. 6.25. As shown in Fig. 6.26, as RBCs pass the diffuser stage of the contraction with the inlet velocity of 10 μm/s, the deceleration of leading cells forms blockage for incoming cells. Therefore, dilation of cells is

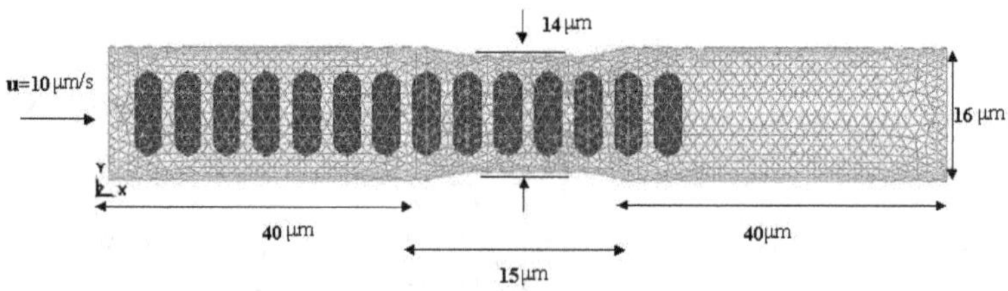

Fig. 6.25. Immersed RBC models passing through a vessel with varying diameters.

coupled with pile-up of cells at the outlet of the vessel constriction. From Fig. 6.26, $U_m > U_w$, thus $H_m < H_w$. From the medical point of view, high hematocrit indicates higher possibility of blood clot, which usually happens at the joints of vessels with different diameters. Also confirmed in Fig. 6.27, under the similar flow conditions, rigid and sticky sickle cells eventually block the micro-vessel entrance, which will certainly result in de-oxygenation of surrounding tissues.

In our study of the behaviors of RBC aggregates and their effects on viscoelastic properties of blood flows, we must also consider cell–cell adhesion. We first give a detailed description of the discrete RBC and aggregation models as shown in Fig. 6.19. In addition, we illustrate the key ingredients of the Immersed Finite Element Method. The first attempt

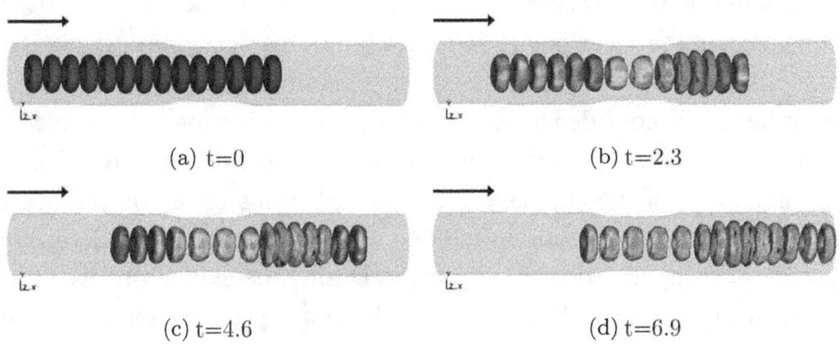

Fig. 6.26. The normal RBC flow with the inlet velocity of 10 μm/s.

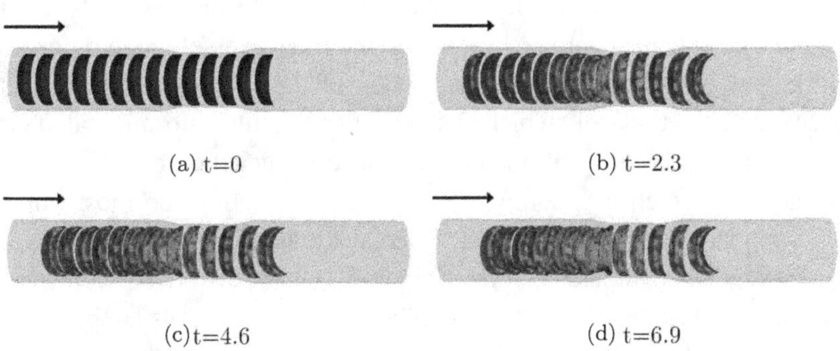

Fig. 6.27. The sickle cell flow with the inlet velocity of 10 μm/s.

to study this phenomenon using three-dimensional continuum models of RBC aggregates with the effects of shear rates, RBC geometries, cell–cell interaction forces, and nonlinear material descriptions was published in Liu *et al.* (2004b). In fact, RBC membranes held together by hydrophobic interactions are chemically and electrostatically active.

In a set of numerical experiments, we subject the RBC and viscous fluid mixer to low shear rates of $0.1 \sim 0.5\ 1/s$ and observe that the RBCs aggregate and rotate as a bulk. With an intermediate shear rate of $0.5\ 1/s$, our numerical demonstrates that the RBC aggregate rotates and aligns with the shear direction and then de-aggregate. At higher shear rate of $0.5 \sim 10\ 1/s$, the RBC aggregate completely disintegrates and starts to orient themselves like parallel layers at $t = 0$, $t = 0.25$, and $t = 0.5$ s. Moreover, the vorticities surrounding deformable cells are clearly captured along with large structural motions and deformations. The disintegration of RBC aggregates with the increase of the shear rate is an indication of the decrease of the macroscopic viscosity, which is consistent with experimental observations.

Furthermore, consider a three-dimensional simulation of a single red blood cell squeezing through a capillary vessel. RBC diameter is 1.2 times larger than that of the capillary vessel, which leads to the divergence of the cytoplasm internal liquid to the two ends of the capsule by deforming into a slug during the squeezing process. During the exiting process, there is a radial expansion of the slug due to the convergence of the cytoplasm, which deforms the capsule into an acaleph or jellyfish shape. In Fig. 6.28, four snapshots illustrate various stages of the red blood cell with respect to the capillary vessel.

At the entrance, divergence of the cytoplasm to the two ends of the capsule by deforming into a slug. Inside the micro-vessel, table deformed slug shape. At the exit, a radial expansion of the slug due to the convergence of the cytoplasm, which deforms the capsule into an acaleph or jellyfish shape. Sickle cell anemia is caused by genetic abnormalities in hemoglobin, which is a complex molecule constituting the most important component of red blood cells. When sickle hemoglobin loses oxygen, the deoxygenated molecule form rigid rods which distort the cell membrane into a sickle or crescent shape. The sickle-shaped cells are both rigid and sticky and tend to block capillary vessels and cause the blood flow blockage to the surrounding tissues and organs. To relate blood rheology

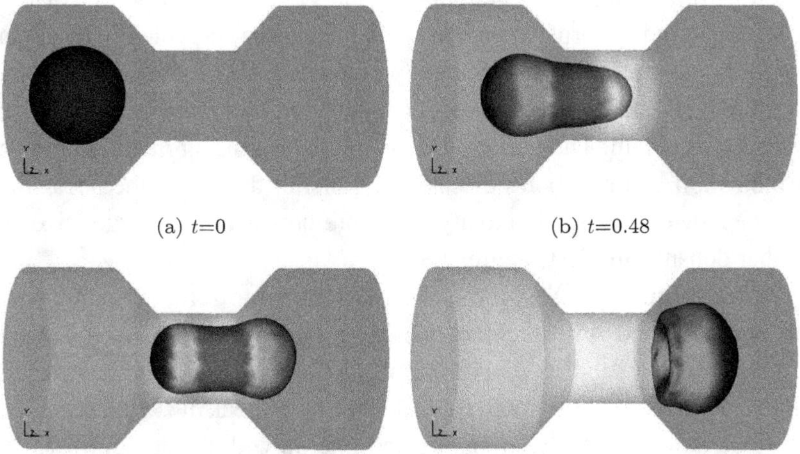

(a) t=0 (b) t=0.48

(c) t=0.96 (d) t=1.44

Fig. 6.28. A single white cell squeezing through a capillary vessel.

to sickle cell anemia, we consider the normal and sickle red blood cells passing through a micro-vessel contraction.

To account for both bending and membrane rigidities, the RBC membrane is modeled with a three-dimensional finite element formulation with 1043 nodes and 4567 tetrahedra elements (Evans and Fung, 1972). For simplicity, we also represent a typical RBC as a flexible three-dimensional thin structure enclosing an incompressible fluid with the same viscosity, i.e., $0.01\ \text{dyn}\cdot\text{s}/\text{cm}^2$. To account for different viscosity of the internal fluid, implicit and compressible immersed continuum methods must be used. The static shape of a normal RBC is a biconcave discoid. Moreover, a function is used to describe the cross section of its shape:

$$\bar{y} = 0.5[1-\bar{x}^2]^{1/2}(a_o + a_1\bar{x}^2 + a_2\bar{x}^4), \quad -1 \leq \bar{x} \leq 1, \tag{6.21}$$

with $a_o = 0.207$, $a_1 = 2.002$, and $a_2 = 1.122$.

The x and y coordinates of the cross section are scaled as

$$x = 5.0\,\bar{x}\ \mu\text{m} \quad \text{and} \quad y = 5.0\,\bar{y}\ \mu\text{m}.$$

Overall, RBC aggregation is one of the main causes of the non-Newtonian behaviors of blood flows. Due to the presence of cross-linking proteins fibrinogen on cell membranes and globulin in blood plasma, red blood cells tend to form aggregates called rouleaus, in which cells adhere loosely like a stack of coins. The presence of massive rouleaus can impair the blood flow through micro- and capillary vessels, causing fatigue and shortness of breath. The variation in the level of RBC aggregation may be an indication of thrombotic disease. In general, the lower the shear rate, e.g., the slower the blood velocity, the more prevalent, i.e., larger size and number density, are RBC aggregates.

As shown in Fig. 6.29, as the shear rate drops to zero, it is anticipated that human blood will become one big aggregate, which then behaves as a viscoelastic solid as illustrated in Section 6.3.2. On the contrary, as the shear rate increases, cell rouleaus tend to break up. Individual red cells also deform slightly, elongate, and line up with the streamlines. The

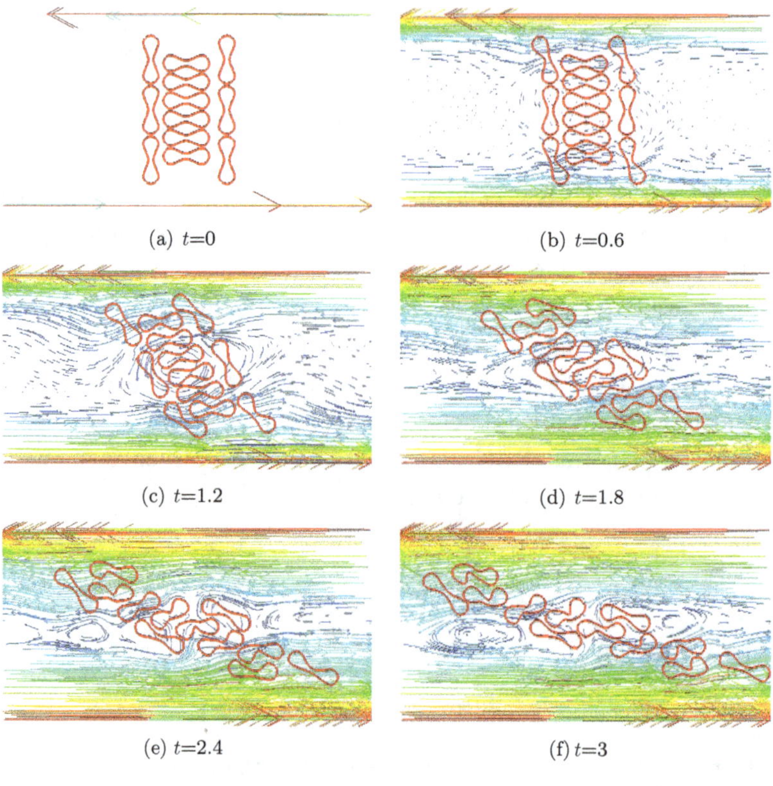

(a) t=0 (b) t=0.6

(c) t=1.2 (d) t=1.8

(e) t=2.4 (f) t=3

Fig. 6.29. Multiple-cell aggregate model with the shear rate of 3.0 1/s.

deformability and the decrease of cell–cell interaction forces combine to reduce blood viscosity with the increase of the shear rate. As shear rate increases above certain level, the whole blood behaves almost like a Newtonian fluid with a nearly constant viscosity. It is important to point out that for rigid particles including hardened red cells, normal leukocytes, etc., bulk viscosity is essentially independent of the shear rate.

In the near future, using the recent diagnostic medical imaging tools, such radiopaque dyes and angiograms, physicians will be able to visualize microcirculation flows through the entire body. The simulation methodology discussed in detail in this section will be an important addition to help understand various physiological phenomena in multi-physics and multi-scale environments.

6.3.2 *Viscoelastic models*

At an even further coarsening, a viscoelastic continuum can be implemented for the modeling of highly deformable biological materials. To further the discussion, we introduce a general Kelvin viscoelastic model as depicted in Fig. 6.30. It is clear that the limiting cases of this model

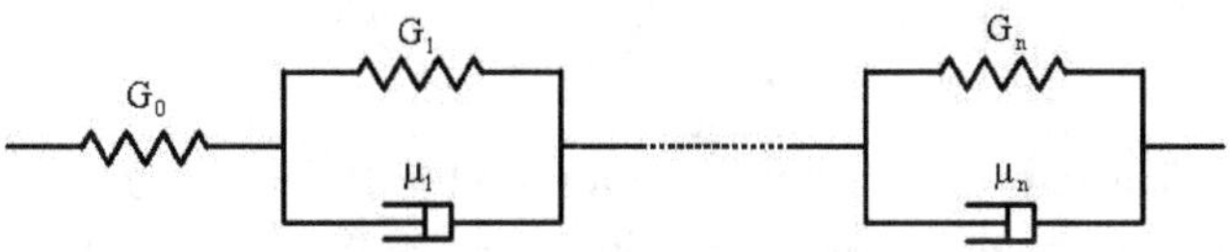

(a)

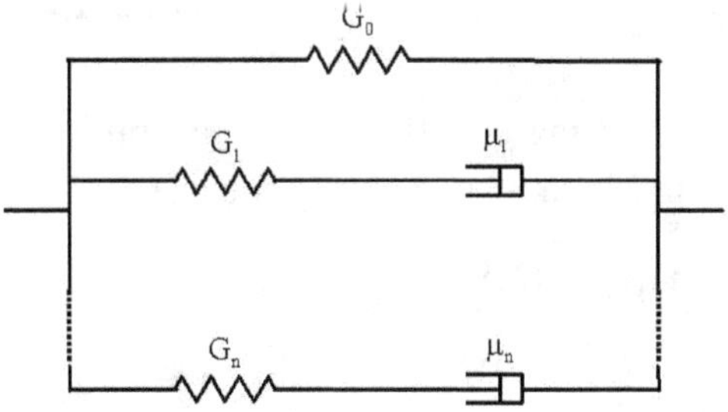

(b)

Fig. 6.30. General viscoelastic models. (a) General Kelvin model and (b) General Maxwell model.

correspond to the familiar Maxwell, Voigt, and Kelvin models (Wang *et al.*, 2019a, 2021).

Designating the shear strains at the spring G_o and the ith spring-dashpot combination (G_i, μ_i) as γ_o and γ_i, respectively, and the stress within the series as τ, we obtain the following governing equations:

$$\tau(t) = G_o \gamma_o(t), \tag{6.22}$$

$$\tau(t) = G_i \gamma_i(t) + \mu_i \dot{\gamma}_i(t), \quad 1 \leq i \leq n. \tag{6.23}$$

Thus, from Eq. (6.23), we obtain

$$\gamma_i(t) = \int_{-\infty}^{t} \frac{1}{\mu_i} e^{-(t-s)/\alpha_i} \tau(s) ds,$$

where the relaxation time α_i with respect to the ith spring-dashpot combination is defined as μ_i / G_i.

Finally, the total shear strain $\gamma(t)$ can be evaluated as

$$\gamma(t) = \frac{\tau(t)}{G_o} + \sum_{i=1}^{n} \gamma_i(t) = \int_{-\infty}^{t} C(t-s) d\tau,$$

where the creep (or compliance) function $C(t)$ is defined as

$$C(t) = \frac{1}{G_o} + \sum_{i=1}^{n} \frac{1}{G_i}(1 - e^{-t/\alpha_i}). \tag{6.24}$$

In a typical creep experiment, with $\tau(s) = 0$ for $s < 0$ and $\tau(s) = \tau$ for $s \geq 0$, the shear strain is simply expressed as $\gamma(t) = \tau C(t)$, which implies that the creep function can be determined experimentally. Naturally, a curve fitting will yield the physical parameters G_o, G_i, and μ_i in Eq. (6.24).

Likewise, the governing equations for a general Maxwell viscoelastic model as depicted in Fig. 6.30 can be expressed as

$$\tau_o(t) = G_o \gamma(t), \tag{6.25a}$$

$$\tau_i(t) = \mu_i \dot{\gamma}(t) - \alpha_i \dot{\tau}_i(t), \quad 1 \leq i \leq n. \tag{6.25b}$$

Thus, from Eq. (6.25b), we obtain

$$\tau_i(t) = \int_{-\infty}^{t} G_i e^{-(t-s)/\alpha_i} \dot{\gamma}(s) ds, \tag{6.26}$$

and the total stress $\tau(t)$ can be evaluated as

$$\tau(t) = G_o\gamma(t) + \sum_{i=1}^{n} \tau_i(t) = \int_{-\infty}^{t} R(t-s)d\gamma, \tag{6.27}$$

where the relaxation (or stiffness) function $R(t)$ is defined as

$$R(t) = G_o + \sum_{i=1}^{n} G_i e^{-t/\alpha_i}. \tag{6.28}$$

In a typical relaxation experiment, with $\gamma(s) = 0$ for $s < 0$ and $\gamma(s) = \gamma$ for $s \geq 0$, the stress is simply expressed as $\tau(t) = \gamma R(t)$, which implies that the relaxation function can also be determined experimentally. Naturally, a curve fitting will yield the physical parameters G_o, G_i, and μ_i in Eq. (6.28).

Equations (6.24) and (6.28) demonstrate that the blood viscoelastic behaviors can be accurately predicted via analytical or numerical experiments. In general, the viscosity of the blood flow depends on hematocrit of blood and the shear rate of blood flow. In particular, RBCs play an important role in the blood viscoelastic behaviors (Fung, 1996, 2000).

6.4 Model Reduction and Noise

With the development of computer hardware and software, it is now becoming feasible to employ so-called monolithic approaches to solve fluids and solids or structures in FSI systems simultaneously as a whole. As a consequence, reduced order modeling techniques can be implemented to derive the dominant coupled FSI models and allow more effective and efficient exploration of phase and parametric spaces. In this section, we employ fluid–structure interaction (FSI) systems with immersed flexible structures with or without free surfaces to explore both singular value decomposition (SVD)-based model reduction methods and mode superposition methods. For acoustoelastic FSI systems, we adopt a three-field mixed finite element formulation with displacement, pressure, and vorticity moment unknowns to effectively enforce the irrotationality constraint. We also propose in this section a new Inf-Sup test based on the lowest non-zero singular value of the coupling matrix for the selection of reliable sets of finite element discretizations for displacement and pressure as well as vorticity moment. Our numerical examples demonstrate that mixed finite element formulations can be effectively used to predict resonance frequencies of fully coupled FSI systems within different ranges of respective physical motions, namely, acoustic, structural, and slosh motions, without the contamination of spurious (non-physical) modes with non-zero frequencies. Our numerical results also confirm that SVD-based model reduction methods can be effectively used to reconstruct from a few snapshots of transient solutions the dominant principal components with moderate level of signal to noise ratio, which may eventually open doors for simulation of long-term behaviors of both linear and nonlinear FSI systems.

Over the past two decades, there have been a surge of research in the modeling of fluid–solid interaction (FSI) systems, many of which are related to the combination of nonlinear and meshless finite element methods with the immersed boundary method concepts (Boffi *et al.*, 2008; Wang, 2006; Wang *et al.*, 2009b; Wang and Liu, 2004; Liu *et al.*, 2006; Zhang *et al.*, 2004). In this section, we explore a further improvement of monolithic FSI modeling techniques by adopting model reduction techniques. More specifically, we employ traditional mixed finite element formulations for very complicated acoustoelastic FSI problems in which there exist three different ranges of coupled modes, namely, sloshing, structural, and acoustic modes. It is important to notice that although we name these three frequency bands as sloshing, structural, and acoustic ranges, the coupled effects are fully incorporated in the formulation. For example, the structural frequencies do depend on added mass effects of the surrounding fluid. In engineering practice, it is essential to focus on particular problems which involve different frequency ranges and to develop corresponding efficient spectrum methods.

In addition to a comprehensive review of mixed formulations for acoustoelastic FSI systems, we also explore two different types of model reduction procedures, namely, mode superposition methods and singular value decomposition (SVD)-based model reduction methods. For mode superposition methods, the advantage is the direct calculation of mode shapes and natural frequencies, whereas the disadvantage is the requirement of an explicit formulation for the FSI system at that particular configuration, which is essential for nonlinear FSI problems such as vocal track acoustic resonance. For SVD-based model reduction methods, snap shots of transient solutions derived from linear or nonlinear FSI formulations or experiments can be used to formulate the dominant principal modes surrounding the physical configuration as illustrated in Schmid (2010).

A number of finite element formulations have been proposed to model an acoustic fluid for the analysis of acoustoelastic FSI problems, namely, the displacement formulation (Belytschko, 1980; Bathe and Hahn, 1979; Hamdi *et al.*, 1978; Belytschko and Kennedy, 1976; Olson and Bathe, 1983), the displacement potential and pressure formulation, and the velocity potential formulation (Morand and Ohayon, 1995; Everstine, 1981; Olson and Bathe, 1985; Felippa and Ohayon, 1990).

Since its inception, the primitive variable formulation has received considerable attention for simple and direct interface conditions no different from element assemblage processes (Kiefling and Feng, 1976; Olson and Bathe, 1983). The key reason is of course the monolithic formulations for both fluid and solid media around their interfaces, which are useful in frequency calculations and response spectrum analysis. Various approaches have been introduced to eliminate spurious *non-zero* frequency circulation modes and to obtain improved formulations (Wilson and Khalvati, 1983; Chen and Taylor, 1990; Bermúdez and Rodríguez, 1994). It has been proposed that the spurious non-zero frequencies are caused by the irrotationality constraint (Wilson and Khalvati, 1983; Hamdi *et al.*, 1978). In 1990s, the true origins of the spurious non-zero frequency rotational modes have been identified (Wang, 2008; Wang and Bathe, 1997a). It was concluded that the reported non-zero frequency spurious modes are caused either by the pure displacement formulation, the use of mixed elements which do not satisfy the Inf-Sup condition, or the improper treatment of the boundary conditions. Therefore, a proper way of eliminating the non-zero frequency spurious modes is to use the displacement/pressure ($\mathbf{u}/p$) based mixed formulation with elements satisfying the Inf-Sup condition. In addition to the traditional $\mathbf{u}/p$ formulation (as used for solids with zero shear modulus) (Wang and Bathe, 1997a), an effective three-field mixed finite element formulation, namely, the displacement/pressure/vorticity moment ($\mathbf{u}-p-\mathbf{\Lambda}$) formulation was presented in Wang (2008).

In this section, we briefly review the $\mathbf{u}/p$ and $\mathbf{u}-p-\mathbf{\Lambda}$ formulations for acoustic fluid media and the related fluid–solid interaction (FSI) models. Moreover, we adopt the same mixed finite element formulations for both fluid and solid domains with mixed finite elements with continuous pressure. Hence, the coupling of fluids and solids in our monolithic fluid–solid is identical to standard finite element assemblage processes. Our numerical examples demonstrate that mixed finite element formulations are feasible for the prediction of coupled frequencies and mode shapes even if primary slosh, structural, and acoustic modes are within separate frequency ranges. A similar yet more straightforward numerical Inf-Sup test based on the lowest non-zero singular value of the coupled matrix is also presented in this section.

The focus of this section however is on issues related to model reduction methods with mixed finite element formulations. One of the most popular model reduction methods is the mode superposition method (Daniel, 1980; Bathe, 1996), the other one is the so-called singular value decomposition (SVD)-based method or principal component analysis (PCA) (Berrar *et al.*, 2004; Jolliffe, 2002). Recent studies include the overview of modeling of large-scale dynamical systems (Antoulas and Sorsensen, 2001; Antoulas, 2005) and interpolating method based on diffeomorphism or differentiable manifold instead of tangent plane (Amsallem *et al.*, 2009).

The mode superposition method is limited to linear systems, whereas various forms of PCA can be extended to nonlinear systems with frequent updates of the principal directions. In order to illustrate the procedures of these two approaches, in this section, we employ a fairly comprehensive two-dimensional acoustoelastic FSI model with both submerged two-dimensional elastic structure and free surface. If we do not know the material properties of the system, with SVD-based principal component analysis, combinations of eigen modes can still be constructed by taking the first few dominant principal components of snapshots of available transient data which could also be derived from experiments or other black box simulation tools (Wang *et al.*, 2019b). In general, excellent agreements are confirmed between the original transient solutions and the data reconstructed with a few dominant principal components. The figures of energy are also plotted in order to verify the realization of this objective, namely, recovering the transient data with a few principal components without losing dominant characteristics. Finally, some temporal coarse-grained techniques are also proposed for the study of long-term behaviors of FSI systems.

6.4.1 *Monolithic formulation for two continua*

In linear analysis, for isentropic and inviscid fluids undergoing small vibrations, the momentum balance and constitutive equations show

$$\rho\ddot{\mathbf{u}} + \nabla p - \mathbf{f}^B = \mathbf{0}, \tag{6.29}$$

$$\nabla\cdot\mathbf{u} + \frac{p}{\beta} = 0, \tag{6.30}$$

where ρ, $\mathbf{u}$, p, $\mathbf{f}^B$, and β represent the density, displacement vector, pressure, body force vector, and fluid bulk modulus, respectively.

Assume the inertia force $-\rho\ddot{\mathbf{u}}$ is included in the body force term and combine Eqs. (6.29) and (6.30), we have

$$\beta\nabla(\nabla\cdot\mathbf{u}) + \mathbf{f}^B = 0. \tag{6.31}$$

In addition, it is known that the inviscid fluid must satisfy the following irrotationality constraint:

$$\nabla\times\mathbf{u} = 0. \tag{6.32}$$

The variational form of Eq. (6.31) can then be expressed as

$$\int_{V_f}\left[\beta(\nabla\cdot\mathbf{u})(\nabla\cdot\delta\mathbf{u}) - \mathbf{f}^B\cdot\delta\mathbf{u}\right]dV + \int_{S_f}\bar{p}\,\delta u_n^s\,dS = 0, \tag{6.33}$$

where V_f and S_f stand for the fluid domain and Neumann boundary, respectively; u_n^s is the displacement normal to S_f; and $\bar{p}$ is the given pressure on S_f corresponding to u_n^s with the following variational indicator

$$\Pi = \int_{V_f}\left[\frac{\beta}{2}(\nabla\cdot\mathbf{u})^2 - \mathbf{f}^B\cdot\mathbf{u}\right]dV + \int_{S_f}\bar{p}u_n^s\,dS.$$

It has been observed that with such a displacement-based formulation circulation modes with non-zero frequencies may emerge with the zero frequency circulation modes (Bathe, 1996). In engineering practice, it is difficult to delineate such non-zero frequencies, often named as spurious or non-physical, from non-zero frequency physical modes. Furthermore, when the fluid acts as almost incompressible, the usual constraint of near incompressibility is shown as Eq. (6.30), with the bulk modulus

$\beta \to \infty$. However, the displacement-based formulation with irrotationality and (almost) incompressibility conditions often lead to a much too stiff response, i.e., 'locking' behavior. In order to eliminate these deficiencies, the penalty-form variational indicator based on constraint (6.30) is introduced

$$\Pi = \int_{V_f} \left[\frac{p^2}{2\beta} - \mathbf{u} \cdot \mathbf{f}^B - \lambda_p \left(\frac{p}{\beta} + \nabla \cdot \mathbf{u} \right) \right] dV + \int_{S_f} \bar{p} u_n^S dS, \qquad (6.34)$$

where λ_p is the Lagrange multiplier.

For this indicator, the first term corresponds to the strain energy with respect to pressure, the second term stands for the external body force potential, which also includes the inertia force, and the last term is the potential due to the boundary pressure. Equation (6.34) is often called $\mathbf{u}/p$ mixed finite element formulation or simply $\mathbf{u}/p$ formulation. Furthermore, because the irrotationality constraint is not imposed in the $\mathbf{u}/p$ formulation, there exist too many exact zero frequencies. In reliable finite element analysis, these zero frequency modes are further eliminated with the following modification of Eq. (6.32)

$$\nabla \times \mathbf{u} = \frac{\mathbf{\Lambda}}{\alpha}, \qquad (6.35)$$

where Λ is the so-called vorticity moment and α stands for a very large number.

With this additional constraint, another variational indicator Π is defined by the Lagrange multiplier method on the basis of the $\mathbf{u}/p$ formulation

$$\int_{V_f} \left[\frac{p^2}{2\beta} - \mathbf{u} \cdot \mathbf{f}^B - \lambda_p \left(\frac{p}{\beta} + \nabla \cdot \mathbf{u} \right) + \frac{\mathbf{\Lambda} \cdot \mathbf{\Lambda}}{2\alpha} - \boldsymbol{\lambda}_\Lambda \cdot \left(\frac{\mathbf{\Lambda}}{\alpha} - \nabla \times \mathbf{u} \right) \right] dV$$
$$+ \int_{S_f} \bar{p} u_n^S dS,$$

where $\boldsymbol{\lambda}_\Lambda$ is a Lagrange multiplier vector corresponding to the vorticity moment.

The fourth term is introduced to statically condense out the degrees of freedom of the vorticity moment $\mathbf{\Lambda}$. From the numerical tests, as discussed in Wang and Bathe (1997a, 1997b), it is found that α can be any numerically reasonable value larger than β, for example, $100\,\beta \le \alpha \le 10^6 \beta$.

In this work, α is defined as 1000 β. Invoke the stationarity of Π with respect to $\mathbf{u}$, p, $\boldsymbol{\Lambda}$ and identify λ_p as p and $\boldsymbol{\lambda}_\Lambda$ as $\boldsymbol{\Lambda}$, we have the three field equations

$$-\mathbf{f}^B + \nabla p + \nabla \times \boldsymbol{\Lambda} = \mathbf{0}, \tag{6.36}$$

$$\frac{p}{\beta} + \nabla \cdot \mathbf{u} = 0, \tag{6.37}$$

$$\nabla \times \mathbf{u} - \frac{\boldsymbol{\Lambda}}{\alpha} = \mathbf{0}, \tag{6.38}$$

with the boundary conditions

$$\mathbf{u} \cdot \mathbf{n} = \bar{u}_n, \quad \text{on } S_u, \tag{6.39}$$

$$p = \bar{p}, \quad \text{on } S_f, \tag{6.40}$$

$$\boldsymbol{\Lambda} = \mathbf{0}, \quad \text{on } S, \tag{6.41}$$

where S_u indicates the Dirichlet boundary with $S = S_u \cup S_f$ and $S_u \cap S_f = \oslash$.

Note that the pressure unknown p is a physical quantity in the formulation as governed by Eq. (6.37) as well as the boundary condition (6.40). However, the vorticity moment $\boldsymbol{\Lambda}$ is an artificial vector introduced to enforce the irrotationality constraint (6.32). Moreover, the governing equation (6.38) implies a small value for the vorticity. It is therefore natural to simplify the problem by imposing a zero value for the vorticity moment on all boundaries as described in (6.41). Detailed descriptions of this point and the relative issues are available in Bao *et al.* (2001) and Wang and Bathe (1997b). Applying the standard Galerkin finite element discretization, and considering a typical element, we have

$$\mathbf{u} = \mathbf{HU}, \quad p = \mathbf{H}_p\mathbf{P}, \quad \boldsymbol{\Lambda} = \mathbf{H}_\Lambda\boldsymbol{\Lambda},$$

$$\nabla \cdot \mathbf{u} = (\nabla \cdot \mathbf{H})\mathbf{U} = \mathbf{BU},$$

$$\nabla \times \mathbf{u} = (\nabla \times \mathbf{H})\mathbf{U} = \mathbf{DU},$$

where $\mathbf{U}$, $\mathbf{P}$, and $\boldsymbol{\Lambda}$ list all the nodal point displacements, pressure, and vorticity moment, respectively in global coordinates and $\mathbf{H}$, $\mathbf{H}_p$, and $\mathbf{H}_\Lambda$ refer to the displacement, pressure, and vorticity moment interpolation matrix, respectively.

Finally, the field equations can be expressed in matrix form with respect to nodal unknown vectors **U**, **P**, and **Λ**

$$\begin{pmatrix} \mathbf{M} & \mathbf{0} & \mathbf{0} \\ \mathbf{0} & \mathbf{0} & \mathbf{0} \\ \mathbf{0} & \mathbf{0} & \mathbf{0} \end{pmatrix} \begin{pmatrix} \ddot{\mathbf{U}} \\ \ddot{\mathbf{P}} \\ \ddot{\mathbf{\Lambda}} \end{pmatrix} + \begin{pmatrix} \mathbf{0} & \mathbf{L} & \mathbf{Q} \\ \mathbf{L}^T & \mathbf{A} & \mathbf{0} \\ \mathbf{Q}^T & \mathbf{0} & \mathbf{G} \end{pmatrix} \begin{pmatrix} \mathbf{U} \\ \mathbf{P} \\ \mathbf{\Lambda} \end{pmatrix} = \begin{pmatrix} \mathbf{R} \\ \mathbf{0} \\ \mathbf{0} \end{pmatrix}, \tag{6.42}$$

with

$$\mathbf{M} = \int_{V_f} \rho \mathbf{H}^T \mathbf{H} dV, \quad \mathbf{L} = -\int_{V_f} \mathbf{B}^T \mathbf{H}_p dV, \quad \mathbf{Q} = \int_{V_f} \mathbf{D}^T \mathbf{H}_\Lambda dV,$$

$$\mathbf{A} = -\int_{V_f} \frac{1}{\kappa} \mathbf{H}_p^T \mathbf{H}_p dV, \quad \mathbf{G} = -\int_{V_f} \frac{1}{\alpha} \mathbf{H}_\Lambda^T \mathbf{H}_\Lambda dV, \quad \mathbf{R} = -\int_{S_f} (\mathbf{H}_n^s)^T \bar{p} dS.$$

Combining the coefficients of different variables, we can systematically rewrite the entries of the mass matrix and the stiff matrix as follows:

$$\begin{pmatrix} \mathbf{M}_{uu} & \mathbf{0} & \mathbf{0} \\ \mathbf{0} & \mathbf{0} & \mathbf{0} \\ \mathbf{0} & \mathbf{0} & \mathbf{0} \end{pmatrix} \begin{pmatrix} \ddot{\mathbf{U}} \\ \ddot{\mathbf{P}} \\ \ddot{\mathbf{\Lambda}} \end{pmatrix} + \begin{pmatrix} \mathbf{0} & \mathbf{K}_{up} & \mathbf{K}_{u\Lambda} \\ \mathbf{K}_{pu} & \mathbf{K}_{pp} & \mathbf{0} \\ \mathbf{K}_{\Lambda u} & \mathbf{0} & \mathbf{K}_{\Lambda\Lambda} \end{pmatrix} \begin{pmatrix} \mathbf{U} \\ \mathbf{P} \\ \mathbf{\Lambda} \end{pmatrix} = \begin{pmatrix} \mathbf{R} \\ \mathbf{0} \\ \mathbf{0} \end{pmatrix}, \tag{6.43}$$

with subscripts indicating their corresponding unknowns.

Consider a structure with isotropic material

$$\begin{aligned} \tau_{ij,j} + f_i^B &= 0, \quad \text{in } V, \\ \tau_{ij} n_j &= f_i^{S_f}, \quad \text{on } S_f, \\ u_i &= u_i^{S_u}, \quad \text{on } S_u, \end{aligned}$$

where $\boldsymbol{\tau}$, $\mathbf{f}^B$, $\mathbf{n}$, and $\mathbf{f}^{S_f}$ represent the stress tensor, body force vector, outward normal vector, and the applied surface traction vector on the Neumann boundary S_f, respectively.

Based on the governing equation and boundary conditions, the principle of virtual displacements can be derived:

$$\int_V \delta\varepsilon_{ij} \tau_{ij} dV = \int_V \delta u_i f_i^B dV + \int_{S_f} \delta u_i^{S_f} f_i^{S_f} dS + \sum_k \delta U_i^k R_{ci}^k, \tag{6.44}$$

where $\delta\boldsymbol{\varepsilon}$ is the virtue strain tensor; $\mathbf{R}_c^k$ is the kth concentrated load vector; $\delta\mathbf{u}$, $\delta\mathbf{u}_{S_f}$, and $\delta\mathbf{U}$ are the virtue displacement over the volume, the Neumann boundary, and isolated points, respectively.

If the material is almost incompressible, the constitutive relations are derived as

$$\tau_{ij} = \kappa\varepsilon_v\delta_{ij} + 2G\varepsilon'_{ij},$$

where κ is the bulk modulus with $\kappa = E/3(1-2\nu)$; G is the shear modulus with $G = E/2(1+\nu)$; ε_v is the volumetric strain with $\varepsilon_v = \varepsilon_{kk}$ in Cartesian coordinates; and ε'_{ij} are the deviatoric strain components with respect to Kronecker delta, $\varepsilon'_{ij} = \varepsilon_{ij} - \varepsilon_v\delta_{ij}/3$.

As the Poisson's ratio ν gradually approaches to 0.5, the bulk modulus κ approaches to infinity and the volumetric strain ε_v approaches to zero. Naturally, it is difficult to evaluate the pressure from the relationship $p = -\kappa\varepsilon_v$, even though the pressure is finite. Therefore, the governing equation should involve another unknown, the pressure p, which is independent of the displacement unknown $\mathbf{u}$. Consequently, with the shear modulus G, the relation can be achieved as

$$\tau_{ij} = -p\delta_{ij} + 2G\varepsilon'_{ij}. \tag{6.45}$$

Now the principle of virtual work is introduced in terms of independent variables $\mathbf{u}$ and p

$$\int_V (\delta\boldsymbol{\varepsilon}'^T : \mathbf{S} - \delta\varepsilon_v p)dV = \delta\mathcal{R},$$

$$\int_V \left(\frac{p}{\kappa} + \varepsilon_v\right)\delta p dV = 0,$$

with $\mathbf{S} = \mathbf{C}'\boldsymbol{\varepsilon}'$, $\mathbf{S} = \boldsymbol{\tau} + p\boldsymbol{\delta}$, and $\boldsymbol{\varepsilon}' = \boldsymbol{\varepsilon} - \varepsilon_v\boldsymbol{\delta}/3$, where $\mathbf{S}$ and $\boldsymbol{\varepsilon}'$ are the deviatoric stress and strain tensors, respectively; the material matrix $\mathbf{C}'$ is defined by Eq. (6.45) and linearly proportional to the shear modulus G; and $\delta\mathcal{R}$ corresponds to the net external virtual work.

Implement the standard Galerkin discretization

$$\boldsymbol{\varepsilon}' = \mathbf{B}_D\mathbf{U}, \quad \varepsilon_v = \mathbf{B}_v\mathbf{U}, \quad \text{and} \quad p = \mathbf{H}_p\mathbf{P},$$

we derive the following governing equation in matrix form with respect to discretized unknowns $\mathbf{U}$ and $\mathbf{P}$

$$\begin{pmatrix} \mathbf{K}_{uu} & \mathbf{K}_{up} \\ \mathbf{K}_{pu} & \mathbf{K}_{pp} \end{pmatrix} \begin{pmatrix} \mathbf{U} \\ \mathbf{P} \end{pmatrix} = \begin{pmatrix} \mathbf{R} \\ \mathbf{0} \end{pmatrix}, \tag{6.46}$$

where $\mathbf{B}_D$ and $\mathbf{B}_v$ are the deviatoric and volumetric strain interpolation matrix, respectively; $\mathbf{R}$ represents the external force corresponding to the net external virtual work $\bar{\mathcal{R}}$; and the components of stiffness matrix are defined as

$$\mathbf{K}_{uu} = \int_{v} \mathbf{B}_D^T \mathbf{C}' \mathbf{B}_D dV \quad \text{and} \quad \mathbf{K}_{pp} = -\int_{v} \frac{1}{\kappa} \mathbf{H}_p^T \mathbf{H}_p dV,$$

$$\mathbf{K}_{up} = \mathbf{K}_{pu} = -\int_{v} \mathbf{B}_v^T \mathbf{H}_p dV.$$

In transient analysis, the inertia forces need to be considered. Using d'Alembert's principle, the element inertia forces can be directly included as part of the body forces

$$\mathbf{R}_B = \int_V \mathbf{H}^T (\mathbf{f}^B - \rho \mathbf{H}\ddot{\mathbf{U}}) dV,$$

where $\mathbf{R}_B$ is the body force vector; ρ is the mass density; and $\ddot{\mathbf{U}}$ lists the nodal point accelerations, i.e., the second time derivative of the nodal point displacements $\mathbf{U}$.

At this point, $\mathbf{f}^B$ no longer includes the inertia forces and the dynamic equilibrium equation becomes

$$\mathbf{M}\ddot{\mathbf{U}} + \mathbf{K}\mathbf{U} = \mathbf{R},$$

where the matrix $\mathbf{M} = \int_V \rho \mathbf{H}^T \mathbf{H} dV$ is the so-called consistent mass matrix of the solid.

In this section, we will explicitly illustrate the coupling procedures for mixed formulations. Note that there is no need to use the so-called mixed formulations with additional pressure and vorticity moment unknowns for the solid domain. Therefore, the additional pressure and vorticity moment unknowns will only show up for the fluid domain. Let's first employ the pure displacement-based formulation for solid

$$\begin{pmatrix} \mathbf{M}_{ss} & \mathbf{M}_{sc} \\ \mathbf{M}_{cs} & \mathbf{M}_{cc}^s \end{pmatrix} \begin{pmatrix} \ddot{\mathbf{U}}_s \\ \ddot{\mathbf{U}}_c \end{pmatrix} + \begin{pmatrix} \mathbf{K}_{ss} & \mathbf{K}_{sc} \\ \mathbf{K}_{cs} & \mathbf{K}_{cc}^s \end{pmatrix} \begin{pmatrix} \mathbf{U}_s \\ \mathbf{U}_c \end{pmatrix} = \begin{pmatrix} \mathbf{R}_s \\ \mathbf{R}_c^s \end{pmatrix}, \tag{6.47}$$

and the $\mathbf{u}/p$ formulation for fluid

$$\begin{pmatrix} \mathbf{M}_{ff} & \mathbf{M}_{fc} & \mathbf{0} \\ \mathbf{M}_{cf} & \mathbf{M}_{cc}^f & \mathbf{0} \\ \mathbf{0} & \mathbf{0} & \mathbf{0} \end{pmatrix} \begin{pmatrix} \ddot{\mathbf{U}}_f \\ \ddot{\mathbf{U}}_c \\ \ddot{\mathbf{P}} \end{pmatrix} + \begin{pmatrix} \mathbf{K}_{ff} & \mathbf{K}_{fc} & \mathbf{K}_{fp} \\ \mathbf{K}_{cf} & \mathbf{K}_{cc}^f & \mathbf{K}_{cp} \\ \mathbf{K}_{pf} & \mathbf{K}_{pc} & \mathbf{K}_{pp} \end{pmatrix} \begin{pmatrix} \mathbf{U}_f \\ \mathbf{U}_c \\ \mathbf{P} \end{pmatrix} = \begin{pmatrix} \mathbf{R}_f \\ \mathbf{R}_c^f \\ \mathbf{0} \end{pmatrix}, \tag{6.48}$$

respectively, where $\mathbf{U}_s$, $\mathbf{U}_f$, and $\mathbf{U}_c$ refer to the nodal displacement unknowns in the structure interior, fluid interior, and fluid–structure interface, respectively; and $\mathbf{R}_s$, $\mathbf{R}_f$, $\mathbf{R}_c^s$, and $\mathbf{R}_c^f$ stand for the nodal forces in the structure interior, fluid interior, structure, and fluid side of the fluid–structure interface, respectively.

It is important to notice that in this case the kinematic matching of the coupling of fluid and solid domains share only the displacement unknown $\mathbf{U}_c$ between the fluid and solid domains, whereas the dynamic matching will cancel the effects of $\mathbf{R}_c^s$ and $\mathbf{R}_c^f$ according to the Newton's third law. A more comprehensive explanation for the coupled fluid–structure matrices are available in Wang (2008) and Wang *et al.* (2019b). In the practical analysis, $\mathbf{U}_c$ might only consist of normal displacement components of fluid elements. In the assemblage process, Eqs. (6.47) and (6.48) are combined and yield the final governing equation for FSI systems

$$\begin{pmatrix} \mathbf{M}_{ss} & \mathbf{0} & \mathbf{M}_{sc} & \mathbf{0} \\ \mathbf{0} & \mathbf{M}_{ff} & \mathbf{M}_{fc} & \mathbf{0} \\ \mathbf{M}_{cs} & \mathbf{M}_{cf} & \mathbf{M}_{cc} & \mathbf{0} \\ \mathbf{0} & \mathbf{0} & \mathbf{0} & \mathbf{0} \end{pmatrix} \begin{pmatrix} \ddot{\mathbf{U}}_s \\ \ddot{\mathbf{U}}_f \\ \ddot{\mathbf{U}}_c \\ \ddot{\mathbf{P}} \end{pmatrix} + \begin{pmatrix} \mathbf{K}_{ss} & \mathbf{0} & \mathbf{K}_{sc} & \mathbf{0} \\ \mathbf{0} & \mathbf{K}_{ff} & \mathbf{K}_{fc} & \mathbf{K}_{fp} \\ \mathbf{K}_{cs} & \mathbf{K}_{cf} & \mathbf{K}_{cc} & \mathbf{K}_{cp} \\ \mathbf{0} & \mathbf{K}_{pf} & \mathbf{K}_{pc} & \mathbf{K}_{pp} \end{pmatrix}$$

$$\times \begin{pmatrix} \mathbf{U}_s \\ \mathbf{U}_f \\ \mathbf{U}_c \\ \mathbf{P} \end{pmatrix} = \begin{pmatrix} \mathbf{R}_s \\ \mathbf{R}_f \\ \mathbf{0} \\ \mathbf{0} \end{pmatrix},$$

with $\mathbf{M}_{cc} = \mathbf{M}_{cc}^s + \mathbf{M}_{cc}^f$ and $\mathbf{K}_{cc} = \mathbf{K}_{cc}^s + \mathbf{K}_{cc}^f$.

In order to choose appropriate mixed elements, we must consider the Inf-Sup condition (Chapelle and Bathe, 1993). It has been proven that $9-4c-4c$ and $9-3-3$ mixed elements are appropriate choices in two-dimensional analysis (Bao *et al.*, 2001; Bathe, 1996). Figure 6.31 depicts the element configurations for these two types of mixed elements schematically along with the so-called full numerical integration, namely, 3×3 Gauss integration. Similarly, for the second example, there are 9 continuous nodal displacements and 4 continuous pressure and vorticity moment nodes in each element. When elements are adjacent to each other. The same assemblage procedures are utilized for the displacement, pressure,

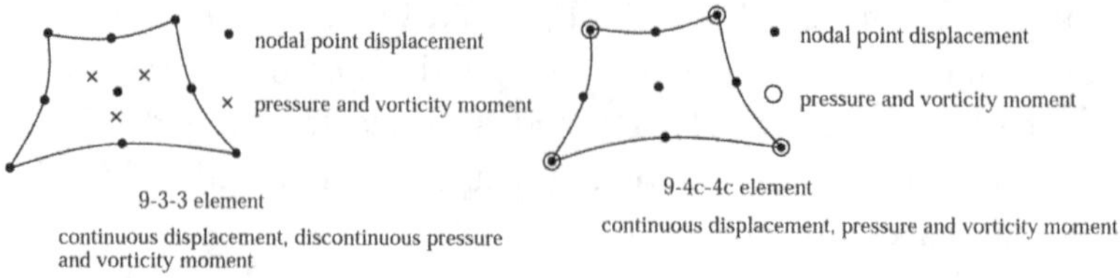

Fig. 6.31. Two mixed elements for $\mathbf{u}$-p-$\boldsymbol{\Lambda}$ formulation.

and vorticity moment unknowns. Letter c indicates that these quantities are continuous among elements. In order to obtain the continuity of the pressure and vorticity moment unknowns, we can also employ $9-4c-4c$ elements for our test problems, namely, FSI Models I and II.

In the mesh generation figure, it shows that each pressure node, except the nodes on the boundary, is shared by four neighboring elements. Based on the configuration of $9-4c-4c$ element, to describe the pressure of a generic element, a bilinear interpolation function is introduced

$$p = p_1 + p_2 r + p_3 s + p_4 rs,$$

where r and s are the natural coordinates for a single element.

Comparing Eq. (6.30) with Eq. (6.32), it can be observed that these two constraints are very similar in nature and are not coupled in the displacement-based formulation. Therefore, it can be assumed that they should be imposed with the same interpolation functions in generic element discretization. That is to say, for the inviscid acoustic fluid formulation, the interpolation for the vorticity moment is

$$\boldsymbol{\Lambda} = \boldsymbol{\Lambda}_1 + \boldsymbol{\Lambda}_2 r + \boldsymbol{\Lambda}_3 s + \boldsymbol{\Lambda}_4 rs.$$

In the following discussion, the stability analysis is based on the $\mathbf{u}-p-\boldsymbol{\Lambda}$ formulation for acoustic fluid and the $\mathbf{u}/p$ formulation for structure in a typical acoustoelastic FSI system. The basic observations and conclusions are also directly applicable to the $\mathbf{u}/p$ formulation for fluid and displacement-based formulation for structure or solid. Assign

$$\mathbf{U}_h = \mathbf{U}, \quad \text{and} \quad \mathbf{S}_h = (\mathbf{P}^T \quad \boldsymbol{\Lambda}^T)^T,$$

$$\mathbf{R}_h = \mathbf{R}, \quad (\mathbf{K}_{us})_h = (\mathbf{K}_{up} \quad \mathbf{K}_{u\Lambda}), \quad (\mathbf{K}_{ss})_h = \begin{pmatrix} \mathbf{K}_{pp} & \mathbf{0} \\ \mathbf{0} & \mathbf{K}_{\Lambda\Lambda} \end{pmatrix},$$

therefore, the coupled formulation with either $\mathbf{u}-p-\mathbf{\Lambda}$ or $\mathbf{u}-p$ formulation for acoustic fluid can be expressed as follows:

$$\begin{pmatrix} \mathbf{M}_{uu} & \mathbf{0} \\ \mathbf{0} & \mathbf{0} \end{pmatrix} \begin{pmatrix} \ddot{\mathbf{U}}_h \\ \ddot{\mathbf{S}}_h \end{pmatrix} + \begin{pmatrix} (\mathbf{K}_{uu})_h & (\mathbf{K}_{us})_h \\ (\mathbf{K}_{us})_h^T & (\mathbf{K}_{ss})_h \end{pmatrix} \begin{pmatrix} \mathbf{U}_h \\ \mathbf{S}_h \end{pmatrix} = \begin{pmatrix} \mathbf{R}_h \\ \mathbf{0} \end{pmatrix},$$

where $\mathbf{U}_h$ contains all the variables of the nodal point displacements and $\mathbf{S}_h$ lists all the pressure and vorticity moment unknowns.

For step-by-step transient analysis, after transforming, the following equation is considered for every time step:

$$\begin{pmatrix} (\mathbf{K}_{uu}^*)_h & (\mathbf{K}_{us})_h \\ (\mathbf{K}_{us})_h^T & (\mathbf{K}_{ss})_h \end{pmatrix} \begin{pmatrix} \mathbf{U}_h \\ \mathbf{S}_h \end{pmatrix} = \begin{pmatrix} \mathbf{R}_h^* \\ \mathbf{0} \end{pmatrix}, \tag{6.49}$$

where $\mathbf{R}_h^*$ is an effective load vector.

Moreover, it is known that at the shear modulus $G = 0$ for invisid fluid, according to Eq. (6.45), we have

$$(\mathbf{K}_{uu}^*)_h = C(\mathbf{M}_{uu})_h,$$

where C is a constant depending on the time integration method, for example, $C = 1/\alpha\Delta t^2$, with $\alpha \geq 0.25$ for Newmark method.

Since $(\mathbf{M}_{uu})_h$ is positive definite, $(\mathbf{K}_{uu}^*)_h$ is always a positive-definite matrix. The stability and effectiveness of the $\mathbf{u}/p$ or $\mathbf{u}-p-\mathbf{\Lambda}$ formulation and corresponding mixed elements which will be employed in the test problems are verified through ellipticity and Inf-Sup conditions, which are necessary and sufficient conditions for well-posedness:

(a) **Ellipticity Condition:**

$$\mathbf{V}_h^T(\mathbf{K}_{uu})_h^*\mathbf{V}_h \geq c_1 \parallel \mathbf{V}_h \parallel_V^2, \quad \forall\ \mathbf{V}_h \in \ker[(\mathbf{K}_{us})_h^T],$$

with $\|v\|_V^2 = \sum_{i,j} \| \frac{\partial v_i}{\partial x_j} \|_{L^2(V)}^2$, $\ker[(\mathbf{K}_{us})_h^T] = \{\mathbf{V}_h | \mathbf{V}_h \in \mathcal{R}^n, (\mathbf{K}_{us})_h^T \mathbf{V}_h = 0\}$, and $c_1 > 0$.

(b) **Inf-Sup Condition:**

$$\inf_{\mathbf{S}_h} \sup_{\mathbf{U}_h} \frac{\mathbf{U}_h^T(\mathbf{K}_{us})_h\mathbf{S}_h}{||\mathbf{U}_h|| \cdot ||\mathbf{S}_h||} \geq c_2 > 0,$$

where the constant c_2 is independent of the mesh size h, the constant α in the irrotationality constraint and the bulk modulus β.

Generally, whether the Inf-Sup condition is satisfied depends on the specific chose of finite element discretization, the boundary condition, and the macromesh generation of the mathematical models (Bao *et al.*, 2001). The Inf-Sup condition or often called Brezzi–Babuška condition, was popularized by Babuška and Brezzi in early 1970s (Babuška, 1973; Brezzi, 1974) and stated in an earlier theoretical work by Necas in 1962. Furthermore, in Brezzi and Fortin (1991) and Ern and Guermond (2003), this condition has been introduced from the viewpoint of Banach space theorems. Although the Inf-Sup condition for mixed formulations was proposed some time ago, the analytical proof of whether the Inf-Sup condition is satisfied by a specific set of elements can be very difficult (Chapelle and Bathe, 1993). In Bathe (1996) the very first numerical proof of the Inf-Sup condition has been proposed based on the matrix equations, which is feasible for engineering applications. As a part of the novelties presented in this section, we will explore a similar yet different type of numerical proof of the Inf-Sup condition which is based on the singular values of the coupling matrix $(\mathbf{K}_{us})_h$, a direct extension or benefit of SVD-based model reduction method with no additional numerical challenge.

6.4.2 *Model reduction*

Since the finite element procedure is realized on the basis of elements, in engineering practice, we often encounter a large number of elements and the degrees of freedom which lead to high-dimensional matrices. To accommodate the challenges in both spatial and temporal resolutions, we must develop efficient methods to construct equivalent low-dimensional dynamic descriptions which can capture the essence of the full fledged finite element model with high fidelity. In this section, two different model reduction methods are introduced, namely, mode superposition methods and SVD-based model reduction methods.

The mode superposition method depends on the explicit construction of mass and stiffness matrices. If the coupled system is well known along with all material properties, this method can be efficient and useful for all linear systems. Suppose we have the following governing dynamic equilibrium equation without damping effects

$$\mathbf{M}\ddot{\mathbf{x}} + \mathbf{K}\mathbf{x} = \mathbf{r}(t). \tag{6.50}$$

We propose to transform this equation into a more effective form for direct integration by using the following transformation $\mathbf{x}(t) = \mathbf{P}\mathbf{u}(t) = u_i(t)\mathbf{p}_i$, where $\mathbf{P}$ is a square modal matrix, each column $\mathbf{p}_i$ of which represents a mode shape, and $\mathbf{u}$ is a time-dependent general coordinate vector. Pre-multiply Eq. (6.50) by $\mathbf{P}^T$, we derive

$$\tilde{\mathbf{M}}\ddot{\mathbf{u}} + \tilde{\mathbf{K}}\mathbf{u} = \tilde{\mathbf{r}}(t), \tag{6.51}$$

with $\tilde{\mathbf{M}} = \mathbf{P}^T\mathbf{M}\mathbf{P}$, $\tilde{\mathbf{K}} = \mathbf{P}^T\mathbf{K}\mathbf{P}$, and $\tilde{\mathbf{r}} = \mathbf{P}^T\mathbf{r}$.

The transformation matrix is established by solving the generalized eigenvalue problem

$$(\mathbf{K} - \omega_i^2\mathbf{M})\mathbf{p}_i = \mathbf{0}. \tag{6.52}$$

The homogeneous solution of Eq. (6.51) can be written as $u_i(t) = c_i \sin \omega_i(t - t_i)$, where c_i is the scaling factor corresponding to the ith eigenvector $\mathbf{p}_i$ and eigenvalue ω_i, t_i is the corresponding time constant related to the phase shifts. Note that both the scaling factor c_i and the time shift t_i depend on the initial conditions. To further simplify the formulation, we can also scale the eigenvectors according to the so-called $\mathbf{M}$-orthonormal condition, i.e., $\mathbf{p}_i^T\mathbf{M}\mathbf{p}_j = \delta_{ij}$, we can finally define the transformation matrix as

$$\mathbf{P} = [\mathbf{p}_1, \mathbf{p}_2, \ldots, \mathbf{p}_n],$$
$$\mathbf{\Lambda} = \text{diag}\{\boldsymbol{\omega}_1^2, \boldsymbol{\omega}_2^2, \ldots, \boldsymbol{\omega}_n^2\}$$

and Eq. (6.52) can be rewritten as

$$\mathbf{K}\mathbf{P} = \mathbf{M}\mathbf{P}\mathbf{\Lambda},$$

with $\mathbf{P}^T\mathbf{K}\mathbf{P} = \mathbf{\Lambda}$ and $\mathbf{P}^T\mathbf{M}\mathbf{P} = \mathbf{I}$, where $\mathbf{I}$ is an $n \times n$ dimensional identity matrix.

Note that in engineering practice, due to the restraints of physical conditions, it is safe to assume the eigenvalue problem in Eq. (6.52) is simple, namely, the matrices are diagonalizable and no Jordan form is needed (Strang, 1988). In comparison with the mode superposition method, SVD-based model reduction methods do not depend on the explicit construction of mass and stiffness matrices. If a few snapshots of the transient solution are available, it is possible to identify the dominant principal directions and the proportionality of the energy content. In fact, this type of model reduction method is also applicable to nonlinear systems in which the principal directions can be altered or updated through time (Gunzburger, 2003). It is also important to recognize that for linear systems, the principal directions are often a linear combination of the mode shapes which depends very much on the transient conditions or snapshots around that particular time.

Consider an $n \times m$ matrix $\mathbf{A}$, each row of which represents a snapshot of all nodal unknowns. We can easily construct two symmetric matrices $\mathbf{A}^T\mathbf{A}$ and $\mathbf{A}\mathbf{A}^T$. Suppose the rank of the matrix $\mathbf{A}$ is r, there will be r non-zero eigenvalues of these two symmetric matrices. Based on the linear algebra theories, the orthogonal subspace of the null space must be the row space while the orthogonal subspace of the left null space is the column space. As a consequence, the eigenvectors of the matrix $\mathbf{A}^T\mathbf{A}$ denoted as $\mathbf{v}_i$ span the row space of the matrix $\mathbf{A}$ and the eigenvectors of the matrix $\mathbf{A}\mathbf{A}^T$ represented with $\mathbf{u}_i$ construct the column space of the matrix $\mathbf{A}$. In fact, $\mathbf{u}_i$ and $\mathbf{v}_i$ are similar to the right and left eigenvectors of the square matrix.

Denote a matrix $\mathbf{U}$, the first r columns of which are the eigenvectors of the symmetric matrix $\mathbf{A}\mathbf{A}^T$, and the last $n-r$ columns are constructed with the left null space vectors. In addition, introduce a matrix $\mathbf{V}$, the first r columns of which are the eigenvectors of the symmetric matrix $\mathbf{A}^T\mathbf{A}$, and the last $m-r$ columns are made of the null space vectors. The singular value decomposition of the matrix $\mathbf{A}$ is then depicted as

$$\mathbf{A} = \mathbf{U}\boldsymbol{\Sigma}\mathbf{V}^T = \sum_{i=1}^{r} \sigma_i \mathbf{A}_i,$$

where $\mathbf{A}_i = \mathbf{u}_i\mathbf{v}_i^T$ are the basis matrices and the first r diagonal entities of the $n \times m$ matrix $\boldsymbol{\Sigma}$ are the square root of the non-zero eigenvalues of the matrix $\mathbf{A}^T\mathbf{A}$ or $\mathbf{A}\mathbf{A}^T$, namely, singular values of the matrix $\mathbf{A}$.

Principal component analysis involves a procedure that transforms a number of correlated variables into a smaller number of uncorrelated variables, i.e., principal components. The objective of PCA is to discover and reduce the dimensionality of the data in a way which best explains the variance in the data. Besides, it provides the most efficient way of capturing the dominant components of many processes with only few instead. Different names such as proper orthogonal decomposition (POD) or Karhunen–Loève decomposition (KLD) are used in different fields. Instead of dealing with infinite dimensional problems in the time domain, it is often more practical to use the temporal discretization in finite dimensions. For example, let us have n time snapshots within the time interval $[t_0, t_1]$. For a typical channel k, denote

$$\bar{a}_k = \frac{1}{n}\sum_{j=1}^{n} a_{jk},$$

the variance of this channel can be expressed as

$$\frac{1}{n}\sum_{j=1}^{n}(a_{jk} - \bar{a}_k)^2.$$

The procedure to maximize the variance in $\mathcal{R}^m$ leads to an optimization problem:
Find $\mathbf{\Phi} \in \mathcal{R}^m$ such that

$$\mathbf{\Phi}^T(\mathbf{a}_k - \bar{a}_k)(\mathbf{a}_k - \bar{a}_k)^T\mathbf{\Phi}$$

is maximized subject to the constraint $\mathbf{\Phi}^T\mathbf{\Phi} = \mathbf{I}$, where $\mathbf{I}$ is an $m \times m$ dimensional identity matrix.

It is then clear that the principal component $\mathbf{\Phi}$ is in fact the eigenvector of the covariance matrix

$$C_{ij} = \frac{1}{n}\sum_{k=1}^{n}(a_{ki} - \bar{a}_k)(a_{kj} - \bar{a}_k), \quad \text{or} \quad \mathbf{C} = \frac{1}{n}\sum_{k=1}^{n}(\mathbf{a}_k - \bar{a}_k)(\mathbf{a}_k - \bar{a}_k)^T$$

and the number of principal components r depends on the rank of the covariance matrix $\mathbf{C}$.

The eigenvalues of the covariance matrix, which are also called principal values, represent the variances of each principal component. In fact, these eigenvalues are the square of the singular values of the shifted matrix $\mathbf{A}$ with respect to the average $\bar{a}_k$, with $1 \le k \le n$. Furthermore the singular values are sorted in descending order. In this way, the directions in which the data set has the most significant amounts of energy can be found. Assume the first l principal components are selected, a $m \times l$ matrix

$\boldsymbol{\Psi} = [\boldsymbol{\Phi}_1, \boldsymbol{\Phi}_2, \ldots, \boldsymbol{\Phi}_l]$ can be built up. This matrix is called feature vectors. Using the feature vectors, the projected data set can be obtained by $\mathbf{x}\boldsymbol{\Psi}$.

PCA involves a procedure that transforms a number of correlated variables into a much smaller set of variables and respective principal components or independent base vectors. The objective of PCA is to discover and to reduce the dimensionality of the data in the most effective way with respect to the variance of the data. Besides, it provides the most efficient way of capturing the dominant components of many processes with only few instead. Another method closely related to PCA and SVD is called proper orthogonal decomposition (POD) or Karhunen–Loeve decomposition (KLD), named after Karl Onni Uolcvi Karbunen (1915–1992) and Michel Loeve (1907–1979). The mathematical proof of the equivalence of these methods is presented in Liang *et al.* (2002).

We consider two acoustoelastic FSI models as shown in Fig. 6.32. In FSI Model I, a square deformable solid is immersed in an open square cavity filled with an acoustic fluid. The immersed solid is connected to the bottom with two springs to compensate the effects of buoyancy and introduce some kind of resistance to rotational motion. In this problem, there exist three distinctive sets of modes, namely, surface wave sloshing modes, structural modes, and acoustoelastic modes.

Table 6.4 lists the first few frequencies of sloshing/structural/acoustic modes based on different formulations and different mesh density. Convergence issues for mixed formulations have be addressed in Bao *et al.* (2001). In general, mixed formulations do not yield monotonic convergence. However, it is clear in Table 6.4 that there exists a clear separation

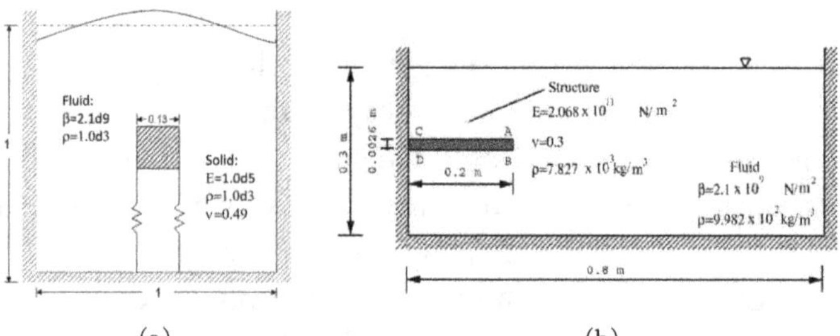

Fig. 6.32. Acoustoelastic FSI (a) Model I and (b) Model II.

Table 6.4. Frequencies of sloshing/structural/acoustic modes of FSI Model I.

Formulation Fluid Structure	Eigenmode	Number of Elements	Number of Zero: Theory	Number of Zero: Result	Frequencies (rad/s): First	Second	Third	Fourth
	Sloshing				1.6742	5.6735	7.7958	10.027
$\mathbf{u}-p-\mathbf{\Lambda}\,\mathbf{u}$	Structure	9	31	31	127.72	156.75	276.62	281.72
	Acoustic				10228	12939	14644	14886
	Sloshing				1.7952	5.5529	7.9521	10.099
$\mathbf{u}-p-\mathbf{\Lambda}\,\mathbf{u}$	Structure	36	159	159	115.43	130.39	140.16	156.83
	Acoustic				10327	11930	12695	13523
	Sloshing				1.7843	5.5405	7.8638	9.6822
$\mathbf{u}-p-\mathbf{\Lambda}\,\mathbf{u}$	Structure	144	703	703	86.456	99.892	99.969	121.25
	Acoustic				10160	10164	10174	10267
	Sloshing				1.8029	5.5529	7.9523	10.099
$\mathbf{u}-p-\mathbf{\Lambda}\,\mathbf{u}/p$	Structure	36	159	159	64.306	73.208	85.608	94.529
	Acoustic				10430	12120	12725	13942
Sloshing					1.7925	5.5405	7.8640	9.6822
$\mathbf{u}-p-\mathbf{\Lambda}\;\mathbf{u}/p$	Structure	144	703	703	107.11	114.25	122.10	130.74
	Acoustic				10161	10296	10304	10670

of sloshing, structure, and acoustic modes and the precise number of zero modes from our implemented program matches exactly with the theoretical prediction based on the same mesh. Furthermore, the first four sloshing, structure, and acoustic modes predicted by both displacement and displacement/pressure mixed formulations with refined meshes match very well with each other. It is also important to note that displacement/pressure formulation is not as accurate as displacement formulation for compressible solid. Displacement/pressure formulation is only needed when the solid is almost incompressible.

To ensure there exists no non-zero spurious (not physical) frequency, we compare the number of zero frequencies with the mathematical prediction (Wang and Bathe, 1997a). Assume that we have n displacement unknowns, m pressure and vorticity moment unknowns, and k free surface nodes, then the total number of zero frequencies is $n-m-(k-1)$. Here, we assume that the physical constant pressure mode arisen from the boundary condition $\mathbf{v}\cdot\mathbf{n}=0$ on S has been eliminated. From the table, we can see the numerical results are identical to the predictions. It is also important to point out that the $\mathbf{u}-p-\mathbf{\Lambda}$ formulation for acoustic fluid works equally well with both the displacement formulation or the $\mathbf{u}/p$ formulation for immersed solid.

Figure 6.33 describes the first four structural modes in addition to the first two sloshing and acoustic modes. Note here for mode shapes, the absolute scale is not relevant since there always exists a scaling constant. The

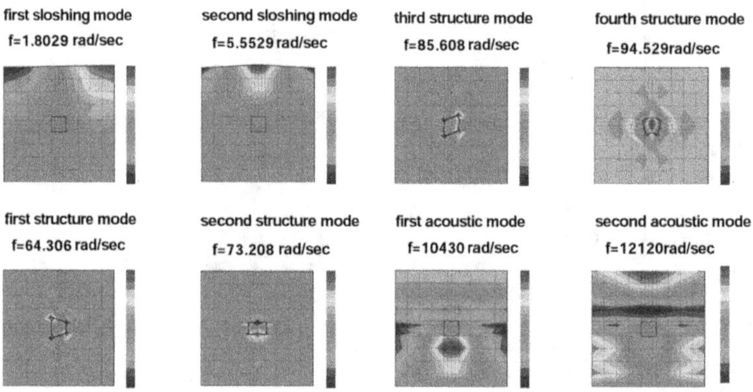

Fig. 6.33. Displacements of sloshing/structural/acoustic modes of FSI Model I.

upward direction is defined as positive. The top two sub-figures depict the sloshing modes, which describe the fluid motion on the free surface. The surface displacements are involved and the free surface is no longer flat. Since the gravitational phenomena depend very much on the gravity, the corresponding coupled frequencies would be within low frequency ranges. The next four sub-figures are structural modes, which involve significant motions of the immersed solid. Naturally, such modes will have higher ranges of frequencies than sloshing modes. Based on the configurations of the immersed solid, it can be recognized that the first and fourth structural modes are flexural or bending modes; the second one is a stretching mode and the third is a shearing mode. The last two sub-figures are acoustic modes, which focus on the coupling of acoustic modes between both immersed solid and the surrounding fluid. Of course, the key characteristics for the acoustic mode is reflected in the constant pressure condition on the free surface and much higher ranges of coupled frequencies.

As for the stability conditions, the ellipticity condition is automatically satisfied, since $(\mathbf{K}^*_{uu})_h$ is always a positive-definite matrix. Therefore, the problem becomes to verify that $9-4c-4c$ elements with corresponding mixed formulation satisfies the Inf-Sup condition. Figure 6.34 shows two different numberical Inf-Sup test values vs. the number of elements N in log−log coordinates. In these numerical Inf-Sup tests, $9-4c-4c$ elements are used along with $\mathbf{u}-p-\mathbf{\Lambda}$ formulation. Moreover, the lowest non-zero singular value of matrix $[\mathbf{K}_{up}\ \mathbf{K}_{ul}]$ is employed in which N represents the square root of the number of the elements.

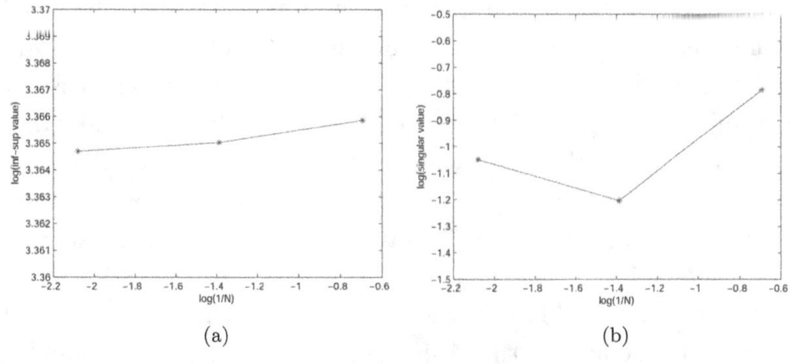

Fig. 6.34. Numerical Inf-Sup tests with (a) eigenvalues and (b) singular values.

It is clear that the Inf-Sup value approaches some value without decaying to zero, which indicates that the $9-4c-4c$ elements pass the test. If an appropriate basis for displacement and pressure unknowns is selected, the $\mathbf{K}_{up}$ matrix can be composed by a diagonal matrix followed by its kernel. Therefore, the minimum non-zero singular value is attempted to generate consistent information with respect to numerical Inf-Sup tests. As shown in Fig. 6.34, the lowest non-zero singular value does not decay as the number of elements employed increases. Since we are using the SVD in our model reduction methods, it is straightforward to employ the singular value of the coupling matrix to replace the standard numerical Inf-Sup test. With the given mesh densities which are self-accommodating, Figure 6.34 demonstrates a possibility of the plateau in the standard numerical Inf-Sup test as described in Bathe (1996); whereas utilizing the lowest singular value of the coupling matrix $[\mathbf{K}_{up}\ \mathbf{K}_{ul}]$, as shown in Fig. 6.34, our proposed new numerical Inf-Sup test yield the same if not more visible conclusion.

Also shown in Fig. 6.32, FSI Model II describes the paradigm of another two-dimensional acoustoelastic FSI model with an elastic slender solid, similar to a structure, immersed in acoustic fluid with the left hand side attached with the cavity boundary, a build-in support. Figure 6.35 depicts the pressure distributions of the first two sloshing, structural, and acoustic modes, respectively. According to the convention, the pressure is assumed positive in compression. As shown in Fig. 6.35, for acoustic modes, the free surface essentially is a surface with a constant pressure which is consistent with traditional assumptions for underwater acoustics. Likewise, for sloshing modes, the pressure distributions match directly with the surface elevations in tank sloshing problems. Since the coupled structural natural frequencies are much higher than the coupled sloshing natural frequencies, it is not difficult to conclude that in this case, surface sloshing waves do not significantly influence the structural dynamical behaviors. Similarly, the structural modes hardly trigger the sloshing motion. Finally, for a test mesh, we have the following theoretical prediction of the zero frequencies as well as the ranges of the sloshing, structural, and acoustic modes in Table 6.5.

In general, if we arbitrarily choose a primary unknown vector $\mathbf{u}$, the Rayleigh–Ritz quotient α, named after Baron Rayleigh John William Strutt

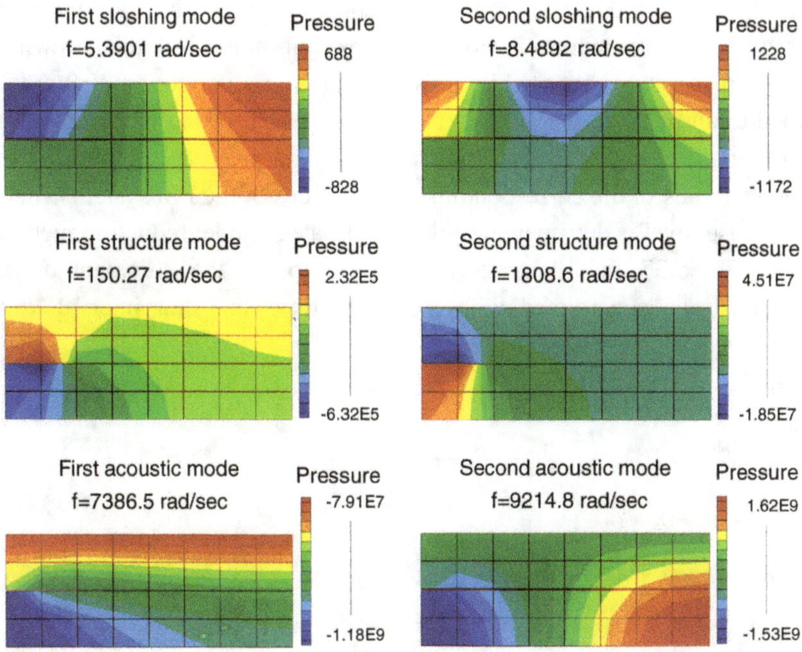

Fig. 6.35. FSI Model II sloshing, structural, and acoustic modes.

Table 6.5. Frequency range of FSI Model II.

Eigen Mode	Number of Elements	Number of Zero: Theory	Number of Zero: Result	Range of Frequency (rad/sec)
Sloshing				5.3889 to 33.468
Structure	40	343	343	149.59 to 1806.7
Acoustic				7384.6 to 103531

(1842–1919) and Walther Heinrich Wilhelm Ritz (1878–1909), as a combination of all natural frequencies ω_i.

$$\alpha^2 = \frac{\mathbf{u}^T\mathbf{K}\mathbf{u}}{\mathbf{u}^T\mathbf{u}}.$$

However, if the vector $\mathbf{u}$ represents a particular eigen mode such as sloshing or structural modes, the Rayleigh–Ritz quotient α will be approaching to the respective natural frequencies as we increase the number of snapshots, which is confirmed by our numerical test shown in Fig. 6.36. Assume that a white noise in Gaussian distribution $N(0;\Delta t)$ is added to all the fluid nodes. Due to the disturbance of the white noise, the first singular vector, on which the transient data projected, could lose the characteristics of the corresponding eigen modes of the FSI system. Therefore, noise level is detrimental to the SVD-based model reduction method.

In Fig. 6.37, we illustrate a procedure to use the dominant principal directions constructed with the small time snapshot interval dt to

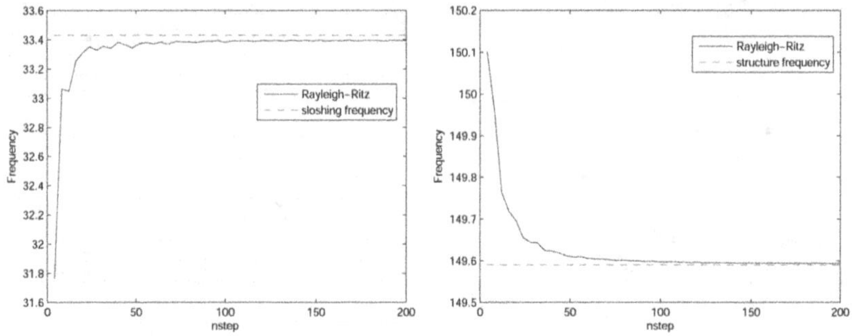

Fig. 6.36. Rayleigh–Ritz quotients for FSI Model II sloshing and structural modes.

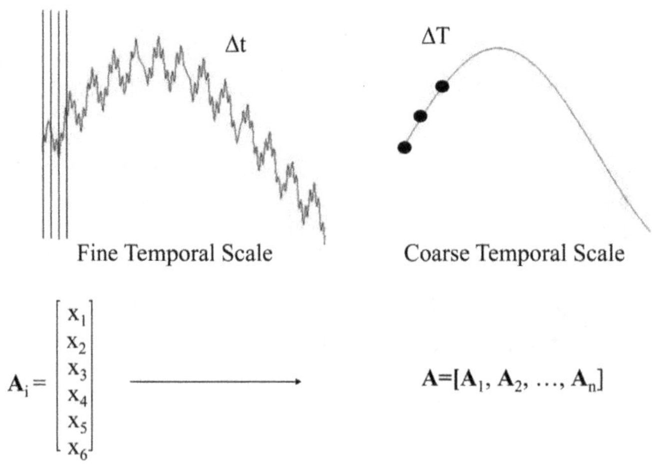

Fig. 6.37. An illustration of coarse graining in time.

extrapolate at a much larger time interval dT. Suppose we start from the initial time and take five snapshots at time step dt and we have m overall degrees of freedom, a $m \times 5$ matrix $\mathbf{A}$ can be assembled. We can recover the transient data $\mathbf{A}$ by a selected number of principal components. Essentially, at every time instance, the coefficients or general coordinates related to these principal components are calculated by solving a normal equation.

Using the recovered transient information, we can then extrapolate the transient system based on a much larger time step dT. At the new time instance $t + dT$, after relaxation within a prescribed time duration which is often smaller than dT. We can then repeat the same process with this type of combination of fine and coarse time step sizes. Figure 6.38 shows the root mean square deviation (RMSD) error between the original and extrapolated results with respect to dT, which represent the energy of original transient displacements and recovered data by the first few principal components. The solid curve depicts the energy at discretized time snapshots while the slash curve describes the energy of recovered data. From the figure, we find that the slopes at the point 40 dt and 60 dt are very small (less than 0.03). Moreover, Fig. 6.39 suggests that as soon as the number of snapshots are sufficiently large, there is no need to further increase the number of snapshots. Therefore, we set $dT = 60$ dt with the number of time snapshots $nstep$ equal to 2000. Similar studies can also be carried out for fine time step size dt. Figure 6.39 shows some results based on different time steps. In this case, considering the memory and numerical flops, dt is set to be 0.0005. Then we go back to the former procedure and relax

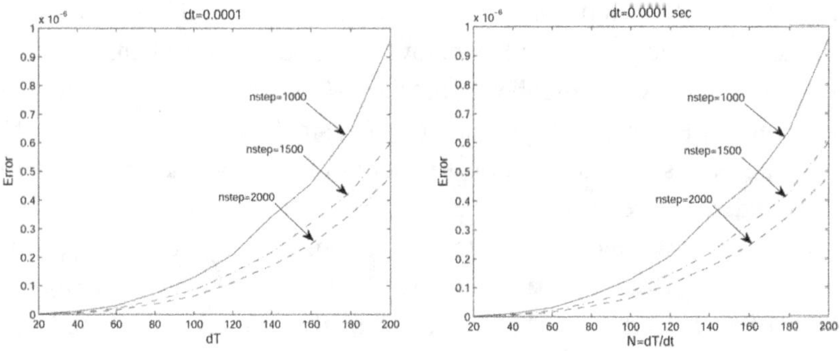

Fig. 6.38. RMSD error vs. dT.

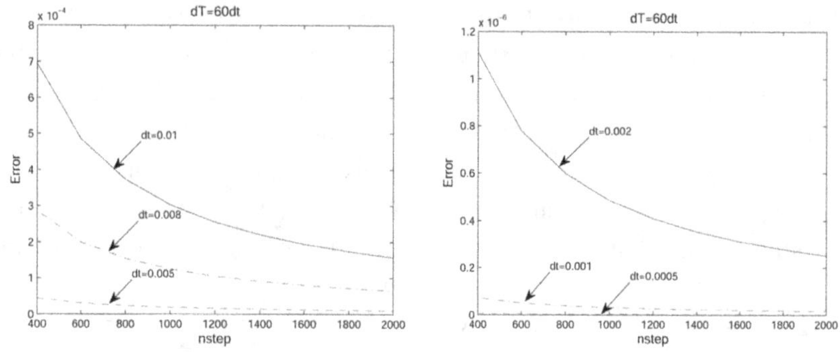

Fig. 6.39. RMSD error vs. nstep.

the extrapolated solution for ever time step dT. Further discussions are also available in the conference publication of earlier results of this work (Yang *et al.*, 2010). In order to verify the effect of PCA, we also employ the first-order Sobolev norm to represent the stored energy in the FSI system denoted with the total volume V

$$(\| \mathbf{v} \|_0)^2 = \sum_i \| v_i \|^2_{L^2(V)} + \sum_{i,j} \left\| \frac{\partial v_i}{\partial x_j} \right\|^2_{L^2(V)}.$$

It is shown in Fig. 6.40 that the sample variance as defined by Eq. (6.56) begins to converge when the number of samples N approaches 50. Figure 6.40 also shows the confidence intervals of displacement and velocity for increasing number of batches M, respectively. It can be seen from the figure that the length of the interval decreases as the number of batches increases. In other words, 90% of the times that upper and lower thresholds are calculated, the true mean lies below the upper threshold and above the lower threshold. As the number of batches M increases, the range between upper and lower thresholds gets smaller and smaller.

Figure 6.41 shows the combined energy of original transient displacements and recovered data by the first few basis matrices. In this case, the white noise is handled with Ito integration. The solid curve depicts the energy at discretized time snapshots while the dashed curve describes the energy of recovered data. This test confirms that even with a limited number of time snapshots the recovered data is fairly close to the original data for the aforementioned acoustoelastic FSI systems.

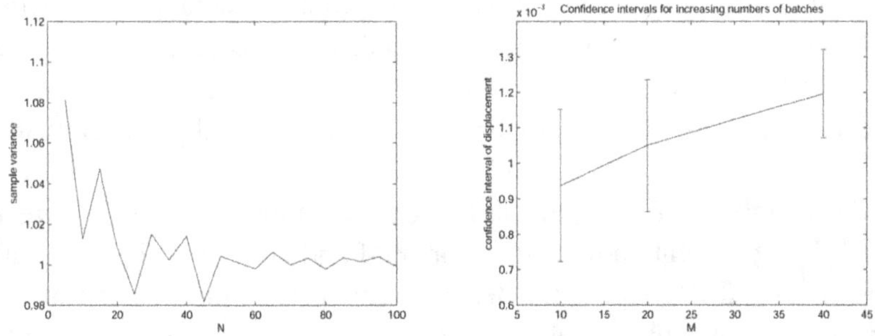

(a) (b)

Fig. 6.40. Sample variance vs. sample number and confidence intervals. (a) Sample number N and (b) Batch number M.

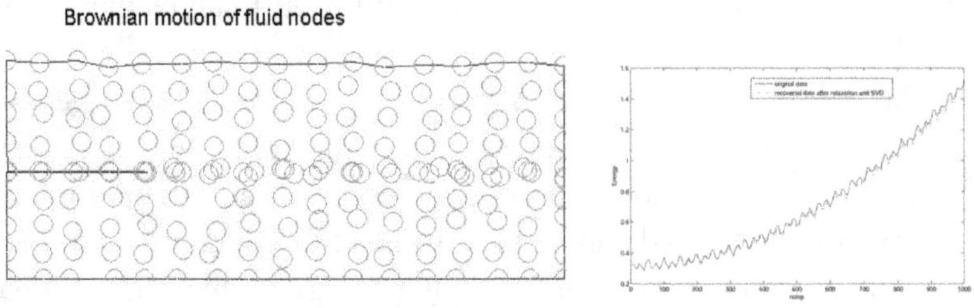

(a) (b)

Fig. 6.41. Transient mode energy and recovered data by SVD and relaxation. (a) Random motion and (b) Recovered energy.

Traditional mode superposition methods can only be applied to linear systems. Furthermore, the entire system behaviors have to be well established. Therefore, it is an effective tool to handle different external loads with different frequencies, for example, the spectrum analysis in civil engineering disciplines. However, SVD based model can be used without the prior knowledge of systems, the selection of dominant response is based on the magnitude of singular values which is a direct presentation of the proportionalities of energies within different response modes. Finally, we must point out the Krylov space constructed with the snapshots of the transient solutions can only be valid for the time vicinity at which the snapshots are recorded. The relationship between the linear Krylov space and the actual nonlinear transient behavior is very much the same as the tangent plane at a particular point of a curved surface. Therefore, for nonlinear problems, such Krylov space must be frequently updated depending on the actual nonlinear behaviors.

The acoustoelastic FSI problems are very challenging, in particular when free surface modes, structural modes, and acoustic modes are fully coupled together. In this section, utilizing the monolithic $\mathbf{u}/p$ and $\mathbf{u}-p-\mathbf{\Lambda}$ mixed finite element formulations for fluid and solid domains, we have demonstrated that coupled frequencies separated in different frequency ranges, can be effectively and efficiently captured in the same implementation. The crucial advantage of $\mathbf{u}-p-\mathbf{\Lambda}$ formulation is that it can solve the acoustoelastic FSI problems without introducing the spurious non-zero frequencies and many zero frequencies. In addition, we have confirmed that similar to the traditional numerical Inf-Sup test the singular values of the coupled matrix can also be used to identify mixed finite elements which satisfy the Inf-Sup condition.

In comparison with the mode superposition methods, we have shown through our numerical experiments that SVD-based model reduction methods do not necessarily depend on the explicit construction of mass and stiffness matrices and can be used to identify the dominant principal components with the proportionality of the energy content. In SVD-based model reduction methods, a few snapshots of the transient solution can be derived from linear or nonlinear FSI formulations as well as experiments. It is also important to recognize that for linear systems, the principal directions are often a linear combination of the mode shapes which depends very much on the initial and transient conditions. Again, unlike the mode superposition method, SVD-based model reduction methods are applicable to nonlinear systems in which the principal directions can be altered through time.

6.4.3 *Noise and stochasticity*

In physical world, random perturbations often present themselves in dynamical modeling of fluid–solid interaction systems. A familiar way to quantify such random perturbations has been elegantly characterized as Einstein's Random Walk (Haw, 2005). In fact, the coupled governing equation can adopt the following general form

$$\mathbf{M}\ddot{\mathbf{x}}+\mathbf{K}\mathbf{x}=\mathbf{R}+\mathbf{R}_r,$$

in which the external forces include a perturbation vector $\mathbf{R}_r$, characterized as the white noise $\mathbf{V}_t$, formal derivative of a Wiener process W_t (Ostoja-Starzewsk and Wang, 1999). Let $\mathbf{y} = \dot{\mathbf{x}}$, we can express the governing equation into a typical dynamical system form,

$$\begin{pmatrix} \dot{\mathbf{x}} \\ \dot{\mathbf{y}} \end{pmatrix} = -\begin{pmatrix} \mathbf{0} & -\mathbf{I} \\ \mathbf{M}^{-1}\mathbf{K} & \mathbf{0} \end{pmatrix} \begin{pmatrix} \mathbf{x} \\ \mathbf{y} \end{pmatrix} + \begin{pmatrix} \mathbf{0} \\ \mathbf{M}^{-1} \end{pmatrix} \mathbf{V}_t. \tag{6.53}$$

In this standard dynamical system form, with $\mathbf{X} = (\mathbf{x}^T \quad \mathbf{y}^T)^T$, so-called Langevin equation, named after Paul Langevin (1872–1946), the governing equation has been abbreviated as

$$\frac{d\mathbf{X}}{dt} = -\mathbf{a}(t,\mathbf{X}) + \mathbf{b}(t,\mathbf{X}), \tag{6.54}$$

in which the drift term $\mathbf{a}(t,\mathbf{X})$ could be simplified as $\mathbf{AX}$ with $\mathbf{A}$ a constant matrix for linear dynamical systems, and the diffusion term $\mathbf{b}(t,\mathbf{X})$ is often characterized as $\mathbf{BV}_t$ with a constant matrix $\mathbf{B}$ multiplied by a vector $\mathbf{V}_t$, every entity of which is represented by a white noise random process (Lemons and Gythiel, 1997).

The initial condition is represented as $\mathbf{X}_{t_0} = \mathbf{X}_0$. Furthermore, the noise perturbation part $\mathbf{V}_t dt$ can be expressed as $d\mathbf{W}_t$, which represents the Brownian motion, named after Robert Brown (1773–1858). Assume the constant matrix $\mathbf{A}$ is simple, namely, diagonalizable, with $\mathbf{A} = \mathbf{\Phi\Lambda\Phi}^{-1}$, the generalized coordinate $\boldsymbol{\xi} = \mathbf{\Phi}^{-1}\mathbf{X}$ is introduced. Following the so-called Ornstein–Uhlenbeck process, named after George Eugene Uhlenbeck (1900–1988) and Leonard Salomon Ornstein (1880–1941), we can then employ the Ito process, named after Kiyosi Ito (1915–2008), for all the generalized coordinate ξ_i satisfying the stochastic differential equation (SDE) with so-called additive noise $c_i dW_t$:

$$d\xi_i = -\lambda_i \xi_i dt + c_i dW_t, \tag{6.55}$$

where for the ith mode the drift term is represented by $\lambda_i \xi_i$ and the diffusive term increment $c_i dW_t$ yields from $\mathbf{\Phi}^{-1}\mathbf{BV}_t dt$ along with the definition of the white noise as the formal derivative of a Wiener process W_t (Ostoja-Starzewsk and Wang, 1999; Higham, 2001).

For the ith mode, if λ_i is zero, the general solution ξ_i for that mode will be Martingale. Consequently, we can express the explicit solution as

$$\xi_i = e^{-\lambda_i t}\left(\xi_i(0) + \int_0^t e^{\lambda_i s} c_i dW_s\right) = e^{-\lambda_i t}\xi_i(0) + e^{-\lambda_i t}\int_0^t e^{\lambda_i s} c_i dW_s,$$

or an implicit approximation as

$$\xi_{i,n+1} = \xi_{i,n} - \lambda_i \xi_{i,n+1}\triangle t_n + c_i \triangle W_n,$$

where $\triangle W_n$ represents a normal distribution with zero mean and variance proportional to $\triangle t_n$.

Assume N is the number of realization, the sample variance can be expressed as

$$S_N^2 = \frac{1}{N-1}\sum_{i=1}^{N}(x_i - \bar{x}_N)^2, \tag{6.56}$$

where $\bar{x}_N$ is the sample mean.

Note that for the entire population, the denominator in Eq. 6.56 is N instead of $N-1$. However, in general, the sample or population number N is much larger than 1, therefore, the difference between $N-1$ and N is negligible. Moreover, based on the Ito calculus, the mean of the solutions for SDE (6.55) for ith mode, can be expressed as

$$E(\xi_i(t)) = E(\xi_0)e^{\lambda_i t}. \tag{6.57}$$

For any given initial value, repeat N different simulations of sample path by implicit Euler method. Then use M different initial values, i.e., generating a series of trajectories with distinct starting points, the mean error of the jth batch by the statistic is

$$\hat{\mu}_j = \frac{1}{N}\sum_{k=1}^{N}\xi_i(t)_{k,j} - E(\xi_i(t)) \tag{6.58}$$

for $j = 1, 2, \ldots, M$, and their average

$$\hat{\mu} = \frac{1}{MN}\sum_{j=1}^{M}\sum_{k=1}^{N}\xi_i(t)_{k,j} - E(\xi_i(t)), \tag{6.59}$$

where $\xi_i(t)_{k,j}$ represents the kth numerical simulation at time t of the ith mode with the jth initial value.

The variance of the batch average $\hat{\mu}_j$ can be estimated by

$$\hat{\sigma}_{\hat{\mu}}^2 = \frac{1}{M-1}\sum_{j=1}^{M}(\hat{\mu}_j - \hat{\mu})^2. \tag{6.60}$$

For the student t-distribution with $M-1$ degree of freedom, a 90% confidence interval $(\alpha = 0.1)$ for the mean error μ is

$$(\hat{\mu} - \Delta\hat{\mu}, \hat{\mu} + \Delta\hat{\mu})$$

with

$$\Delta\hat{\mu} = t_{1-\alpha,M-1}\sqrt{\frac{\hat{\sigma}_{\hat{\mu}}^2}{M-1}}.$$

To reiterate, suppose there are m degree of freedoms of the system after condensation and the maximum number of time snapshots is $nstep$, a m-by-$nstep$ incident matrix can be built up. We perform the SVD of this matrix and obtain the first principal singular vector $\mathbf{u}$. If we excite the FSI system with a single mode, we can recover through the transient data by taking the first few principal components. It is anticipated that by varying time step dt, we may obtain different recovered transient results. Moreover, using the recovered transient information, we can then extrapolate the system based on a much larger time step dT. At the new time instance $t + dT$, after relaxation within a prescribed time duration which is often smaller than dT. We can then repeat the same process with this type of combination of fine and coarse time step sizes. Finally, to further verify the effect of PCA, the stored energy in the FSI system with the first Sobolev norm can also be employed.

Bibliography

Alipour, F., Berry, D., and Titze, I. (2000). A finite-element model of vocal-fold vibration, *Journal of the Acoustical Society of America* 14, pp. 442–454.

Alipour, F. and Scherer, R. (2000). Vocal fold bulging effects on phonation using a biophysical computer model, *Journal of Voice* 14(4), pp. 470–483.

Alipour, F. and Titze, I. (1991). Elastic models of vocal fold tissues, *The Journal of the Acoustical Society of America* 90(3), pp. 1326–1331.

Amsallem, D., Cortial, J., Carlberg, K., and Farhat, C. (2009). A method for interpolating on manifolds structural dynamics reduced-order models, *International Journal for Numerical Methods in Engineering* 80(9), pp. 1241–1258.

Antoulas, A. (2005). *Approximation of Large-Scale Dynamical Systems (Advances in Design and Control)*, SIAM.

Antoulas, A. and Sorsensen, D. (2001). Approximation of large-scale dynamical systems: An overview, *International Journal of Applied Mathematics and Computer Science* 11(5), pp. 1093–1121.

Axelsson, O. (1996). *Iterative Solution Methods*, Cambridge University Press.

Babuška, I. (1973). The finite element method with Lagrangian multipliers, *Numerische Mathematik* 20, pp. 179–192.

Babuška, I. and Suri, M. (1994). The *p* and *h-p* versions of the finite element method, basic principles and properties, *SIAM Review* 36, pp. 578–632.

Bao, W., Wang, S., and Bathe, K. (2001). On the Inf-Sup condition of mixed finite element formulations for acoustic fluids, *Mathematical Models & Methods in Applied Sciences* 11, pp. 883–901.

Bathe, K. (1996). *Finite Element Procedures*, Prentice Hall, Englewood Cliffs, New Jersey.

Bathe, K. (1998). Fluid-structure interactions, *Journal of Mechanical Engineering* 120, pp. 66–68.

Bathe, K. and Dvorkin, E. (1986). A formulation of general shell elements-The use of mixed interpolation of tensorial components, *International Journal for Numerical Methods in Engineering* 22, pp. 697–722.

Bathe, K. and Hahn, W. (1979). On transient analysis of fluid-structure systems, *Computers & Structures* 10, pp. 383–391.

Bathe, K., Nitikitpaiboon, C., and Wang, S. (1995a). A mixed displacement-based finite element formulation for acoustic fluid-structure interaction, *Computers & Structures* 56(2/3), pp. 225–237.

Bathe, K. and Sonnad, V. (1980). On effective implicit time integration in analysis of fluid-structure problems, *International Journal for Numerical Methods in Engineering* 15, pp. 943–948.

Bathe, K. and Zhang, H. (2004). Finite element developments for general fluid flows with structural interactions, *International Journal for Numerical Methods in Engineering* 60, pp. 213–232.

Bathe, K., Zhang, H., and Wang, M. (1995b). Finite element analysis of incompressible and compressible fluid flows with free surfaces and structural interactions, *Computers & Structures* 56(2/3), pp. 193–213.

Belytschko, T. (1980). Fluid-structure interaction, *Computers & Structures* 12, pp. 459–469.

Belytschko, T. and Kennedy, J. (1976). A fluid-structure finite element method for the analysis of reactor safety problems, *Nuclear Engineering and Design* 38, pp. 71–81.

Belytschko, T., Liu, W., and Moran, B. (2000). *Nonlinear Finite Elements for Continua and Structures*, John Wiley & Sons.

Benek, J., Buning, P., and Steger, J. (1985). A 3-d chimera grid embedding technique, *7th Computational Physics Conference*, p. 1523.

Benek, J., Steger, J., and Dougherty, F. (1983). A flexible grid embedding technique with application to the euler equations, *6th Computational Fluid Dynamics Conference Danvers*, p. 1944.

Berg, J. V. D., Zatema, J., and Doomenbal, P. (1957). On the air resistance and Bernoulli effect in the human larynx, *Journal of the American Statistical Association* 29, pp. 626–631.

Bermúdez, A. and Rodríguez, R. (1994). Finite element computation of the vibration modes of a fluid-solid system, *Computer Methods in Applied Mechanics and Engineering* 119, pp. 355–370.

Berrar, D., Dubitzaky, W., and Granzow, M. (2004). *A Practical Approach to Microarray Data Analysis*, Kluwer Academic Publishers.

Beyer, R. (1992). A computational model of the cochlea using the immersed boundary method, *Journal of Computational Physics* 98, pp. 145–162.

Beyer, R. and LeVeque, R. (1992). Analysis of a one-dimensional model for the immersed boundary method, *SIAM Journal of Numerical Analysis* 29, pp. 332–364.

Bisplinghoff, R. and Ashley, H. (1962). *Principles of Aeroelasticity*, Dover.

Bisplinghoff, R., Ashley, H., and Halfman, R. L. (1983). *Aeroelasticity*, Dover.

Blevins, R. (1994). *Flow-Induced Vibration*, 2nd (edn.), Krieger.

Boffi, D. and Gastaldi, L. (2003). A finite element approach for the immersed oboundary method, *Computers & Structures* 81, pp. 491–501.

Boffi, D., Gastaldi, L., and Heltai, L. (2007). On the CFL condition for the finite element immersed boundary method, *Computers & Structures* 85, pp. 775–783.

Boffi, D., Gastaldi, L., Heltai, L., and Peskin, C. (2008). On the hyperelastic formulation of the immersed boundary method, *Computer Methods in Applied Mechanics and Engineering* 197, pp. 2210–2231.

Brezzi, F. (1974). On the existence, uniqueness, and approximation of saddle point problems arising from Lagrangian multipliers, *Revue d'Automatique, d'Informatique et de Recherche Opérationnelle, Analyse Numérique* 2, pp. 129–151.

Brezzi, F., Bathe, K., and Fortin, M. (1989). Mixed-interpolated elements for Reissner-Mindlin plates, *International Journal for Numerical Methods in Engineering* 28, pp. 1787–1801.

Brezzi, F. and Fortin, M. (1991). *Mixed and Hybrid Finite Element Methods*, Springer-Verlag.

Briscoe, B. and Panesar, S. (1991). The effect of surface topography on the adhesion of poly(urethane)-metal contacts, *International Conference on Frontiers of Tribology*, pp. A20–A27.

Briscolini, M. and Santangelo, P. (1989). Development of the mask method for incompressible unsteady flows, *Journal of Computational Physics* 84, pp. 57–75.

Campbell, J. (1961). The in-plane elastic constants of paper, *Australian Journal of Applied Science* 12, pp. 356–357.

Campo, E. (2012). *The Effect of Vocal Fold Geometry on the Fluid Structure Acoustic Interactions in an Experimental Model of the Human Airway*, Master's thesis, The Pennsylvania State University.

Cao, J., Donell, B., Deaver, D., Lawrence, M., and Dong, C. (1998). In Vitro side-view imaging technique and analysis of human T-leukemic cell adhesion to ICAM-1 in shear flow, *Microvascular Research* 55, 124–137.

Chapelle, D. and Bathe, K. (1993). The inf-sup test, *Computers & Structures* 47(4/5), pp. 537–545.

Chen, H. and Taylor, R. (1990). Vibration analysis of fluid-solid systems using a finite element displacement formulation, *International Journal for Numerical Methods in Engineering* 29, pp. 683–698.

Chessa, J. and Belytschko, T. (2003). The extended finite element method for two-phase fluids, *Journal of Applied Mechanics* 70, pp. 10–17.

Chhetri, D., Zhang, Z., and Neubauer, J. (2011). Measurement of young's modulus of vocal folds by indentation, *Journal of Voice* 25, 1, pp. 1–7.

Chien, S. (1976). Electrochemical interactions between erythrocyte surfaces, *Thrombosis Research* 8, pp. 189–202.

Chorin, A. (1967). A numerical method for solving incompressible viscous flow problems, *Journal of Computational Physics* 2, pp. 12–26.

Chorin, A. and Marsden, J. (1998). *A Mathematical Introduction to Fluid Mechanics*, Springer-Verlag.

Clift, R., Grace, J., and Weber, M. (1978). *Bubbles, Drops, and Particles*, Academic Press.

Cortez, R. and Minion, M. (2000). The blob projection method for immersed boundary problems, *Journal of Computational Physics* 161, pp. 428–453.

Cranen, B. (1985). *The Acoustic Impedance of the Glottis*, Ph.D. thesis, University of Nijmegen.

Cremer, L. and Heckl, M. (1973). *Structure-Borne Sound*, Springer-Verlag, translated and Revised by E.E. Ungar.

Daniel, W. (1980). Performance of reduction methods for fluid-structure and acoustic eigenvalue problems, *International Journal for Numerical Methods in Engineering* 15, pp. 1585–1594.

Deverge, M., Pelorson, X., Vilain, C., Lagree, P., Chentouf, F., Willems, J., and Hirschberg, A. (2003). Influence of collision on the flow through in-vitro rigid models of the vocal folds, *Journal of the Acoustical Society of America* 114(6), pp. 3354–3362.

Diamond, S. (2001). Reaction complexity of flowing human blood, *Biophysical Journal* 80, pp. 1031–1032.

Dillon, R., Fauci, L., Fogelson, A., and Gaver III, D. (1996). Modeling biofilm processes using the immersed boundary method, *Journal of Computational Physics* 129, pp. 57–73.

Dowell, E., Crawley, E., Curtiss, H. C. Jr., Perters, D. A., Scanlan, R. H., and Sisto, F. (1995). *A Modern Course in Aeroelasticity*, Kluwer Academic.

Dowell, E. and Ilgamov, M. (1988). *Studies in Nonlinear Aeroelasticity*, Springer-Verlag.

Dugundji, J., Dowell, E., and Perkin, B. (1963). Subsonic flutter of panels on continuous elastic foundations, *AIAA Journal* 1, pp. 1146–1154.

Dvorkin, E. and Bathe, K. (1984). A continuum mechanics based four-node shell element for general nonlinear analysis, *Engineering Computation* 1, pp. 77–88.

Dzwinel, W., Boryczko, K., and Yuen, D. A. (2003). A discrete-particle model of blood dynamics in capillary vessels, *Journal of Colloid and Interface Science* 258, pp. 163–173.

Ern, A. and Guermond, J. (2003). *Theory and Practice of Finite Elements*, Vol. 159, Springer-Verlag, New York.

Evans, E. and Fung, Y. (1972). Improved measurements of the erythrocyte geometry, *Microvascular Research* 4, p. 335.

Everstine, G. (1981). A symmetric potential formulation for fluid-structure interaction, *Journal of Sound and Vibration* 79, pp. 157–160.

Fadlun, E., Verzicco, R., Orlandi, P., and Mohd-Yusof, J. (2000). Combined immersed-boundary finite-difference methods for three-dimensional complex flow simulations, *Journal of Computational Physics* 161, pp. 35–60.

Farhat, C., Geuzaine, P., and Grandmont, C. (2001). The discrete geometric conservation law and the nonlinear stability of ALE schemes for the solution of flow problems on moving grids, *Journal of Computational Physics* 174, pp. 669–694.

Fauci, L. and Peskin, C. (1988). A computational model of aquatic animal locomotion, *Journal of Computational Physics* 77, pp. 85–108.

Felippa, C. and Ohayon, R. (1990). Mixed variational formulation of finite element analysis of acoustoelastic/slosh fluid-structure interaction, *Journal of Fluids and Structures* 4, pp. 35–57.

Fogelson, A. and Peskin, C. (1988). A fast numerical method for solving the three-dimensional stokes equations in the presence of suspended particles, *Journal of Computational Physics* 79, pp. 50–69.

Fogelson, A., Wang, S., and Liu, W. (2008). Special issue: Immersed boundary method and its extensions — Preface, *Computer Methods in Applied Mechanics and Engineering* 197, pp. 25–28.

Freitas, C. and Runnels, S. (1999). Simulation of fluid-structure interaction using patched-overset grids, *Journal of Fluids and Structures* 13, pp. 191–207.

Fung, Y. (1955). *An Introduction to the Theory of Aeroelasticity*, John Wiley & Sons.

Fung, Y. (1996). *Biomechanics: Circulation*, Springer-Verlag.

Fung, Y. (2000). *Biomechanics: Motion, Flow, Stress, and Growth*, Springer-Verlag.

Gay, M., Zhang, L., and Liu, W. (2006). Stent modeling using immersed finite element method, *Computer Methods in Applied Mechanics and Engineering* 195, pp. 4358–4370.

Glimm, J., Grove, J., Li, X., and Tan, D. (2000). Robust computational algorithms for dynamic interface tracking in three dimensions, *SIAM Journal on Scientific Computing* 21, pp. 2240–2256.

Glowinski, R., Pan, T., Hesla, T. I., and Joseph, D. (1999). A distributed Lagrange multiplier/fictitious domain method for particulate flows, *International Journal of Multiphase Flow* 25, pp. 755–794.

Glowinski, R., Pan, T., Hesla, T. I., Joseph, D., and Périaux, J. (2001). A fictitious domain approach to the direct numerical simulation of incompressible viscous flow past moving rigid bodies: Application to particulate flow, *Journal of Computational Physics* 169, pp. 363–426.

Goldstein, D., Handler, R., and Sirovich, L. (1993). Modeling a no-slip flow boundary with an external force field, *Journal of Computational Physics* 105, pp. 354–366.

Goriely, A. and Tabor, M. (1997a). Nonlinear dynamics of filaments i: Dynamical instabilities, *Physica D* 105, pp. 20–44.

Goriely, A. and Tabor, M. (1997b). Nonlinear dynamics of filaments i: Nonlinear analysis, *Physica D* 105, pp. 45–61.

Griffith, B., Hornung, R., McQueen, D., and Peskin, C. (2007). An adaptive, formally second order accurate version of the immersed boundary method, *Journal of Computational Physics* 223(1), pp. 10–49.

Griffith, B. and Luo, X. (2017). Hybrid finite difference/finite element immersed boundary, *International Journal for Numerical Methods in Biomedical Engineering* 33(11), pp. 1–31.

Griffith, B. and Patankar, N. (2020). Immersed methods for fluid-structure interaction, *Annual Review of Fluid Mechanics* 52, pp. 421–448.

Griffith, B. and Peskin, C. (2005). On the order of accuracy of the immersed boundary method: Higher order convergence rates for sufficiently smooth problems, *Journal of Computational Physics* 208(1), pp. 75–105.

Gunzburger, M. (2003). *Perspectives in Flow Control and Optimization*, SIAM.

Guo, X., Takai, E., Liu, K., and Wang, S. (2001). An exploration of cell stress and deformation under shear flow, *ASME Winter Bioengineering Conference*, BED-51.

Häggblad, B. and Bathe, K. (1990). Specifications of boundary conditions for Reissner/Mindlin plate bending finite elements, *International Journal for Numerical Methods in Engineering* 30, pp. 981–1011.

Hamdi, M., Ousset, Y., and Verchery, G. (1978). A displacement method for the analysis of vibrations of coupled fluid-structure systems, *International Journal for Numerical Methods in Engineering* 13, pp. 139–150.

Haszpra, O. (1979). *Modelling Hydroelastic Vibrations*, Pitman.

Haw, M. (2005). Einstein's random walk, *Physics World* 18(1), pp. 19–22.

Heinrich, J. and Connolly, D. (1992). Three-dimensional finite element analysis of self-acting foil bearings, *Computer Methods in Applied Mechanics and Engineering* 100, pp. 31–43.

Hencky, H. (1947). Über die berücksichtigung der schubverzerrung in ebenen platten, *Ingenieur Archiv* 16, pp. 72–76.

Higham, D. (2001). An algorithmic introduction to numerical simulation of stochastic differential equations, *SIAM Review* 43, pp. 525–546.

Hildebrand, F. (1976). *Advanced Calculus for Applications*, Prentice Hall.

Howe, M. (1998). *Acoustics of Fluid-Structure Interactions*, Cambridge University Press.

Hu, H., Patankar, N., and Zhu, M. Y. (2001). Direct numerical simulations of fluid-solid systems using the arbitrary Lagrangian-Eulerian technique, *Journal of Computational Physics* 169, pp. 427–462.

Huang, Y., Doerschuk, C. M., and Kamm, R. D. (2001). Computational modeling of RBC and neutrophil transit through the pulmonary capillaries, *Journal of Applied Physiology* 90, pp. 545–564.

Huerta, A. and Liu, W. (1988). Viscous flow with large free surface motion, *Computer Methods in Applied Mechanics and Engineering* 69, pp. 277–324.

Hughes, T., Franca, L., and Balestra, M. (1986). A new finite element formulation for computational fluid dynamics: V. Circumventing the Babuška-Brezzi condition: A stable Petrov-Galerkin formulation of the Stokes problem accommodating equal-order interpolations, *Computer Methods in Applied Mechanics and Engineering* 59, pp. 85–99.

Hughes, T., Liu, W., and Zimmerman, T. (1981). Lagrangian-Eulerian finite element formulations for incompressible viscous flows, *Computer Methods in Applied Mechanics and Engineering* 29, pp. 329–349.

Hughes, T. and Tezduyar, T. (1981). Finite elements based upon Mindlin plate theory with particular reference to the four-node bilinear isoparametric element, *Journal of Applied Mechanics* 48, pp. 587–596.

Iosilevich, A., Bathe, K., and Brezzi, F. (1997). On evaluating the Inf-Sup condition for plate bending elements, *International Journal for Numerical Methods in Engineering* 40, pp. 3639–3663.

Jiang, J. J. and Titze, I. R. (1994). Measurement of vocal fold intraglottal pressure and impact stress, *Journal of Voice* 8(2), pp. 132–144.

Johnson, A. and Tezduyar, T. (1999). Advanced mesh generation and update methods for 3d flow simulations, *Computational Mechanics* 23, pp. 130–143.

Jolliffe, I. (2002). *Principal Component Analysis*, Springer.

Junger, M. and Feit, D. (1972). *Sound, Structures, and their Interaction*, MIT Press.

Kamakoti, R. and Shyy, W. (2004). Fluid-structure interaction for aeroelastic applications, *Progress in Aerospace Sciences* 40, pp. 535–558.

Kay, D., Loghin, D., and Wathen, A. (2002). A preconditioner for the steady-state Navier-Stokes equations, *SIAM Journal of Scientific Computing* 24(1), pp. 237–256.

Kiefling, L. and Feng, G. (1976). Fluid-structure finite element vibrational analysis, *AIAA Journal* 14, pp. 199–203.

Kim, D. and Liu, W. (2006). Maximum principle and convergence analysis for the meshfree point collocation method, *SIAM Journal on Numerical Analysis* 44(2), pp. 515–539.

Kim, Y. and Peskin, C. (2007). Penalty immersed boundary method for an elastic boundary with mass, *Physics of Fluids* 19, p. 053103.

Kim, Y., Zhu, L., Wang, S., and Peskin, C. (2003). On various techniques for computer simulation of boundaries with mass, *Second MIT Conference on Computational Fluid and Solid Mechanics* 193, pp. 1746–1750.

Knoll, D. and Keyes, D. (2004). Jacobian-free Newton-Krylov methods: A survey of approaches and applications, *Journal of Computational Physics* 193, pp. 357–397.

Krane, M., Barry, M., and Wei, T. (2007). Unsteady behavior of flow in a scaled-up vocal folds model, *Journal of the Acoustical Society of America* 122(6), pp. 3659–3670.

Krane, M., Wei, T., and Barry, M. (2004). Flow measurements in a scaled-up vocal folds model, *International Conference on Voice Physiology and Biomechanics*.

Kreiss, B. (1983). Construction of a curvilinear grid, *SIAM Journal on Scientific and Statistical Computing* 4, pp. 270–279.

Lacey, C. and Talke, F. (1990). A tightly coupled numerical foil bearing solution, *IEEE Transactions on Magnetics* 26, pp. 3039–3043.

Lai, M. and Peskin, C. (2000). An immersed boundary method with formal second-order accuracy and reduced numerical viscosity, *Journal of Computational Physics* 160, pp. 705–719.

Lemons, D. and Gythiel, A. (1997). Paul mlangevin 1908 paper on the theory of brownian motion, *American Journal of Physics* 65, p. 1079.

LeVeque, R. and Li, Z. (1997). Immersed interface methods for Stokes flow with elastic boundaries or surface tension, *SIAM Journal on Scientific Computing* 18, pp. 709–735.

Li, S. and Liu, W. (1996). Moving least square reproducing kernel method part II: Fourier analysis, *Computer Methods in Applied Mechanics and Engineering* 139, pp. 159–194.

Li, S. and Liu, W. (1999). Reproducing kernel hierarchical partition of unity part I: Formulation and theory, *International Journal for Numerical Methods in Engineering* 45, pp. 251–288.

Li, X., Jin, B., and Glimm, J. (1996). Numerical study for the three-dimensional Rayleigh-Taylor instability through the TVD/AC scheme and parallel computation, *Journal of Computational Physics* 126, pp. 343–355.

Li, Z. and LeVeque, R. (1994). The immersed interface methods for elliptic equations with discontinuous coefficients and singular sources, *SIAM Journal on Scientific Computing* 31, pp. 1019–1044.

Lian, Y. and Shyy, W. (2005). Numerical simulations of membrane wing aerodynamics for micro air vehicle applications, *Journal of Aircraft* 42, pp. 865–873.

Liang, Y., Lin, W., Lee, H. P., Lim, S. P., and Lee, K. H. (2002). Proper orthognonal decomposition and its applications - part ii: model reduction for mems dynamical analysis, *Journal of Sound and Vibration* 256, pp. 512–532.

Lim, S., Ferent, A., Wang, S., and Peskin, C. (2008). Dynamics of a closed rod with twist and bend in fluid, *SIAM Journal on Scientific Computing* 31, pp. 273–302.

Liu, W. (1981). Finite element procedures for fluid-structure interactions and application to liquid storage tanks, *Nuclear Engineering and Design* 65, pp. 221–238.

Liu, W., Belytschko, T., and Chang, H. (1986). An arbitrary Lagrangian-Eulerian finite element method for path-dependent materials, *Computer Methods in Applied Mechanics and Engineering* 58, pp. 227–245.

Liu, W., Chang, H., Chen, J., and Belytschko, T. (1988). Arbitrary Lagrangian-Eulerian Petrov-Galerkin finite elements for nonlinear continua, *Computer Methods in Applied Mechanics and Engineering* 68, pp. 259–310.

Liu, W., Chen, Y., Chang, C., and Belytschko, T. (1996a). Advances in multiple scale kernel particle methods, *Computational Mechanics* 18, pp. 73–111.

Liu, W., Chen, Y., Uras, R. A., and C. Chang, C. (1996b). Generalized multiple scale reproducing kernel particle methods, *Computer Methods in Applied Mechanics and Engineering* 139, pp. 91–157.

Liu, W. and Gvildys, J. (1986). Fluid-structure interaction of tanks with an eccentric core barrel, *Computer Methods in Applied Mechanics and Engineering* 58, pp. 51–77.

Liu, W., Jun, S., and Zhang, Y. (1995). Reproducing kernel particle methods, *International Journal for Numerical Methods in Fluids* 20, pp. 1081–1106.

Liu, W., Kim, D., and Tang, S. (2007a). Mathematical foundations of the immersed finite element, *Computational Mechanics* 39, pp. 211–222.

Liu, W., Liu, Y., Farrell, D., Zhang, L., Wang, S., Fukui, Y., Patankar, N., Zhang, Y., Bajaj, C., Chen, X., and Hsu, H. (2006). Immersed finite element method and its applications to biological systems, *Computer Methods in Applied Mechanics and Engineering* 195, pp. 1722–1749.

Liu, W., Liu, Y., Zhang, L., Wang, S., Gerstenberger, A., and Farrell, D. (2004a). *Immersed finite element method and applications to biological systems. Finite Element Methods: 1970's and Beyond.* (International Center for Numerical Methods and Engineering).

Liu, W. and Ma, D. (1982). Computer implementation aspects for fluid-structure interaction problems, *Computer Methods in Applied Mechanics and Engineering* 31, pp. 129–148.

Liu, W. and Chen, Y. J. (1995). Wavelet and multiple scale reproducing kernel methods, *International Journal for Numerical Methods in Fluids* 21, pp. 901–932.

Liu, Y. and Liu, W. (2006a). Rheology of red blood cell aggregation in capillary by computer simulation, *Journal of Computational Physics* 220, pp. 139–154.

Liu, Y., Liu, W., Belytschko, T., Patankar, N., To, A., Kopacz, A., and Chung, J. (2007b). Immersed electrokinetic finite element method, *International Journal for Numerical Methods in Engineering* 71, pp. 379–405.

Liu, Y. and Liu, W. K. (2006b). Rheology of red blood cell aggregation by computer simulation, *Journal of Computational Physics* 220, pp. 139–154.

Liu, Y., Zhang, L., Wang, S., and Liu, W. (2004b). Coupling of Navier-Stokes equations with protein molecular dynamics and its application to hemodynamics, *International Journal for Numerical Methods in Fluids* 46, pp. 1237–1252.

Luo, H., Mittal, R., and Bielamowicz, S. (2009). Analysis of flow-structure interaction in the larynx during phonation using an immersed-boundary method, *Journal of the Acoustical Society of America* 126(2), pp. 816–824.

Luo, H., Mittal, R., Zheng, X., Bielamowicz, S., Walsh, R., and Hahn, J. (2008). An immersed-boundary method for flow-structure interaction in biological systems with application to phonation, *Journal of Computational Physics* 227, pp. 9303–9332.

Mayo, A. and Peskin, C. (1993). An implicit numerical method for fluid dynamics problems with immersed elastic boundaries, *Contemporary Mathematics* 141, pp. 261–277.

McCormick, S. (1987). *Multigrid Methods*, SIAM.

McQueen, D. and Peskin, C. (1985). Computer-assisted design of butterfly bileaflet valves for the mitral position, *Scandinavian Journal of Thoracic and Cardiovascular Surgery*, 19, pp. 139–148.

McQueen, D. and Peskin, C. (1997). Shared-memory parallel vector implementation of the immersed boundary method for the computation of blood flow in the beating mammalian heart, *The Journal of Supercomputing* 11, pp. 213–236.

McQueen, D. and Peskin, C. (2001). Heart simulation by an immersed boundary method with formal second-order accuracy and reduced numerical viscosity. In *Mechanics for a New Millennium, Proceedings of the International Conference on Theoretical and Applied Mechanics (ICTAM).*

Mindlin, R. (1951). Influence of rotatory inertia and shear on flexural motions of isotropic, elastic plates, *Journal of Applied Mechanics* 18, pp. 31–38.

Mittal, R., Zheng, X., Bhardwaj, R., Seo, J., Xue, Q., and Bielamowicz, S. (2011). Toward a simulation-based tool for the treatment of vocal fold paralysis, *Frontiers in Physiology* 2(19), pp. 1–5.

Mohd-Yusof, J. (1998). Development of immersed boundary methods for complex geometries, *Center for Turbulence Research Annual Research Briefs*, pp. 325–336 NASA Ames/Stanford University.

Morand, H. and Ohayon, R. (1995). *Fluid Structure Interaction - Applied Numerical Methods*, John Wiley & Sons, translated by C.A. James.

Mori, Y. and Peskin, C. (2008). aimplicit second-order immersed boundary methods with boundary mass, *Computer Methods in Applied Mechanics and Engineering* 197, pp. 2049–2067.

Müftü, S. and Benson, R. (1995). Modelling the transport of paper webs including the paper permeability effects, *Proceedings of ISPS at the ASME - International Congress and Exposition in San Fransisco, CA.*

Munkres, J. (1991). *Analysis on Manifolds*, Addison-Wesley Publishing.

Newren, E., Fogelson, A., Guy, R., and Kirby, R. (2007). Unconditionally stable discretizations of the immersed boundary equations, *Journal of Computational Physics* 222, pp. 702–719.

Olson, L. and Bathe, K. (1983). A study of displacement-based fluid finite elements for calculating frequencies of fluid and fluid-structure systems, *Nuclear Engineering and Design* 76, pp. 137–151.

Olson, L. and Bathe, K. (1985). Analysis of fluid-structure interactions. A direct symmetric coupled formulation based on the fluid velocity potential, *Computers & Structures* 21(1/2), pp. 21–32,

Osher, S. and Fedkiw, R. (2002). *Level Set Methods and Dynamic Implicit Surfaces*, Springer.

Ostoja-Starzewsk, M. and Wang, S. (1999). Stochastic finite elements as a bridge between random material microstructure and global response, *Computer Methods in Applied Mechanics and Engineering* 168(1/4), pp. 35–49.

Paidoussis, M. (1998). *Fluid-Structure Interactions - Slender Structures and Axial Flow*, Vol. I, Academic Press.

Paidoussis, M. (2004). *Fluid-Structure Interactions - Slender Structures and Axial Flow* Vol. II, Academic Press.

Patankar, N., Wang, S., and Liu, W. (2003). A note on immersed boundary formulation for elastic bodies in fluids, *Preprint*.

Pedley, T. (1980). *The Fluid Mechanics of Large Blood Vessels*, Cambridge University Press.

Perlman, A., Titze, I., and Cooper, D. (1984). Elasticity of canine vocal fold tissue, *Journal of Speech, Language, and Hearing Research* 27(2), pp. 212–219.

Peskin, C. (1977). Numerical analysis of blood flow in the heart, *Journal of Computational Physics* 25, pp. 220–252.

Peskin, C. (2002). The immersed boundary method, *Acta Numerica* 11, pp. 479–517.

Peskin, C. (2012). Immersed boundary method, *NYU Courant Lecture Notes*.

Peskin, C. and Printz, B. F. (1993). Improved volume conservation in the computation of flows with immersed elastic boundaries, *Journal of Computational Physics* 105, pp. 33–46.

Pivkin, I., Caswell, B., and Karniadakis, G. (2011). Dissipative particle dynamics, in K. Lipkowitz (ed.), *Review in Computational Chemistry*, John Wiley & Sons, 27, pp. 85–110.

Quarteroni, A. (2001). Modeling the cardiovascular system - A mathematical adventure: Part I, *SIAM News* 34.

Reissner, E. (1979). Note on the effect of transverse shear deformation in laminated anisotropic plates, *Computer Methods in Applied Mechanics and Engineering* 20, pp. 203–209.

Roma, A., Peskin, C., and Berger, M. (1999). An adaptive version of the immersed boundary method, *Journal of Computational Physics* 153, pp. 509–534.

Rosa, M., Pereira, J., Grellet, M., and Alwan, A. (2003). A contribution to simulating a three-dimensional larynx model using the finite element method, *Journal of the Acoustical Society of America* 114(5), pp. 2893–2905.

Saad, Y. and Schultz, M. (1986). GMRES: A generalized minimal residual algorithm for solving nonsymmetric linear systems, *SIAM Journal on Scientific and Statistical Computing* 7, pp. 856–869.

Saiki, E. and Biringen, S. (1996). Numerical simulation of a cylinder in uniform flow: Application of a virtual boundary method, *Journal of Computational Physics* 123, pp. 450–465.

Scherer, R. and Shinwari, D. (2001). Intraglottal pressure profiles for a symmetric and oblique glottis with a divergence angle of 10 degrees, *Journal of the Acoustical Society of America* 109(4), pp. 1616–1630.

Scherer, R. and Titze, I. (1983). Pressure-flow relationships in two models of the larynx having rectangular glottal shapes, *The Journal of the Acoustical Society of America* 73(2), pp. 668–676.

Schey, H. (1992). *Div, Grad, Curl, and all that* (Norton).

Schmid, P. (2010). Dynamic mode decomposition of numerical and experimental data, *Journal of Fluid Mechanics* 656, pp. 5–28.

Sethian, J. (1996). *Level Set Methods and Fast Marching Methods*, Cambridge University Press.

Shu, C. (2020). Essential non-oscillatory and weighted essentially non-oscillatory schemes, *Acta Numerica* 29, pp. 701–762.

Shyy, W., Berg, M., and Ljungqvist, D. (1999). Flapping and flexible wings for biological and micro air vehicles, *Progress in Aerospace Sciences* 35, pp. 455–505.

Shyy, W., Francois, M., Udaykumar, H., N. N'dri, and Tran-Son-Tay, R. (2001). Moving boundaries in micro-scale biofluid dynamics, *Applied Mechanics Review* 54, pp. 405–453.

Shyy, W., Udaykumar, H., Rao, M., and Smith, R. (1996). *Computational Fluid Dynamics with Moving Boundaries*, Taylor & Francis.

Stenberg, R. and Suri, M. (1997). An *hp* error analysis of MITC plate elements, *SIAM Journal of Numerical Analysis* 34, pp. 544–568.

Stockie, J. and Wetton, B. (1999). Analysis of stiffness in the immersed boundary method and implications for time-stepping schemes, *Journal of Computational Physics* 154, pp. 41–64.

Strang, G. (1988). *Linear Algebra and its Applications*, (3rd edn.), Saunders HBJ.

Sulsky, D. and Brackbill, J. U. (1991). A numerical method for suspension flow, *Journal of Computational Physics* 96, pp. 339–368.

Sulsky, D. and Kaul, A. (2004). Implicit dynamics in the material-point method, *Computer Methods in Applied Mechanics and Engineering* 193, pp. 1137–1170.

Suri, M., Babuška, I., and Schwab, C. (1995). Locking effects in the finite element approximation of plate models, *Mathematics of Computation* 64(210), pp. 461–482.

Tanaka, H. and Araki, T. (2000). Simulation method of colloidal suspensions with hydrodynamic interactions: Fluid particle dynamics, *Physical Review Letters* 85, pp. 1338–1341.

Tao, C. and Jiang, J. (2006). Anterior-posterior biphonation in finite element model of vocal fold vibration, *Journal of the Acoustical Society of America* 120(3), pp. 1570–1577.

Tao, C., Jiang, J., and Zhang, Y. (2006). Simulation of vocal fold impact pressure with a self-oscillating finite-element model, *Journal of the Acoustical Society of America* 119(6), pp. 3987–3994.

Tao, C., Zhang, Y., Hottinger, D., and Jiang, J. (2007). Asymmetric airflow and vibration induece by the coanda effect in a symmetric model of the vocal folds, *Journal of the Acoustical Society of America* 122(4), pp. 2270–2278.

Taylor, C., Hughes, T., and Zarins, C. (1996). Computational investigations in vascular disease, *Computers in Physics* 10, pp. 224–232.

Tezduyar, T. (1991). Stabilized finite element formulations for incompressible-flow computations, *Advances in Applied Mechanics* 28, pp. 1–44.

Tezduyar, T. (2001). Finite element methods for flow problems with moving boundaries and interfaces, *Archives of Computational Methods in Engineering* 8, pp. 83–130.

Thomson, S., Mongeau, L., and Frankel, S. (2005). Aerodynamic transfer of energy to the vocal folders, *Journal of the Acoustical Society of America* 118(3), pp. 1689–1700.

Tidriri, M. (1997). Preconditioning techniques for the Newton-Krylov solution of compressible flows, *Journal of Computational Physics* 132, pp. 51–61.

Timoshenko, S. (1922). On the transverse vibrations of bars of uniform cross section, *Philosophical Magazine* 43(A), pp. 125–131.

Timoshenko, S. and Woinowsky-Krieger, S. (1987). *Theory of Plates and Shells* (2nd edn.), McGraw-Hill Publishing Company.

Titze, I. and Durham, P. (1987). Passive mechanisms influencing fundamental frequency control, *Laryngeal Function in Phonation and Respiration*. College-Hill Press, Boston, pp. 304–319.

Titze, I. R. (1989). On the relation between subglottal pressure and fundamental frequency in phonation, *Journal of the Acoustical Society of America* 85(2), pp. 901–906.

Tornberg, A. and Shelley, M. (2004). Simulating the dynamics and interactions of flexible fibers in stokes flows, *Journal of Computational Physics* 196, pp. 8–40.

Triantafyllou, M., Triantafyllou, G., and Yue, D. (2000). Hydrodynamics of fishlike swimming, *Annual Review of Fluid Mechanics* 32, pp. 33–53.

Tryggvason, G. (1988). Numerical simulations of the Rayleigh-Taylor instability, *Journal of Computational Physics* 75, pp. 253–282.

Uflyand, Y. (1948). The progagation of waves in the transverse vibrations of bars and plates, *Applied Mathematics and Mechanics* 12, pp. 287–300.

Van Loon, R., Anderson, P., Van de Vosse, F. N., and Sherwin, S. (2007). Comparison of various fluid-structure interaction methods for deformable bodies, *Computers & Structures* 85, pp. 833–843.

Veldin, T., Brank, B., and Brojan, M. (2019). Computational finite element model for surface wrinkling of shells on soft substrates, *Communications in Nonlinear Science and Numerical Simulation* 78(104863), pp. 1–17.

Vos, P., Van Loon, R., and Sherwin, S. (2008). A comparison of fictitious domain methods appropriate for spectral/hp element discretisations, *Computer Methods in Applied Mechanics and Engineering* 197, pp. 25–28.

Wagner, G. (2001). *A Numerical Investigation of Particulate Channel Flow*, Ph.D. thesis, Northwestern University, Evanston.

Wagner, G., Moës, N., Liu, W., and Belytschko, T. (2001). The extended finite element method for rigid particles in stokes flow, *International Journal for Numerical Methods in Engineering* 51, pp. 293–313.

Wagner, G., Ghosal, S., and Liu, W. (2003). Particulate flow simulations using lubrication theory solution enrichment, *International Journal for Numerical Methods in Engineering* 56, pp. 1261–1289.

Wang, C. and Zhang, L. (2015). Numerical modeling of gas–liquid–solid interactions: Gas–liquid free surfaces interacting with deformable solids, *Computer Methods in Applied Mechanics and Engineering* 286, pp. 123–146.

Wang, S. (1997). Finite element analysis of air-sheet interactions and flutter suppression devices, *Computers & Structures* 64(5/6), pp. 983–994.

Wang, S. (1999). Numerical analysis of moving orthotropic thin plates, *Computers & Structures* 70, pp. 467–486.

Wang, S. (2000). Velocity/Pressure mixed finite element and finite volume formulation with ALE descriptions for nonlinear fluid-structure interaction problems, *Advances in Engineering Software* 31, pp. 35–44.

Wang, S. (2003). On the discretized delta function and force calculation in the immersed boundary method. In *Computational Fluid and Solid Mechanics*, Elsevier, pp. 2164–2169.

Wang, S. (2006). From immersed boundary method to immersed continuum method, *International Journal for Multiscale Computational Engineering* 4, pp. 127–145.

Wang, S. (2007). An iterative matrix-free method in implicit immersed boundary continuum methods, *Computers & Structures* 85, pp. 739–748.

Wang, S. (2008). *Fundamentals of Fluid-Solid Interactions-Analytical and Computational Approaches*, Elsevier Science.

Wang, S. (2021). A revisit of implicit monolithic algorithms for compressible solids immersed inside a compressible liquid, *Fluids* 6(8), 273, pp. 1–25.

Wang, S. (2022). *Essential Mathematical Tools for Engineers*, Sentia Publishing.

Wang, S. (2025). Multi-scale modeling of proteins and cells: A protocol for modeling of complex systems, *Journal of Vibration Testing and System Dynamics* 9(1), pp. 21–46.

Wang, S. and Bathe, K. (1997a). Displacement/pressure based finite element formulations for acoustic fluid-structure interaction problems, *International Journal for Numerical Methods in Engineering* 40, pp. 2001–2017.

Wang, S. and Bathe, K. (1997b). On mixed elements for acoustic fluid-structure interactions, *Mathematical Models & Methods in Applied Sciences* 7, pp. 329–343.

Wang, S. and Gordnier, R. (2009). A combination of immersed methods and compact scheme based computational fluid dynamics for fluid-structure interactions, *Air Force ASEE Summer Faculty Fellow Report.*

Wang, S., Gordnier, R., Ji, S., and Bathe, K. (2009a). A study of 2d+ airfoil fsi models with different fsi approaches, *50th AIAA/ASME/ASCE/AHS/ASC Structures, Structural Dynamics, and Materials Conference* SDM-54, p. 2573.

Wang, S. and Liu, W. (2004). Extended immersed boundary method using FEM and RKPM, *Computer Methods in Applied Mechanics and Engineering* 193, pp. 1305–1321.

Wang, S. and Lua, J. (2010). Private Communication.

Wang, S. and Renner, G. (1998). Computational simulation of three-dimensional pretensioned moving paper webs, *Nineteenth Southeastern Conference on Theoretical and Applied Mechanics.*

Wang, S., Rowlan, L., Elsharafi, M., Ermila, M., Grejtak, T., and Taylor, C. (2019a). On leakage issues of sucker rod pumping systems, *ASME Journal of Fluids Engineering* 141(11), p. 111201, 7(pages).

Wang, S., Rowlan, L., Henderson, A., Aleman, S. T., Creacy, T., and Taylor, C. (2021). Viscoelastic representation of the operation of sucker rod pumps, *Fluids* 7(70), pp. 1–12, https://doi.org/10.3390/fluids7020070.

Wang, S., Yang, Y., and Wu, T. (2019b). Model studies of fluid-structure interaction problems, *Computer Modeling in Engineering and Sciences* 119(1), pp. 5–34.

Wang, S., Zhang, L., and Liu, W. (2009b). On computational issues of immersed finite element methods, *Journal of Computational Physics* 228, pp. 2535–2551.

Wang, X. and Zhang, L. (2010). Interpolation functions in the immersed boundary and finite element methods, *Computational Mechanics* 45(4), pp. 321–334.

Wei, T., Mark, R., and Hutchison, S. (2014). The fluid dynamics of competitive swimming, *Annual Review of Fluid Mechanics* 46, pp. 547–565.

Wilson, E. and Khalvati, M. (1983). Finite elements for the dynamic analysis of fluid-solid systems, *International Journal for Numerical Methods in Engineering* 19, pp. 1657–1668.

Wu, T., Wang, S., and Cohn, B. (2011). Modeling of proteins and their interactions with solvent, in S. Li and B. Sun (eds.), *Advances in Cell Mechanics*, Springer, pp. 55–116.

Wu, T., Wang, S., Cohn, B., and Ge, H. (2010). Molecular modeling of normal and sickle hemoglobins, *International Journal for Multiscale Computational Engineering* 8, pp. 237–244.

Yang, J., Wang, X., Krane, M., and Zhang, L. (2017). Fully-coupled aeroelastic simulation with fluid compressibility—for application to vocal fold vibration, *Computer Methods in Applied Mechanics and Engineering* 315, pp. 584–606.

Yang, Y., Wu, T., and Wang, S. (2010). Reduced order models for fluid-solid interaction systems with mixed finite element formulations, *International Symposium on Computational Mechanics* 1233(1), pp. 1303–1308.

Yu, Z. (2005). A DLM/FD method for fluid/flexible-body interactions, *Journal of Computational Physics* 207, pp. 1–27.

Zhang, J., Childress, S., Libchaber, A., and Shelley, M. (2000). Flexible filaments in a flowing soap film as a model for one-dimensional flags in a two-dimensional wind, *Nature* 408, pp. 835–838.

Zhang, L. (2014). Modeling of soft tissues interacting with fluid (blood or air) using the immersed finite element method, *Journal of Biomedical Science and Engineering* 7, pp. 130–145.

Zhang, L. (2017). Immersed methods for high reynolds number fluid-structure interactions, *International Journal of Computational Methods*, p. 1750068.

Zhang, L. and Gay, M. (2007). Immersed finite element method for fluid-structuure interactions, *Journal of Fluids and Structures* 23, pp. 839–857.

Zhang, L., Gerstenberger, A., Wang, S., and Liu, W. (2004). Immersed finite element method, *Computer Methods in Applied Mechanics and Engineering* 193, pp. 2051–2067.

Zhang, L., Wagner, G., and Liu, W. (2002). A parallized meshfree method with boundary enrichment for large-scale CFD, *Journal of Computational Physics* 176, pp. 483–506.

Zhang, L., Wagner, G., and Liu, W. (2003). Modeling and simulation of fluid structure interaction by meshfree and FEM, *Communications in Numerical Methods in Engineering* 19, pp. 615–621.

Zhang, Z., Neubauer, J., and Berry, D. A. (2006). The influence of subglottal acoustics on laboratory models, *The Journal of the Acoustical Society of America* 120(3), pp. 1558–1569.

Zheng, X., Bielamowicz, S., Luo, H., and Mittal, R. (2009). A computational study of the effect of false vocal folds on glottal flow and vocal fold vibration during phonation, *Annals of Biomedical Engineering* 37(3), pp. 625–642.

Zhou, Y. and Wei, G. (2006). On the fictitious-domain and interpolation formulations of the matched interface and boundary (mib) method, *Journal of Computational Physics* 219, pp. 228–246.

Index

www.ingramcontent.com/pod-product-compliance
Lightning Source LLC
LaVergne TN
LVHW010329220926
848484LV00004BB/865
* 9 7 8 9 8 1 3 2 3 4 5 0 5 *